INTERMEDIATE FINITE ELEMENT METHOD

Series in Computational and Physical Processes in Mechanics and Thermal Sciences

W. J. Minkowycz and E. M. Sparrow, *Editors*

Anderson, Tannehill, and Pletcher, Computational Fluid Mechanics and Heat Transfer

Aziz and Na, Perturbation Methods in Heat Transfer

Baker, Finite Element Computational Fluid Mechanics

Beck, Cole, Haji-Shiekh, and Litkouhi, Heat Conduction Using Green's Functions

Carey, Computational Grids: Generation, Adaptation, and Solution Strategies

Chung, Editor, Numerical Modeling in Combustion

Comini, Del Giudice, and Nonino, Finite Element Analysis in Heat Transfer: Basic Formulation and Linear Problems

Heinrich and Pepper, Intermediate Finite Element Method: Fluid Flow and Heat Transfer Applications

Jaluria, Computer Methods for Engineering

Jaluria and Torrance, Computational Heat Transfer

Koenig, Modern Computational Mechanics

Patankar, Numerical Heat Transfer and Fluid Flow

Pepper and Heinrich, The Finite Element Method: Basic Concepts and Application

Shih, Numerical Heat Transfer

Shyy, Udaykumar, Rao, and Smith, Computational Fluid Dynamics with Moving Boundaries

Tannehill, Pletcher, and Anderson, Computational Fluid Mechanics and Heat Transfer, Second Edition

Willmott, Modeling and Dynamics of Regenerative Heat Transfer

PROCEEDINGS

Chung, Editor, Finite Elements in Fluids: Volume 8

Haji-Sheikh, Editor, Integral Methods in Science and Engineering-90

Shih, Editor, Numerical Properties and Methodologies in Heat Transfer: Proceedings of the Second National Symposium

FORTHCOMING

Averill and Reddy, Computational Methods in Damage Mechanics

INTERMEDIATE FINITE ELEMENT METHOD

Fluid Flow and Heat Transfer Applications

Juan C. Heinrich
University of Arizona
Tucson, Arizona

Darrell W. Pepper
University of Nevada
Las Vegas, Nevada

USA	Publishing Office:	TAYLOR & FRANCIS 325 Chestnut Street Philadelphia, PA 19106 Tel: (215) 625-8900 Fax: (215) 625-2940
	Distribution Center:	TAYLOR & FRANCIS 47 Runway Road, Suite G Levittown, PA 19057 Tel: (215) 269-0400 Fax: (215) 269-0363
UK		Taylor & Francis Ltd. 1 Gunpowder Square London EC4A 3DE Tel: 171 583 0490 Fax: 171 583 0581

INTERMEDIATE FINITE ELEMENT METHOD: Fluid Flow and Heat Transfer Applications

1 2 3 4 5 6 7 8 9 0

Printed by Hamilton Printing Co., Castleton, NY, 1999. Cover design by Michelle Fleitz.

A CIP catalog record for this book is available from the British Library.
∞ The paper in this publication meets the requirements of the ANSI Standard Z39.48-1984 (Permanence of Paper).

Library of Congress Cataloging-in-Publication Data

Heinrich, Juan C.
 Intermediate finite element method: fluid flow and heat transfer applications/Juan C. Heinrich, Darrell W. Pepper.
  p.   cm.
 Includes bibliographical references (p.).

 1. Fluid mechanics——Mathematical models. 2. Finite element method. 3. Heat——Transmission——Mathematical models. I. Pepper, D. W. (Darrell W.) II. Title.
TA357.H45
620.1'06'015118——dc21   97-16149

ISBN: 1-56032-309-4 (case)

To our parents:

Carlos and Ruby Heinrich

Weldon and Marjorie Pepper

CONTENTS

Foreword   xv
Preface   xvii
Disclaimer   xxi

1   INTRODUCTION   1
1.1   Overview   1
1.2   Short History of the Finite Element Method   3
1.3   Finite Element Concept   5
1.4   Present Text   6
1.5   Closure   7
        References   8

2   BASIC EQUATIONS OF FLUID DYNAMICS
2.1   Overview   11
2.2   Substantial Derivative   11
2.3   Mass Conservation Equation   12
2.4   Navier-Stokes Equations   13
2.5   Equation of Energy Conservation   15
2.6   Mass Transport   16
2.7   Wave Equation   17
2.8   Boundary Conditions   18
2.9   Closure   19
        References   19

3	**FUNDAMENTAL CONCEPTS**	21
3.1	Overview	21
3.2	Linear Heat Conduction	21
3.3	Linear Operators and Linear Function Spaces	25
3.4	Weighted Residuals Formulation	28
3.5	Galerkin Method	37
	3.5.1 Galerkin Method in One Dimension	39
	3.5.1.1 Approximation 1	40
	3.5.1.2 Approximation 2	43
	3.5.2 Galerkin Method in Two Dimensions	45
3.6	Finite Element Method in One Dimension	46
	3.6.1 Basic Piecewise Linear Spaces	46
	3.6.2 Heat Conduction in One Dimension	54
	3.6.3 Error in the Finite Element Approximation	59
	3.6.4 Boundary Fluxes	62
	3.6.5 Interelement Conditions	66
3.7	Finite Element Method in Two Dimensions	73
	3.7.1 Basic Piecewise Bilinear Space	73
	3.7.2 Heat Conduction in Two Dimensions	77
	3.7.3 Error in Two-dimensional Approximation	87
	3.7.4 Interelement Conditions	88
3.8	Finite Element Method in Three Dimensions	95
	3.8.1 Trilinear Element	95
	3.8.2 Heat Conduction in Three Dimensions	96
3.9	Closure	97
	References	98
	Exercises	99
4	**HIGHER ORDER ELEMENTS**	109
4.1	Overview	109
4.2	One-Dimensional Elements	110
4.3	Two-Dimensional Elements	116
	4.3.1 Triangular Elements	117
	4.3.2 Rectangular Elements	124
	4.3.3 Error Bounds for Two-dimensional Elements	127
4.4	Isoparametric Elements	130
	4.4.1 Difficulty 1	130
	4.4.2 Difficulty 2	132
4.5	Blending Function Interpolation	147

4.6	Three-Dimensional Elements	163
4.7	Closure	165
	References	166
	Exercises	169

5	**NUMERICAL INTEGRATION**	175
5.1	Overview	175
5.2	Quadrature Formulae	175
	5.2.1 Newton-Cotes Formulae	178
	5.2.2 Gaussian Quadrature	180
5.3	Multiple Integrals	182
5.4	Minimum and Optimal Order of Integration	184
5.5	Reduced Integration, Evaluating Gradients	194
	5.5.1 Reduced and Selective Integration	194
	5.5.2 Evaluating Gradients	196
5.6	Closure	204
	References	204
	Exercises	205

6	**NONLINEARITY**	209
6.1	Overview	209
6.2	Basic Methods for Nonlinear Equations	210
	6.2.1 The Newton-Raphson Method	210
	6.2.2 Direct Iteration Methods	220
6.3	Nonlinear Examples	225
	6.3.1 Heat Transfer with Temperature-Dependent Conductivity	225
	6.3.2 Stationary Navier-Stokes Equations	228
	6.3.3 Steady State Natural Convection	243
6.4	Closure	249
	References	250
	Exercises	251

7	**TIME DEPENDENCE**	257
7.1	Overview	257
7.2	Diffusion Equations	257
	7.2.1 Semidiscrete Galerkin Method	258
	7.2.2 The θ Method	260
	7.2.3 Accuracy and Stability of the θ Method	265

	7.2.4	Mass Lumping	268
7.3		Runge-Kutta Methods	275
7.4		Generalized Newmark Algorithms	283
	7.4.1	Newmark Method for Second-Order Hyperbolic Equations	284
	7.4.2	Generalized Newmark Method for Parabolic Equations	297
7.5		Closure	305
		References	305
		Exercises	307

8		**STEADY STATE CONVECTIVE TRANSPORT**	**313**
8.1		Overview	313
8.2		One-Dimensional Convection-Diffusion	314
8.3		Petrov-Galerkin Method	321
8.4		Petrov-Galerkin Method in Two Dimensions	331
8.5		Petrov-Galerkin Method in Three Dimensions	340
8.6		Nonlinear Equations	342
8.7		Closure	346
		References	347
		Exercises	350

9		**TIME-DEPENDENT CONVECTION-DIFFUSION**	**357**
9.1		Overview	357
9.2		Time-Dependent Convection	357
	9.2.1	Numerical Damping	361
	9.2.2	Phase Error	368
9.3		Petrov-Galerkin Method for Time-Dependent Convection-Diffusion	370
	9.3.1	Quadratic in Time, Linear in Space Weights	371
	9.3.2	Stability Analysis	377
9.4		Multidimensional Time-Dependent Convection-Diffusion	384
9.5		Closure	395
		References	395
		Exercises	397

10		**VISCOUS INCOMPRESSIBLE FLUID FLOW**	**403**
10.1		Overview	403
10.2		Basic Form of the Navier-Stokes Equations	404

10.3	Constant-Density Flows in Two Dimensions	409
	10.3.1 Mixed Formulation	409
	10.3.2 Fractional Step Method	422
	10.3.3 Penalty Function Formulation	424
10.4	Stratified Flows	434
	10.4.1 Finite Element Approximations	436
	10.4.2 Calculation of the Pressure	446
	10.4.3 Open Boundaries	448
10.5	Free Surface Flows	452
10.6	Closure	456
	References	458
	Exercises	461
11	**MESH GENERATION**	465
11.1	Overview	465
11.2	Introduction	466
	11.2.1 Types of Meshes	466
	11.2.2 Popular Mesh Generation Schemes	470
	11.2.2.1 Manual Generation	470
	11.2.2.2 Semi-Automatic Generation	470
	11.2.2.3 Transport Mapping	471
	11.2.2.4 Explicit Solution of PDEs	471
	11.2.2.5 Overlapping and Deformation	472
	11.2.2.6 Advancing-front Method	473
	11.2.2.7 Combination	473
11.3	Mesh Generation Techniques	474
	11.3.1 Structured Meshes	475
	11.3.1.1 One Dimension	475
	11.3.1.2 Two Dimensions	476
	11.3.1.3 Three Dimensions	476
	11.3.1.4 Boundary Fitted Coordinates	477
	11.3.2 Unstructured Meshes	484
	11.3.2.1 General Types of Elements	485
	11.3.2.2 One-dimensional Elements	485
	11.3.2.3 Two-dimensional Elements	486
	11.3.2.4 Three-dimensional Elements	488
	11.3.3 Mesh Generation Guidelines	490
11.4	Bandwidth	495
	11.4.1 Nodal Renumbering Schemes	498

11.4.1.1 Gibbs Method   499
11.4.1.2 Groom's Method   500
11.4.1.3 Lipton-Tajan Method   500
11.4.1.4 Akhras and Dhatt Method   500
11.4.1.5 Frontal Method   501
11.4.1.6 Element Colorization Method   501
11.4.1.7 Nested Dissection Method   501
11.4.2 Simple Bandwidth Reduction Algorithm   501
11.4.3 Delaunay Triangulation   505
11.5 Adaptation   508
11.5.1 Types of Adaptation   508
11.5.2 Error Estimates and Adaptation Criteria   511
11.5.3 Simple h-adaptive Technique   517
11.5.3.1 Mesh Regeneration   517
11.5.3.2 Element Subdivision   520
11.5.3.3 Adaptation Parameters   521
11.5.3.4 Adaptation Rules   524
11.5.4 Mesh Adaptation Example   525
11.6 Closure   527
References   529
Exercises   533

12 FURTHER APPLICATIONS   535
12.1 Overview   535
12.2 Time-Dependent Flows and Flows in Rotating Systems   535
12.2.1 Isothermal Flow Past a Circular Cylinder   540
12.2.2 Natural Convection in a Horizontal Circular Cylinder   542
12.2.3 Lubricant Flow in a Microgap   545
12.3 Turbulent Flow   547
12.4 Compressible Flow   552
12.4.1 Supersonic Flow Impinging on a Cylinder   558
12.4.2 Chemically Reacting Supersonic Flow   560
12.5 Three-dimensional Flow   564
12.5.1 Natural Convection Within a Sphere   564
12.5.2 Transonic Flow Through a Rectangular Nozzle   566
12.6 Closure   568
References   568

APPENDIX A: LINEAR OPERATIONS 571
    A.1   Linear Vector Spaces 571
    A.2   Linear Operators 575

APPENDIX B: UNITS 577

APPENDIX C: NOMENCLATURE 579

INDEX 585

FOREWORD

The finite element first became important in the 1960s as a powerful new approach for solving problems in structural analysis. As the mathematical foundation of the scheme became better understood, it was realized that it could be applied quite generally in other application domains, such as fluid mechanics, and the first conference on finite elements applied to fluids was held in Swansea, UK, in 1974. Since that time we have witnessed a rapid expansion in this area, and today several commercial computer programs based on the finite element approach are available for simulation of compressible or incompressible flows and coupled transport processes. The present volume provides an intermediate-level approach to this applications area. The early chapters summarize the basic methodology and the main equations to be considered later. Some topics such as nonlinear solution algorithms and time integrators are included to give insight into the algorithmic issues without overburdening the reader with details. Other important related computational issues such as grid generation and adaptation are summarized and a welcome inclusion. Finally, the concluding section provides a glimpse of the power of the method in treating several interesting real flow and transport applications.

Graham F. Carey
Director, CFD Laboratory
University of Texas, Austin

PREFACE

One of the most interesting and practical numerical techniques for finding approximate solutions to partial differential equations in engineering and science is the finite element method (FEM). The FEM is used to solve a wide variety of problems and, in many instances, is the only viable method for obtaining solutions. While the FEM is built on a rich mathematical basis, it is still one of the most practical numerical schemes yet devised for solving complex problems. It is becoming increasingly important for engineers and designers today to be knowledgeable of the FEM, and to know how to apply it using either in-house or commercial codes now on the market.

In this book, we advance the concept of the FEM to address more difficult problem areas, with emphasis on fluid flow with heat transfer. This book follows from the introductory text, *The Finite Element Method: Basic Concepts and Applications*, by D. W. Pepper and J. C. Heinrich, first published in 1992. The 1992 publication has been used in numerous finite element short courses given by the authors for ASME, and Home Study courses for AIAA. The present book is an outgrowth of advanced finite element courses given over the last few years to both ASME and AIAA participants. As in the 1992 book, we have attempted to keep the subject information at an applied level, with numerous examples. However, when one introduces advanced topics in a numerically based subject, more mathematical detail is typically required. We have tried to keep this detail to a minimum but still provide enough mathematical basis to understand the principles.

We focus here on the solutions of linear and nonlinear partial differential equations, with emphasis on fluid motion. Fluid flow is a difficult subject; many textbooks have been written on the myriad aspects of fluid motion, including recent texts on computational fluid dynamics. Understanding fluid flow problems is difficult enough, let alone when one must numerically solve these complex equations. We first reacquaint the reader with the governing equations for fluid flow, then proceed to show how the FEM is ideally suited for finding numerical approximations to the solutions of the equations, especially when problem geometries are complex. We leave many aspects of fluid motion and their mathematical analyses to the existing excellent books and technical papers on fluid dynamics.

Each chapter describes particular applications of the FEM and illustrates the concepts through example problems. Chapter 1 gives a brief history of the FEM and an overview of the text. The basic equations for fluid flow, with emphasis on the continuity, energy, and Navier-Stokes equations, are reviewed in Chapter 2. In Chapter 3, the basis of the Galerkin formulation is discussed, and the reader is reacquainted with the FEM using linear heat conduction. Chapter 4 discusses shape functions for one, two, and three dimensions, and the concept of isoparametric elements. In Chapter 5, various numerical integration schemes are examined, including the use of reduced integration. A brief overview of nonlinearity is given in Chapter 6. Integration in time, including implicit and explicit time stepping, several iteration schemes, and lumping of the mass matrix are explained in Chapter 7. Petrov-Galerkin weighting, which is applied to convective terms, is introduced in Chapter 8, with emphasis on solving steady state convection-diffusion problems. Chapter 9 deals with the time-dependent aspects of the convective-diffusion equation. Applications to viscous incompressible fluid flow are discussed in Chapter 10, which includes sections on stratified and free surface flows, along with a comprehensive review on the state-of-the-art in finite element modeling of fluid flow, illustrating the use of penalty and mixed approximations. The various aspects of mesh generation are described in Chapter 11, along with optimization of the bandwidth and adaptive meshing techniques; a set of simple algorithms is included for two-dimensional mesh generation and optimization. In Chapter 12, issues dealing with flow in rotating systems, turbulence, and compressible flows are discussed.

Example problems are provided within each chapter to help illustrate application of the FEM for the respective type of problem.

Appendices are included, along with a floppy disk, which contains both the source and an executable version of a FORTRAN code (FLOW2D) for two-dimensional fluid flow with heat transfer using the penalty approach. This code runs on 486 and Pentium based PCs with at least 8 MB RAM, and should port easily to UNIX workstations.

We wish to thank Mrs. Lisa Ehmer, former senior acquisitions editor, for her support and assistance from the beginning of this book. We also wish to thank Mrs. Carolyn Ormes, former development editor, for her valuable insight and assistance in organizing the text, and patience in preparing the manuscript. We are especially thankful to Mrs. Jeannie Pepper for typing the manuscript, revising the chapters when needed, and preparation of the final copy. We also wish to thank Mr. Vernon Lau and Mr. Jason Mulvey, undergraduate mechanical engineering students at UNLV, for their efforts in generating the figures.

Juan C. Heinrich
Darrell W. Pepper

DISCLAIMER

The code FLOW2D, enclosed on a disk attached to the back of this text, is a two-dimensional finite element program based on the penalty approach for fluid flow. The program is written in FORTRAN 77. The source code is denoted by the extension .FOR; the executable version is denoted by the extension .EXE, and was obtained using Microsoft FORTRAN 4.0. Example data sets are listed with .DAT extensions. FLOW2D has been tested on 486 and Pentium class PC's, and has been ported to SGI and SUN workstations.

When using the program, the user understands and acknowledges that no warranty is expressed or implied by the authors or by the publishers as to the accuracy or reliability of the program. It is up to the user to ensure that the basic assumptions of the program are understood and that verification of results from user defined applications are left to the user. A README.TXT file is included on the disk with additional instructions as to the uses and limitations of FLOW2D.

ONE

INTRODUCTION

1.1 OVERVIEW

The finite element method (FEM) is a unique numerical approach used to solve partial differential equations that describe engineering and scientific problems. The method is particularly effective in problems requiring the numerical solution of transient or steady state convection-diffusion transport in regions of complex geometry.

In this book, we advance the concept of the FEM to solution of these more difficult, but typically more realistic, types of problems. This book is a follow-on to the introductory text *The Finite Element Method: Basic Concepts and Applications*, by Pepper and Heinrich (1992); the introductory text has been used for many years by the authors in the AIAA Home Study Course on the Introduction to the Finite Element Method and for numerous ASME Finite Element Short Courses. We strongly encourage readers of this intermediate book to reacquaint themselves with the fundamentals of the FEM. We recommend reviewing either our introductory book or one of the many fundamental texts available in the literature before delving too deeply into these more advanced topics.

Our primary emphasis will be on the solution of linear and nonlinear partial differential equations, with particular concentration on the equations of viscous fluid motion. The equations of fluid motion and accompanying boundary conditions are generally difficult to properly formulate for most problems; however, once the equation set and boundary conditions are established, one must face the formidable task of numerically solving these complex equations. It is important that one fully comprehend the nature of the equations that describe the physics of the problem–a great numerical scheme that solves the wrong set of equations or improperly constrained boundary conditions still produces wrong answers. Our word of advice is to first examine the problem, understand the underlying physics and appropriate equation set, and then begin to formulate the numerical approach.

There are so many numerical schemes now available that one can rapidly become overwhelmed by the plethora of numerical intricacies and algorithms for solving the intended equation set. This is not surprising with the current interest in computing. The continuing development of faster and more powerful computers has allowed engineers and scientists to solve sophisticated problems at their desks using numerical methods once deemed available only on mainframe computers.

As one quickly discovers, the types of numerical methods used by the majority of researchers and applications oriented engineers for solution of partial differential equations fall into three categories: 1) finite element methods, 2) finite difference (or finite volume) methods (FDM/FVM), and 3) other approaches (boundary integrals, hybrids, analytical, spectral, etc.). The FDM has been used for a wide variety of problems; nearly all of the early numerical simulations dealing with heat transfer and fluid flow revolved around various solution strategies for the FDM. There is still a large group of FDM/FVM advocates who continue to use such methods for exceedingly complex problem geometries.

The difficulties associated with irregular geometries have been resolved by incorporating Boundary Fitted Coordinates (BFC) into the solution algorithm for FVM. However, the use of BFC and its inherent difficulties in formulating "good grids" (i.e., a mesh that will produce accurate results) exact a high price just to use the FDM/FVM. There is some work now being done on using the FVM with unstructured grids (triangular elements); one commercial code company now sells an unstructured FVM model for transient compressible flows. Such hybrid codes are still rather difficult to formulate, and can become computationally costly.

There appears to be a "converging" of FVM and FEM. Advocates of the FVM have begun to appreciate the benefits of methods that utilize unstructured meshing. Fortunately, many researchers are becoming more familiar with the FEM and its ease of application. We know of very few individuals who ever ventured back into the FDM/FVM after discovering the power and versatility of the FEM.

1.2 SHORT HISTORY OF THE FINITE ELEMENT METHOD

The beginning of the FEM does not start with the beginning of the digital computer age. As one traces the development of the mathematical foundations and theory behind the FEM, much of the development of the early theory occurred in the late 19th century.

One of the earliest purveyors of the FEM was Lord Rayleigh, although it was not evident at the time. Rayleigh (1877) would obtain answers to many practical problems related to structural design, especially railway bridges and trusses, using minimization of energy principles. While the minimization concept was not new, Rayleigh was able to expand upon the original concept. However, geometry problems were difficult to overcome, and problems had to be reduced to simple geometric concepts to achieve analytical solution.

Following Rayleigh's early efforts, Ritz (1909) extended the principle of energy minimization by including the use of multiple independent functions. Ritz came close to establishing the concept of finite elements; however, it would be nearly 40 years before the first true mathematical basis of the FEM appeared. Independent of Ritz's work, Galerkin (1915) published his work regarding the integral concept of the method of weighted residuals (MWR). This work would become the theoretical basis for the FEM approach now used almost universally.

In 1941, Hrenikoff began work on the framework method, which was essentially a collection of beams and bars to analyze plane elasticity. In 1943, Courant published a paper that demonstrated the use of triangular regions and piecewise approximations within the triangular regions. It was Courant who put everything together (without the benefit of a computer). He examined different geometric regions, used separate approximating functions, and assembled separate solutions into a global solution, all based around the principle of total potential energy previously formulated by Ritz.

The first appearance of a more modern form of the FEM is in the work by Turner et al. (1956). Their work centered on structural analysis of aircraft sections being built at Boeing Aircraft. One of the co-authors, R. W. Clough, coined the name "finite element" in a paper published in 1960. Likewise, Argyris and Kelsey published a text in 1960 describing the development of this rather unique mathematical approach to structural problems. By the end of the 1960s, the term "finite element" had become the accepted name for the method.

The need of matrix algebra became evident when dealing with the finite element technique. A conference held by the Air Force in the mid-1960s emphasized the application of the finite element approach and matrix algebra techniques to structural problems, and the need for large computational resources (Wright-Patterson Air Force Base, 1966). At this time the development (and ensuing competition) of large mainframe computers for calculating science and engineering problems began to appear. As a result of the availability of these mainframe computers and their continued improvement, along with numerous papers appearing throughout the 1960s on the FEM, it was not long before the first commercial FEM program – NASA STRuctural ANalysis (NASTRAN) – became available. Soon after, the Structural Analysis Program (SAP) followed, which was an FEM program developed by a student of Clough's. Likewise, the ANSYS finite element program was developed for use in the nuclear industry. Interestingly, these programs, along with the proliferation of many others, are advertised today in many trade and technical journals.

In the early 1970s, the automotive industry became very interested in the use of the FEM for structural analysis related to the design and safety of automobiles. The FEM was no longer confined to just the aircraft industry. The textbook published by R. D. Cook in 1974 made the finite element principles and capabilities more readily understandable to a vast number of engineers. Today, one can find numerous textbooks that discuss the basic theory and history of the FEM; most are focused on the application of the FEM to structural problems.

As structural analyses using FEM became more routine, researchers began to apply the method to more difficult problem areas, particularly those fields dealing with fluid flow. Some of the earliest work in fluid simulation with finite elements can be traced to the mid-1960s; a comprehensive review is given in Zienkiewicz and Taylor (1989, 1991). Based on this early pioneering work, the numerical simulation of fluid flow with finite elements began to proliferate by the 1970s. Today, the FEM is a strong contender for simulating all modes of fluid flow processes, rivaling performance standards associated with FDM/FVM.

1.3 FINITE ELEMENT CONCEPT

The two most often used ways to formulate the FEM are the Rayleigh-Ritz variational method and the Galerkin MWR. Both approaches rely on using a linear combination of appropriate functions $\varphi_i(x,y,z)$ to approximate the solution in the form $\sum \varphi_i a_i$. The coefficients a_i are determined using the integral statements in such a way as to approximately satisfy the original differential equations.

There is a major difference between the Rayleigh-Ritz method and the Galerkin method. The Rayleigh-Ritz method finds the unknown coefficients through an energy minimization process; this process requires a minimum principle. The Galerkin method is based on making the projection of the error in the approximating functions φ_i vanish in the finite dimensional space spanned by the functions in order to determine the unknown coefficients a_i. This approach allows the Galerkin method to be used in situations when minimum principles do not exist. Such cases occur when convection is the dominant transport mechanism in a fluid system. The Galerkin method is therefore the method of choice in problems involving fluid flow.

The next important step lies in the choice of the approximating functions φ_i. In early developments φ_i values were chosen from analytical functions defined over the entire domain, e.g., trigonometric functions. However, this is very limiting because of the impossibility to enforce boundary conditions on such globally defined functions except for very simple geometries. In the FEM we use locally defined functions, which are typically piecewise polynomials defined over small geometric subregions of the domain called *elements*. This allows us to achieve virtually unlimited freedom in terms of the geometric complexity of the domain and the type of boundary conditions that can be imposed.

Once a solution to a problem has been obtained, the question of error estimation and accuracy naturally arises. Sources of error can generally be attributed to the way in which the problem domain and the solution to the governing equations have been approximated. If the domain is discretized with a coarse mesh, we may never get close enough to the right answer; likewise, the mesh may be considered suitable, but the approximation functions may be too low in order. Furthermore, one must face the inherent computational errors associated with round-off, numerical differentiation, and numerical integration.

In general, the following steps are needed in any finite element approximation to the solution of a differential equation: 1) the equation (or system of equations) and its boundary and initial conditions must be carefully defined to ensure that a well-posed problem is formulated, 2) an element type must be chosen to define the approximation functions to be used in the solution, 3) a mesh must be created that adequately refines regions where large changes in the solution are expected, and that allows the boundary conditions to be properly imposed, 4) the finite element algorithm used to produce the system of equations from the unknown coefficients a_i must be formulated and used to solve the linear system of algebraic equations, and 5) the error in the approximation must be assessed to determine if the solution is converged or if a more refined solution is needed.

These five steps are very basic and concisely describe the procedure for obtaining solutions with the FEM. However, the requisite substeps within each major step are typically not trivial. A similar set of steps and the details residing within each step are discussed by Reddy (1993).

1.4 PRESENT TEXT

This intermediate text begins with a discussion and review of the underlying principles of the FEM, including solution techniques for linear and nonlinear steady state and transient problems. Application of the FEM begins with an examination of the convective-diffusion equation and introduction of the Petrov-Galerkin technique for reducing numerical dispersion errors. An in-depth analysis of incompressible fluid flow follows, with emphasis on penalty and mixed schemes. Mesh generation is discussed in some depth, as is application of the finite element scheme to compressible flows, turbulence, heat and mass transfer, lubrication, unsteady flows, and free surfaces.

Chapter 1 gives a brief history of the FEM and an overview of the text, as described above. Chapter 2 reviews the basic equations for fluid flow, with emphasis on the continuity, energy, and Navier-Stokes equations. The Galerkin formulation and its application to the FEM are reviewed in Chapter 3, utilizing a set of simple examples from heat transfer. The formulation of shape functions for one, two, and three dimensions are described in Chapter 4, along with the concept of isoparametric elements. Numerical integration schemes are reviewed in Chapter 5, along with the impact of using reduced integration (as some commercial finite element codes now employ). An introduction into nonlinearity is given in Chapter 6. Time integration, including implicit and explicit time stepping, iteration schemes, and lumping of the mass matrix are discussed in Chapter 7. The application of Petrov-Galerkin weighting to steady state convection-diffusion problems is introduced in Chapter 8; this concept is becoming widely accepted as the defacto standard for dealing with convective terms in both linear and nonlinear equations–this chapter is particularly important in the way one handles first-order (convective type) derivatives. Time-dependent solutions to the convection-diffusion equation are examined in Chapter 9, including use of Petrov-Galerkin methods for transient problems. Chapter 10 gives a comprehensive review on the state of the art, finite element modeling of viscous incompressible fluid flow. Sections are included on resolving stratified as well as free surface flows; illustrations of the uses of penalty and mixed approximations are given. In Chapter 11 the most common approaches to mesh generation are described for both structured (finite volume) and unstructured (finite element) meshes, along with optimization of the bandwidth and adaptive meshing techniques. Chapter 12 includes examples on flows in rotating systems, finite element approaches to modeling turbulence, and the use of FEM for compressible flows.

Example problems are provided within each chapter to help illustrate application of the FEM for the various types of problems. A set of appendices is included, along with a computer code for solving two-dimensional fluid flow using the penalty approach (FLOW2D). A 3.5" diskette accompanies the text and includes both source (FORTRAN) and executable versions of the software.

1.5 CLOSURE

One wonders how engineers designed and analyzed many of the complex systems we take for granted today before the availability of commercial FEM codes. Recall that commercial aircraft prior to the jet age were designed and built, then tested and examined for structural integrity after being put into service; companies could not wait until exhaustive testing and analyses were completed before going into production. Our automobiles and aircraft of today have become much more sophisticated and streamlined. The simplistic approaches of the past can no longer be used to accommodate budget constraints and deadlines. Today, the analyses must be performed first–it is fairly simple to change loads, materials, and geometry and then recompute to obtain new results. Our reliance and confidence in many of these commercial finite element codes are well grounded, particularly in structural analysis.

The use of finite elements in fluid flow analysis is becoming more widespread; many finite element software companies now include codes for calculating fluid flow. However, the field is still fairly young. With the recent advances and future prospects for even faster computers, the use of FEM for fluid analysis as well as other difficult nonlinear problems will proliferate rapidly. There is considerable interest in new aircraft designs and suborbital spacecraft utilizing airbreathing propulsion, i.e., a craft that takes off from a ground based runway and ultimately climbs to the upper reaches of the atmosphere. While still a concept, considerable work has been done using finite element techniques to analyze thermal, structural, and fluid dynamic problems, and their interactions. New flight vehicles are now being designed and evaluated using specifications and data obtained by computational fluid dynamics (CFD) methods.[*] Even marginal success of these vehicles will demonstrate the necessity and value of the FEM, and its use for a wide range of projects yet to be developed.

REFERENCES

Argyris, J. H. and Kelsey, S. (1960). *Energy Theorems and Structural Analysis.* London: Butterworth Scientific Publications.

Clough, R. W. (1960). "The Finite Element Method in Plane Stress Analysis." *J. Struct. Div., ASCE, Proc. Conf. Electronic Comput.,* 2nd, pp. 345–378.

Cook, R. D. (1974). *Concepts and Applications of Finite Element Analysis.* New York: John Wiley and Sons.

Courant, R. (1943). "Variational Methods for the Solution of Problems of Equilibrium and Vibration," *Bull. Amer. Math. Soc.,* Vol. 49, pp. 1–43.

Galerkin, B. G. (1915). "Series Occurring in Some Problems of elastic Stability of Rods and Plates," *Eng. Bull.,* Vol. 19, pp. 897–908.

Hrenikoff, A. (1941). "Solution of Problems in Elasticity by the Framework Method," *Trans. ASME, J. Appl. Mech.,* Vol. 8, pp. 169–175.

Pepper, D. W. and Heinrich, J. C. (1992). *The Finite Element Method:*

---

[*] Current ground test facilities are limited to Mach 8 for testing times up to several minutes. Higher Mach number testing usually lasts only milliseconds.

Basic Concepts and Applications. Washington, D.C.: Hemisphere (Taylor and Francis).

Rayleigh, J. W. S. (1877). *Theory of Sound,* 1st rev. ed., New York: Dover.

Reddy, R. N. (1993). *An Introduction to the Finite Element Method,* 2nd ed., New York: McGraw-Hill.

Ritz, W. (1909). "Uber eine Neue Methode zur Losung Gewisses Variations-Probleme der Mathematischen Physik." *J. Reine Angew. Math.*, Vol. 135, pp. 1–61.

Turner, M., Clough, R. W., Martin, H. H., and Topp, L. (1956). "Stiffness and Deflection Analysis of Complex Structures." *J. Aero. Sci.*, Vol. 23, pp. 805–823.

Wright-Patterson Air Force Base (1966). *Proc. Matrix Meth. Struct. Mech. Conf.* AFFDL-TR-66-80.

Zienkiewicz, O. C. and Taylor, R. L. (1989). *The Finite Element Method,* vol. 1, *Basic Formulation and Linear Problems*. London: McGraw-Hill.

Zienkiewicz, O. C. and Taylor, R. L. (1991). *The Finite Element Method,* Vol. 2, *Solid and Fluid Mechanics, Dynamics and Non-linearity*. London: McGraw-Hill.

TWO

BASIC EQUATIONS OF FLUID DYNAMICS

2.1 OVERVIEW

The FEM can be characterized as a method to find numerical approximations to the solution of differential equations–it requires a mathematical model that represents the physical problem of interest before it can be applied. The finite element solutions that we generate, even if they are exact, will be a good solution only if the mathematical model that we use is a good representation of the physical system.

In fluid mechanics and heat transfer, the basic conservation equations are second-order partial differential equations that can sometimes be reduced to systems of first-order equations or transformed into single higher order equations by means of mathematical manipulations. In this section, we will introduce the basic conservation equations for Newtonian laminar flow that we will use later in the textbook. For more detailed derivations and discussions, the reader is referred to the texts of Batchelor (1967), Bird et al. (1960), Malvern (1969), and White (1974), among many others.

2.2 SUBSTANTIAL DERIVATIVE

The derivative operator

$$\frac{D}{Dt} = \frac{\partial}{\partial t} + u \cdot \nabla \tag{2.1}$$

is called the material derivative, particle derivative, or substantial derivative of a function and it reflects the fact that we are using a representation based on an Eulerian system, which is appropriate for fluid flow.

In Eq. (2.1), the velocity vector field in three-dimensional Cartesian form is

$$\mathbf{u}\,(\mathbf{x}, t) = u\,(x, y, z, t)\,\mathbf{i} + v\,(x, y, z, t)\,\mathbf{j} + w\,(x, y, z, t)\,\mathbf{k} \quad (2.2)$$

where u, v, w are the velocity components in the x-, y- and z-direction respectively, t denotes time, and $\mathbf{i}$, $\mathbf{j}$, and $\mathbf{k}$ are unit vectors in the direction of the x, y, and z axis, respectively.

The divergence operator ∇ in Cartesian coordinates is given by

$$\nabla \equiv \mathbf{i}\frac{\partial}{\partial x} + \mathbf{j}\frac{\partial}{\partial y} + \mathbf{k}\frac{\partial}{\partial z} \quad (2.3)$$

Throughout this book we will also use indicial notation whenever this is more convenient. In this case the spatial coordinates will be denoted by x_1, x_2, x_3. The velocity components become u_1, u_2, u_3, and the unit coordinate vectors will be denoted by $\mathbf{e}_1$, $\mathbf{e}_2$, $\mathbf{e}_3$. Furthermore, repeated indices assume a summation in that index unless stated otherwise. For example, in indicial notation the operator ∇ becomes

$$\nabla \equiv \mathbf{e}_i\frac{\partial}{\partial x_i} \quad (2.4)$$

2.3 MASS CONSERVATION EQUATION

For flows that are nonreactive and do not undergo changes of phase, the most general form of the mass conservation or *continuity* equation is given by

$$\frac{D\rho}{Dt} + \nabla \cdot \rho u = 0 \quad (2.5)$$

where ρ is the fluid density. If the fluid is incompressible, i.e., the density ρ is constant, the rate of change of density vanishes and Eq. (2.5) becomes

$$\nabla \cdot \mathbf{u} = 0 \qquad (2.6)$$

The term $\nabla \cdot \mathbf{u}$ is also called the *dilatation* term and corresponds to the change of volume of a fluid particle. Hence, Eq. (2.6) is equivalent to requiring particles to remain at constant volume.

2.4 NAVIER-STOKES EQUATIONS

The dynamic equations for a fluid particle can be obtained directly from Newton's second law and written in the form

$$\rho \frac{D u_i}{D t} = \frac{\partial}{\partial x_j} \sigma_{ij} + \rho B_i \qquad (2.7)$$

where B_i are the components of applied body forces $\mathbf{B}$ and σ_{ij} are the components of the stress tensor.

For a Newtonian fluid, the stress tensor can be related to the velocity field through the constitutive relation

$$\sigma_{ij} = - p \delta_{ij} + \mu \left(\frac{\partial u_i}{\partial x_j} + \frac{\partial u_j}{\partial x_i} \right) + \lambda \frac{\partial u_k}{\partial x_k} \delta_{ij} \qquad (2.8)$$

In Eq. (2.8), p is the thermodynamic pressure that must reduce to the hydrostatic pressure under conditions of no fluid motion. The function δ_{ij} is the Kronecker delta, i.e., $\delta_{ij} = 1$ if $i = j$ and $\delta_{ij} = 0$ if $i \neq j$. The coefficient μ is the first coefficient of viscosity or *dynamic viscosity*, or simply, viscosity; it represents the stresses arising from the shearing motion of the fluid. The coefficient λ is the second coefficient of viscosity and is associated with volumetric changes; it gives a measure of the energy dissipation that results from a uniform expansion or contraction of the fluid and it is usually called the *coefficient of bulk viscosity*.

In addition, we define the *mean pressure* $\bar{p}$ as the negative of the average of the normal stresses, i.e.,

$$\bar{p} = -\frac{1}{3}\sigma_{ii} = p - \left(\lambda + \frac{2}{3}\mu\right)\nabla\cdot\mathbf{u} \tag{2.9}$$

In order to make the mean pressure equal to the thermodynamic pressure, Stokes (1845) assumed that $\lambda + 2/3\mu = 0$. This is known as Stokes' hypothesis and we will come back to it later when we discuss finite element algorithms for incompressible fluids.

The Navier-Stokes equations for a Newtonian viscous fluid, also known as the momentum equations, can now be written in the form

$$\rho\left(\frac{\partial u}{\partial t} + u\frac{\partial u}{\partial x} + v\frac{\partial u}{\partial y} + w\frac{\partial u}{\partial z}\right) = -\frac{\partial p}{\partial x} + \frac{\partial}{\partial x}\left(2\mu\frac{\partial u}{\partial x}\right) + \frac{\partial}{\partial y}\left[\mu\left(\frac{\partial u}{\partial y} + \frac{\partial v}{\partial x}\right)\right]$$

$$+ \frac{\partial}{\partial z}\left[\mu\left(\frac{\partial u}{\partial z} + \frac{\partial w}{\partial x}\right)\right] + \frac{\partial}{\partial x}\left[\lambda\left(\frac{\partial u}{\partial x} + \frac{\partial v}{\partial y} + \frac{\partial w}{\partial z}\right)\right] + \rho B_x$$

$$\tag{2.10a}$$

$$\rho\left(\frac{\partial v}{\partial t} + u\frac{\partial v}{\partial x} + v\frac{\partial v}{\partial y} + w\frac{\partial v}{\partial z}\right) = -\frac{\partial p}{\partial y} + \frac{\partial}{\partial x}\left[\mu\left(\frac{\partial u}{\partial y} + \frac{\partial v}{\partial x}\right)\right] + \frac{\partial}{\partial y}\left(2\mu\frac{\partial v}{\partial y}\right)$$

$$+ \frac{\partial}{\partial z}\left[\mu\left(\frac{\partial v}{\partial z} + \frac{\partial w}{\partial y}\right)\right] + \frac{\partial}{\partial y}\left[\lambda\left(\frac{\partial u}{\partial x} + \frac{\partial v}{\partial y} + \frac{\partial w}{\partial z}\right)\right] + \rho B_y$$

$$\tag{2.10b}$$

$$\rho\left(\frac{\partial w}{\partial t} + u\frac{\partial w}{\partial x} + v\frac{\partial w}{\partial y} + w\frac{\partial w}{\partial z}\right) = -\frac{\partial p}{\partial z} + \frac{\partial}{\partial x}\left[\mu\left(\frac{\partial u}{\partial z} + \frac{\partial w}{\partial x}\right)\right]$$

$$+ \frac{\partial}{\partial y}\left[\mu\left(\frac{\partial v}{\partial z} + \frac{\partial w}{\partial y}\right)\right] + \frac{\partial}{\partial z}\left(2\mu\frac{\partial w}{\partial z}\right) + \frac{\partial}{\partial z}\left[\lambda\left(\frac{\partial u}{\partial x} + \frac{\partial v}{\partial y} + \frac{\partial w}{\partial z}\right)\right] + \rho B_z$$

$$\tag{2.10c}$$

or using indicial notation

$$\rho\frac{Du_i}{Dt} = -\frac{\partial p}{\partial x_i} + \frac{\partial}{\partial x_j}\left[\mu\left(\frac{\partial u_i}{\partial x_j} + \frac{\partial u_j}{\partial x_i}\right) + \lambda\frac{\partial u_k}{\partial x_k}\delta_{ij}\right] + \rho B_i \quad (2.11)$$

If we replace λ by $-2/3\mu$ in the above equation, we obtain the usual form for compressible flow. If the flow is incompressible, and using Eq. (2.6), Eq. (2.11) reduces to

$$\rho\frac{Du_i}{Dt} = -\frac{\partial p}{\partial x_i} + \frac{\partial}{\partial x_j}\left(\mu\frac{\partial u_i}{\partial x_j}\right) + \rho B_i \qquad (2.12)$$

Further simplifications are possible if more assumptions are made, such as constant viscosity, and unidirectional flow. However, this is not desirable at this point. We will use Eq. (2.11) and Eq. (2.12) as our starting point when dealing with compressible and incompressible flows, respectively.

For compressible flows an equation of state that relates the pressure and the density is required. Throughout this book we will assume that the ideal gas equation is valid, as given by the relation

$$p = \rho RT \qquad (2.13)$$

where R is the gas constant and T denotes temperature.

2.5 EQUATION OF ENERGY CONSERVATION

The energy equation follows from the first law of thermodynamics. In terms of the internal energy of a fluid particle, conservation of energy is

$$\rho\frac{De}{Dt} = -\nabla\cdot\mathbf{q} + \sigma_{ij}\frac{\partial u_i}{\partial x_j} + f \qquad (2.14)$$

In Eq. (2.14), e denotes the internal energy per unit mass. The heat transfer $\mathbf{q}$ is assumed to be given by Fourier's law

$$\mathbf{q} = -k\,\nabla T \qquad (2.15)$$

where k is the material's conductivity and f is a volumetric heat source or sink. For a calorically perfect fluid, we can relate the internal energy to temperature through the thermodynamic equation

$$de = c_V \, dT \qquad (2.16)$$

where c_V is the specific heat at constant volume. With this assumption, Eq. (2.14) becomes[*]

$$\rho c_v \frac{DT}{Dt} = \nabla \cdot k \nabla T + \sigma_{ij} \frac{\partial u_i}{\partial x_j} + f \qquad (2.17)$$

For an incompressible fluid, it is customary to rewrite Eq. (2.17) in the form

$$\rho c_v \frac{DT}{Dt} = \nabla \cdot k \nabla T + \phi + f \qquad (2.18)$$

where ϕ is the viscous heat dissipation function and is defined as

$$\phi = \left[\mu \left(\frac{\partial u_i}{\partial x_j} + \frac{\partial u_j}{\partial x_i} \right) + \lambda \frac{\partial u_k}{\partial x_k} \delta_{ij} \right] \frac{\partial u_i}{\partial x_j} \qquad (2.19)$$

In many applications involving incompressible fluids, the dissipation term is small and is neglected. However, there are some incompressible flows, such as many lubricating film flows in which viscous heat dissipation is important.

2.6 MASS TRANSPORT

In a great variety of practical situations, we need to consider multicomponent systems in which the transport of individual components must be described. If we consider only the case of

---

[*] For gases flowing with constant density, c_V must be replaced by c_p, the heat capacity at constant pressure.

incompressible flow and ignore interactive diffusion effects (Dufour and Soret effects), as is often the case in atmospheric dispersion, sediment transport, and double-diffusive mixtures, the conservation equations for the transported quantity can be written as

$$\frac{Dc}{Dt} = -\nabla \cdot \mathbf{G} + S \tag{2.20}$$

where c is the mass fraction of the material under consideration and $\mathbf{G}$ is the mass flux vector. Using Fick's law of diffusion, the mass flux vector can be expressed in terms of the local concentration as

$$\mathbf{G} = -D\nabla c \tag{2.21}$$

where D is the mass diffusion coefficient. The function S represents sources and/or sinks.

For a particular component c, in a binary system, Eq. (2.20) can be written as[*]

$$\frac{\partial c}{\partial t} + u_i \frac{\partial c}{\partial x_i} = \frac{\partial}{\partial x_i}\left(D \frac{\partial c}{\partial x_i}\right) + S \tag{2.22}$$

In general, one equation of the form of Eq. (2.22) is needed for each component in the fluid. The coupling between the equations can be quite complicated, particularly in the case of flows with chemical reactions. However, the single form given by Eq. (2.22), or slight modification to it, usually suffices to correctly model a great variety of practical situations involving mass transport.

2.7 WAVE EQUATION

There is a variety of physical processes that are governed by wave propagation equations, such as acoustic, free surface, electromagnetic waves, and structural vibrations. We will write it in the form

---

[*] External forces, pressure gradients and temperature gradients also contribute to mass diffusion, but their effect is usually small, and we will neglect it. More complete expressions for the diffusion flux are given by Bird et al. (1960).

$$\rho \frac{\partial^2 \phi}{\partial t^2} + d\, \frac{\partial \phi}{\partial t} = \nabla \cdot \mathbf{K} \nabla \phi + F(\phi, \dot{\phi}) \qquad (2.23)$$

where d is a viscous damping coefficient and $\mathbf{K}$ has a different meaning, depending on the type of problem. For example, $\mathbf{K}$ is the tension if Eq. (2.23) models the vibrations of a membrane, and is the structural stiffness in the case of structural vibrations. F is a forcing term and is nonlinear in general.

Even though our emphasis will be on the solution of parabolic equations, the importance of the wave equation is significant. We feel that is it important to discuss how the numerical methods that we develop can be applied to the solution of Eq. (2.23), expanding the range of applications of the algorithms considerably.

2.8 BOUNDARY CONDITIONS

The conservation equations described in the previous sections, with the exception of the continuity equation, Eq. (2.5), are all second-order partial differential equations that require boundary conditions in order to make the formulations mathematically closed. The boundary conditions must also be physically realistic and are therefore dependent on the particular geometry, the materials involved and the values of the pertinent parameters. For example, at a solid boundary it is appropriate to require a liquid to have the same velocity as the solid. This is the so-called, *no-slip* boundary condition. However, the no-slip boundary condition is only valid when the continuum hypothesis is justified; it is not a realistic boundary condition at a solid boundary if the fluid under consideration is a gas with a large mean free path. In this case there is a slip velocity of the gas relative to the solid boundary.

In general, the pressure is not required to satisfy boundary conditions, but we must prescribe a reference value for it to be determined uniquely. In some flow problems the known pressure at a free surface provides the equilibrium condition along that interface, or the prescribed pressure at an open fluid boundary is the driving force in the system. In these cases, we may prescribe the pressure along these boundaries.

The importance of the use of physically meaningful boundary conditions in numerical simulations cannot be overstressed. Lack of attention to properly defining boundary conditions is an important source of error. We will pay special attention to this issue throughout this book. However, owing to the large variety of situations that can be encountered, we can only discuss the boundary conditions in the context of each particular situation as they arise. White (1974) treats a variety of boundary conditions for fluid flow problems.

2.9 CLOSURE

In this chapter, we have established the basic equations of fluid dynamics around which this book is centered. There are many variations and simplifications of these equations that lead to other equations no less important than the ones considered here. However, as we develop finite element methods to deal with these basic equations, it should also become clear how these methods can be applied to many other differential equations or sets of differential equations with appropriate modifications.

The equations presented in this section are quite formidable. In general, they are nonlinear coupled equations; their solutions are by no means simple. To be able to develop good universal methods for their solutions, a great deal of help can be obtained from the general theory of partial differential equations and an understanding of the behavior of the solutions of model equations such as Laplace's equation, the wave equation and the heat equation. For the reader who is not familiar with the basic character of partial differential equations, we strongly encourage some fundamental reading on the subject. The works by Carrier and Pearson (1976) and Greenberg (1978) are among the many in this area.

REFERENCES

Batchelor, G. K. (1967) *An Introduction to Fluid Dynamics.* Cambridge, UK: Cambridge University Press.

Bird, R. B., Stewart, W. E., and Lightfoot, E. N. (1960) *Transport Phenomena.* New York: John Wiley and Sons.

Carrier, G. F. and Pearson, C. E. (1976) *Partial Differential Equations.* New York: Academic Press.

Greenberg, M. D. (1978) *Foundations of Applied Mathematics.* Englewood Cliffs, N.J.: Prentice-Hall.

Malvern, L. E. (1969) *Introduction to the Mechanics of a Continuous Medium*. Englewood Cliffs, N.J.: Prentice-Hall, Inc.

Stokes, G. G. (1845) "On the Theories of Internal Friction of Fluids in Motion," *Trans. Cambridge Phil. Soc.*, Vol. 8, pp. 287–295.

White, F. W. (1974) *Viscous Fluid Flow*. New York: McGraw-Hill.

THREE

FUNDAMENTAL CONCEPTS

3.1 OVERVIEW

In this chapter we develop the fundamental concepts that lead to the form of the FEM we will apply to model fluid flow and heat transfer, and many other physical phenomena. The concepts are established using model problems in heat conduction taken from practical applications.

The mathematical theory is kept to the minimum necessary in order to put the FEM on a rigorous footing without neglecting the fundamental results on convergence. The emphasis in this book lies in the physical aspects of the problem and how they relate to the mathematical concepts. It is well known that the FEM has a strong mathematical basis. However, our premise is that in numerical simulations of engineering problems it is more important to be able to determine whether the physics of the problem at hand are captured correctly by the numerical model than to know how to prove convergence of the method. The latter we leave to any of the excellent books that have already been written about the mathematical aspects of the FEM.

3.2 LINEAR HEAT CONDUCTION

Let us consider the following model problem. We want to calculate the temperature distribution on an electronic package consisting of a heat-producing silicon chip sitting on a ceramic substrate with its bottom in

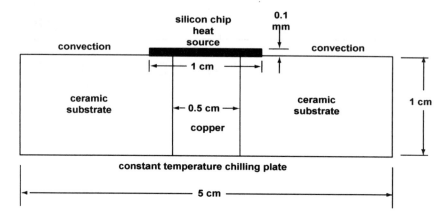

Figure 3.1 Heat conduction in an electronic package.

contact with a chilling plate. In the center of the substrate a copper plate is added to enhance heat extraction to the chill plate. Figure 3.1 shows the geometry, boundary conditions and typical dimensions. This configuration is very common in modern electronic components.

We will make several simplifying assumptions:

1. Gradients in the z direction are small; thus we may assume that the problem is two-dimensional.

2. Different materials are in perfect contact, i.e., we will not take into account the effect of bonding materials or imperfect interfaces.

3. The bottom plate is at a constant temperature.

4. The vertical boundaries are perfectly insulated.

5. The top boundary air interface is at a constant (ambient) temperature.

6. No radiation effects are included.

7. It is a steady state situation.

The equations governing the example problem can be written for one half of the region using symmetry about the mid-vertical line as depicted in Fig. 3.2. The basic equation is

$$-\nabla \cdot k\nabla T = f \tag{3.1}$$

over the open region Ω called the *domain* of the differential equation, which is bounded by the lines $\overline{AB}, \overline{BD}, \overline{DE}, \overline{EF}, \overline{FG},$ and $\overline{GA}$ which

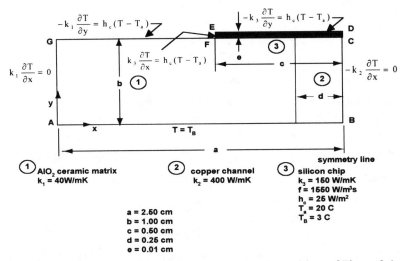

Figure 3.2 Mathematical model for the heat transfer problem of Figure 3.1.

together constitute the *boundary* Γ of Ω. The source term f represents the heat generated by the silicon chip, but f = 0 in the copper and the ceramic components.

The boundary conditions associated with Eq. (3.1) are

$$T = T_B \text{ on } \overline{AB} \qquad (3.2)$$

$$-k_i \frac{\partial T}{\partial x} = 0 \text{ on } \overline{BC}, \ \overline{CD} \text{ and } \overline{GA} \qquad (3.3)$$

$$-k_i \frac{\partial T}{\partial y} = h_c(T - T_a) \text{ on } \overline{DE} \text{ and } \overline{FG} \qquad (3.4)$$

and

$$k_i \frac{\partial T}{\partial x} = h_c(T - T_a) \text{ on } \overline{EF} \qquad (3.5)$$

where i = 1 for $\overline{FG}$ and $\overline{GA}$; i = 2 for $\overline{BC}$ and i = 3 for $\overline{CD}$, $\overline{DE}$ and $\overline{EF}$.

Conditions of the type of Eq. (3.2), in which the dependent variable is assigned a prescribed value, are called *Dirichlet*, or *essential*, boundary conditions, and we will denote by Γ_1 the union of all those portions of the boundary where a condition of this type is prescribed. In this case, the only place where a fixed temperature is given is in that part of the component in contact with the chilling plate; hence $\Gamma_1 = \overline{AB}$.

A condition of the type of Eq. (3.3), in which the heat flux normal to the boundary is prescribed, is called a *Neumann*, or *natural*, boundary condition. In this case the heat flux normal to the left wall $\overline{GA}$ must be zero because the wall was assumed to be perfectly insulated and along the symmetry line no heat flux is possible either. The union of all boundary segments where a Neumann boundary condition is prescribed will be denoted by Γ_2; in this case, $\Gamma_2 = \overline{BC} \cup \overline{CD} \cup \overline{GA}$, where $\cup$ is the union set operator.

Boundary conditions of the type of Eq. (3.4) and Eq. (3.5), in which both the heat flux normal to the boundary and the unknown temperature at the boundary appear are called *mixed*, or Robbins, boundary conditions. This type of boundary condition models the normal heat flux from the solid body to a fluid medium at a temperature T_a through a convective heat transfer coefficient h_c. Also, radiation boundary conditions can sometimes be linearized and modeled in this fashion with an appropriate heat transfer coefficient h_{rad}. The union of all segments over which a mixed boundary condition is applied will be denoted by Γ_3. In this case, $\Gamma_3 = \overline{DE} \cup \overline{EF} \cup \overline{FG}$. It should be clear that the union $\Gamma_1 \cup \Gamma_2 \cup \Gamma_3 = \Gamma$ and the intersections $\Gamma_i \cap \Gamma_j = \phi$, $i \neq j$ where ϕ is the empty set. Here the symbol $\cap$ denotes the intersection set operator; because the intersections are empty when $i \neq j$, the boundary segments cannot overlap but must cover the whole boundary.

We will also require a simpler one-dimensional model, particularly when dealing with theoretical aspects of the FEM. Consider a thin metal wire. We assume that the metal wire (or rod) is surrounded by insulating material so that heat can only conduct in the axial direction. We further assume that we can run an electrical current through the wire that acts as a heat source f. Figure 3.3 depicts the model problem. The governing equation for this problem is

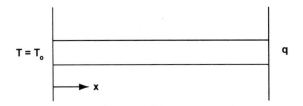

Figure 3.3 Heat conduction in a thin rod.

$$-\frac{d}{dx}\left(k\frac{dT}{dx}\right) = f \qquad \Omega = \{x/0 < x < L\} \qquad (3.6)$$

The left end is kept at a fixed temperature T_0, i.e.,

$$T(0) = T_0 \qquad (3.7)$$

and hence in this case, $\Gamma_1 = \{0\}$. At the right end we have a prescribed heat flux q,

$$-k\frac{\partial T}{\partial x}\bigg|_{x=L} = q \qquad (3.8)$$

and we have $\Gamma_2 = \{L\}$. Clearly, in this case, $\Gamma_3 = 0$.

These well-defined examples will be used to introduce and discuss the basis of the FEM. We will modify them as the need arises to incorporate other aspects, such as time dependency, or transport modes such as radiation and advection.

3.3 LINEAR OPERATORS AND LINEAR FUNCTION SPACES

In a more general mathematical notation, we interpret the differential equations Eqs. (3.1) and (3.6) as linear operators that act on functions $u(x)$ defined over the corresponding domain Ω, and that satisfy a set of boundary conditions on Γ; see Appendix A. We can then write

$$Lu = f \qquad x \in \Omega \qquad (3.9)$$

$$u = g_1 \qquad x \in \Gamma_1 \qquad (3.10)$$

$$k\frac{\partial u}{\partial n}(x) = g_2(x) \quad x \in \Gamma_2 \tag{3.11}$$

$$k\frac{\partial u}{\partial n}(x) = h(x) \cdot u(x) + g_3(x) \quad x \in \Gamma_3 \tag{3.12}$$

where L is typically a second-order differential operator.

Clearly, Eq. (3.11) is a special case of Eq. (3.12), where $h \equiv 0$. However, each of these kinds of boundary conditions is important enough in practice to deserve individual treatment.

The normal derivative $\partial u/\partial n$ in Eqs. (3.11) and (3.12) is defined as

$$\frac{\partial u}{\partial n} = \mathbf{n} \cdot \nabla u = n_i \frac{\partial u}{\partial x_i} \tag{3.13}$$

where $\mathbf{n}$ denotes the unit vector normal to the boundary Γ and pointing away from the region Ω. For example, in Eq. (3.4) the boundary segment $\overline{DE}$ has unit outward normal

$$\mathbf{n} = \mathbf{j}$$

Hence

$$\frac{\partial T}{\partial n} = \frac{\partial T}{\partial y}$$

and Eq. (3.12) becomes Eq. (3.4) if we set

$$h = h_c$$

and

$$g_3 = h_c T_a$$

Example 3.1

In the case of our one-dimensional example, the operator L is given by

$$L \equiv -\frac{d}{dx}\left(k\frac{d}{dx}\right) \tag{3.14}$$

The operator L is linear because of the linearity of the derivative. Furthermore, Eq. (3.10) must be satisfied on Γ_1 with $g_1 = T_o$ and Eq. (3.11) on Γ_2 with $g_2 = q/k$.

In our two-dimensional example we have

$$L \equiv -\nabla \cdot k\nabla = -\frac{\partial}{\partial x}\left(k\frac{\partial}{\partial x}\right) - \frac{\partial}{\partial y}\left(k\frac{\partial}{\partial y}\right) \tag{3.15}$$

The boundary conditions can be easily expressed in the form of Eqs. (3.10) – (3.12) and the linearity of the operator L is again clear.   ∎

We now introduce the classes of real functions over which the linear operators act and those that will be needed to find our finite element solutions. In this section we will only define the necessary function spaces. We will explain the rationale behind the choice of these vector spaces as the need arises.

1. Given a domain Ω, we say that a function is in $C^n(\Omega)$ if it has continuous derivatives of up to order n in Ω. $C^0(\Omega)$ denotes the space of functions that are continuous in Ω, and $C^{-1}(\Omega)$ denotes the space of piecewise continuous functions in Ω.

2. The space $L^2(\Omega)$ is the space of functions defined in Ω that are square integrable over Ω, i.e.,

$$L^2(\Omega) = \left\{ f(x) \,/\, \int_\Omega (f(x))^2 \, d\Omega < \infty \right\} \tag{3.16}$$

Example 3.2

As an example, in our one-dimensional problem, $\Omega = \{x \,/\, 0 < x < L\}$ and for any integer n, the function $f_n(x) = x^n$ is in L^2 only if $n \geq 0$. Moreover, for $n \geq 0$ the function $f_n(x)$ is in C^∞ because all its derivatives are continuous in Ω.   ∎

3. The Sobolev space H^1 is the space of functions defined over Ω such that both the function and all its first partial derivatives are in $L^2(\Omega)$. For example, in a two-dimensional problem we have

$$H^1(\Omega) = \left\{ f(x,y) / \int_\Omega \left[f^2 + \left(\frac{\partial f}{\partial x} \right)^2 + \left(\frac{\partial f}{\partial y} \right)^2 \right] d\Omega < \infty \right\} \quad (3.17)$$

On the other hand, the Sobolev space $H^1(\Omega)$ associated with the one-dimensional equation, Eq. (3.6), is given by

$$H^1(\Omega) = \left\{ f(x) / \int_0^L \left[f^2 + \left(\frac{d f}{d x} \right)^2 \right] dx < \infty \right\} \quad (3.18)$$

4. Finally, we will need a subspace of the Sobolev space $H^1(\Omega)$ defined for each particular problem by the functions of H^1 that vanish in the portion Γ_1 of the boundary where Dirichlet boundary conditions are imposed. This subspace is denoted by $H_0^1(\Omega)$ and is characterized as

$$H_0^1(\Omega) = \left\{ f / f \in H^1(\Omega) \text{ and } f(\mathbf{x}) = 0 \text{ if } \mathbf{x} \in \Gamma_1 \right\} \quad (3.19)$$

In other words, $H_0^1(\Omega)$ is the set of functions that satisfy homogeneous Dirichlet conditions in Γ_1. In the problem shown in Fig. 3.2, the space $H_0^1(\Omega)$ consists of those functions defined in Ω that vanish along the line $\overline{AB}$.

We will provide some more illustrative examples through exercises later in the text.

3.4 WEIGHTED RESIDUALS FORMULATION

We will now reformulate the basic problem given by Eqs. (3.9) – (3.12) in a way appropriate for the application of the FEM. Given a differential equation of the form of Eq. (3.9), we define the *residual* function as

$$R(u,x) \equiv Lu(x) - f(x) \tag{3.20}$$

It then follows that if u* is the solution to the differential equation given by Eq. (3.9), we must have $R(u^*, x) \equiv 0$. However, if u is only an approximation to the solution of Eq. (3.9), the residual provides a measure of the error in the satisfaction of the equation.

The residual function for Eq. (3.1) becomes

$$R(u,x) = -\frac{\partial}{\partial x}\left(k\frac{\partial u}{\partial x}\right) - \frac{\partial}{\partial y}\left(k\frac{\partial u}{\partial y}\right) - f(x) \tag{3.21}$$

and similarly for Eq. (3.6).

A function u*(**x**) that has second-order derivatives in Ω and satisfies Eqs. (3.9) – (3.12) is called a *classical* solution to the differential equation.

If we now multiply Eq. (3.20) by a weighting function w defined over Ω, integrate over Ω, and set the integral equal to zero, i.e.,

$$\int_\Omega w(x)R(u,x)d\Omega = \int_\Omega w(Lu - f)d\Omega = 0 \tag{3.22}$$

we obtain what is called the *weighted residuals* form of Eq. (3.9). It can be proved (Finlayson, 1972) that if for a fixed u, Eq. (3.22) is satisfied for all w in $H_0^1(\Omega)$, then we must have u = u*, i.e., u must be the solution to Eq. (3.9). However, Eq. (3.9) requires the second derivative of u to exist in Ω. This condition is very restrictive, and if we enforce it, we cannot find solutions to many important equations. For example, examine the solution to

$$-\frac{d^2u}{dx^2} = \delta(x - \zeta) \qquad a < x, \quad \zeta < b \tag{3.23}$$

where ζ is a parameter. Clearly, the second derivative of u does not exist at x = ζ; however, solutions to equations of the form of Eq. (3.23) exist and are particularly important in mechanics (see Exercise 3.7).

To explain how we resolve these difficulties, let us first consider our one-dimensional problem given by Eqs. (3.6) – (3.8). The weighted residual form is

$$\int_0^L w \left(-\frac{d}{dx}\left(k\frac{dT}{dx}\right) - f \right) dx = 0 \tag{3.24}$$

If we integrate the second derivative term by parts, we obtain

$$\int_0^L \left(k\frac{dw}{dx}\frac{dT}{dx} - wf \right) dx - \left[w\left(-k\frac{dT}{dx}\right)\right]_{x=0}^{x=L} = 0 \tag{3.25}$$

Notice that Eq. (3.24) and Eq. (3.25) express the exact same condition. However, Eq. (3.25) contains only first-order derivatives of the unknown function. Hence the differentiability requirements have been relaxed or weakened. For this reason Eq. (3.25) is called the *weak form* of Eq. (3.6).

So far we have not introduced any conditions for the weighting functions w, which remain very general. If we set w = T in Eq. (3.25), we see that in order for Eq. (3.25) to be meaningful, the integrals

$$\int_0^L k\left(\frac{dw}{dx}\right)^2 dx \qquad \int_0^L k\left(\frac{dT}{dx}\right)^2 dx$$

must exist. Hence both w and T must be in $H^1(\Omega)$. Therefore the Sobolev space $H^1(\Omega)$ is obtained naturally from the condition that the integrals be well defined. We mentioned before that w must be further restricted to be in the subspace $H_0^1(\Omega)$. To show this, we have to work a little bit harder. Let T = T*, the exact solution to Eqs. (3.6) – (3.8), and replace it in Eq. (3.25), which we rewrite in the form

$$\int_0^L \left(k\frac{dw}{dx}\frac{dT^*}{dx} - wf \right) dx - \left[w\left(-k\frac{dT^*}{dx}\right)\right]_{x=L} + \left[w\left(-k\frac{dT^*}{dx}\right)\right]_{x=0} = 0$$

Using Eq. (3.8), to incorporate the natural boundary condition at the right end, we get

$$\int_0^L \left(k \frac{dw}{dx} \frac{dT^*}{dx} - wf \right) dx - w(L) \, q + \left[w \left(-k \frac{dT^*}{dx} \right) \right]_{x=0} = 0 \quad (3.26)$$

Now we choose a family of functions $w_n(x)$ in $H^1(\Omega)$ such that $w_n(x)$ does not vanish at $x = 0$. A convenient set of functions is given by

$$w_n(x) = e^{-nx^2} \qquad\qquad (3.27)$$

The functions $w_n(x)$ and their derivatives $w'_n(x)$ are depicted in Fig. 3.4 for the interval $0 \le x \le 1$. We can observe in Fig. 3.4a that $w_n(0) = 1$ for all values of n and that as n increases in value, the functions decay very rapidly to zero. In fact, we note that

$$w_\infty(x) = \lim_{n \to \infty} w_n = \begin{cases} 1 & \text{if } x = 0 \\ 0 & \text{if } x > 0 \end{cases}$$

We also observe from Fig. 3.4b that the derivatives are bounded and approach zero rapidly as n increases, i.e.,

$$\lim_{n \to \infty} w'(x) = 0 \qquad \forall \, x \ge 0$$

It is clear that the functions w' and w_∞ are in the space $H^1(\Omega)$ and are admissible functions (although a formal proof is quite cumbersome).

If we now make $w = w_n$ in Eq. (3.26), we can choose n large enough that the first two terms in Eq. (3.26) can be made arbitrarily small. In fact, in the limit when $w = w_\infty$, which is also an admissible function those terms are zero, and Eq. (3.26) reduces to

$$-k \frac{dT^*}{dx} \Big|_{x=0} = 0$$

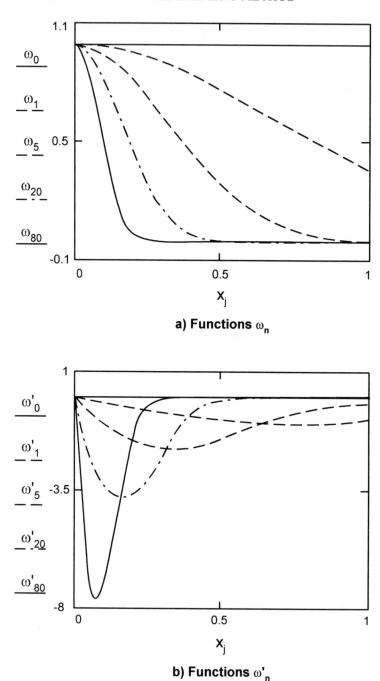

Figure 3.4 Test functions of Eq. (3.27).

This is not necessarily true, however, since $dT*/dx|_{x=0}$ can assume any value, depending on the boundary conditions. Moreover, the left-hand side of Eq. (3.26) must be identically zero because T* is the exact solution to the problem. This contradiction shows that the weighting functions w(x) must vanish at x = 0 and hence must be in the subspace $H_0^1(\Omega)$.

It is not difficult to extend the above results to two- and three-dimensional situations and to show that the weighting functions must vanish over the portion Γ_1 of the boundary where Dirichlet conditions are imposed. We now state the above results for the two-dimensional problem defined by Eqs. (3.9) – (3.12) in the case when the operator L is given by Eq. (3.15).

The weak formulation for the above boundary value problem is as follows. Find a function u* in the space $H^1(\Omega)$ such that it satisfies the Dirichlet boundary condition, $u(x) = g_1(x) \text{ in } \Gamma_1$ and

$$\int_\Omega \left[k \frac{\partial w}{\partial x} \frac{\partial u}{\partial x} + k \frac{\partial w}{\partial y} \frac{\partial u}{\partial y} - wf \right] d\Omega$$

(3.28)

$$+ \int_{\Gamma_2} wg_2 d\Gamma_2 + \int_{\Gamma_3} w(hu + g_3) d\Gamma_3 = 0$$

for all functions w in $H_0^1(\Omega)$.

The function u* that satisfies the above formulation[*] is called a *weak* or *generalized* solution to Eqs. (3.9) – (3.12). We can prove (Hughes, 1987) that the weak solution in Eq. (3.28) satisfies Eqs. (3.9) – (3.12) and vice versa. Notice that the conditions on u* have been weakened as compared to the classical solution. We no longer require a second derivative to exist everywhere in Ω but only that the first partial derivatives be in the $L^2(\Omega)$ space.

To illustrate the procedure, let us derive the form of Eq. (3.28) for the boundary value problem of Fig. 3.2. Set the residual function

---

[*]It is easy to prove that if L is linear and the only solution to the equation Lu = 0 is u = 0, then the solution u* is unique; see Greenberg (1978).

given by Eq. (3.1), which is multiplied by a weighting function w($\mathbf{x}$) and integrated over Ω, to zero, i.e.,

$$-\int_\Omega \left[w \left(\frac{\partial}{\partial x} \left(k \frac{\partial T}{\partial x} \right) + \frac{\partial}{\partial y} \left(k \frac{\partial T}{\partial y} \right) \right) + wf \right] d\Omega = 0 \qquad (3.29)$$

Using the differentiation formula (no summation implied)

$$\frac{\partial}{\partial x_i} \left(v \frac{\partial u}{\partial x_i} \right) = v \frac{\partial^2 u}{\partial x_i^2} + \frac{\partial v}{\partial x_i} \frac{\partial u}{\partial x_i} \qquad (3.30)$$

Eq. (3.29) can be rewritten in the form

$$\int_\Omega \left[k \frac{\partial w}{\partial x} \frac{\partial T}{\partial x} + k \frac{\partial w}{\partial y} \frac{\partial T}{\partial y} - wf \right] d\Omega$$

$$\qquad (3.31)$$

$$-\int_\Omega \left[\frac{\partial}{\partial x} \left(kw \frac{\partial T}{\partial x} \right) + \frac{\partial}{\partial y} \left(kw \frac{\partial T}{\partial y} \right) \right] d\Omega = 0$$

We now apply Green's theorem in the plane* to the second integral in Eq. (3.31). The theorem states that if Ω is a *connected* region with boundary Γ and M(x,y), N(x,y) are differentiable functions defined in Ω, then

$$\int_\Omega \left(\frac{\partial N}{\partial x} - \frac{\partial M}{\partial y} \right) d\Omega = \int_\Gamma (N\,dy + M\,dx) \qquad (3.32)$$

The second integral in Eq. (3.31) can then be rewritten in the form

$$\int_\Omega \left[\frac{\partial}{\partial x} \left(kw \frac{\partial T}{\partial x} \right) + \frac{\partial}{\partial y} \left(kw \frac{\partial T}{\partial y} \right) \right] d\Omega = \int_\Gamma \left(kw \frac{\partial T}{\partial x} dy - kw \frac{\partial T}{\partial y} dx \right)$$

$$\qquad (3.33)$$

---

*For three-dimensional problems, we will invoke the Gauss divergence theorem to obtain an equivalent result. Those readers needing to review the basic theorems of vector calculus are referred to the book by Greenberg (1978).

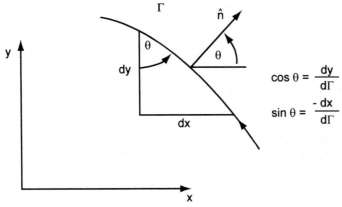

Figure 3.5 Unit normal to surface.

Using the fact that the components of the unit outward normal to Γ, as depicted in Fig. 3.5, are

$$n_x = \frac{dy}{d\Gamma} \qquad n_y = -\frac{dx}{d\Gamma}$$

the line integral in Eq. (3.33) becomes

$$\int_\Gamma \left(kw\frac{\partial T}{\partial x}dy - kw\frac{\partial T}{\partial y}dx \right) = \int_\Gamma kw\frac{\partial T}{\partial n}d\Gamma \qquad (3.34)$$

Hence, Eq. (3.32) can be rewritten as

$$\int_\Omega \left[k\frac{\partial w}{\partial x}\frac{\partial T}{\partial x} + k\frac{\partial w}{\partial y}\frac{\partial T}{\partial y} - wf \right]d\Omega + \int_\Gamma w\left(-k\frac{\partial T}{\partial n} \right)d\Gamma = 0 \qquad (3.35)$$

Example 3.3

Recall that (Fig. 3.2) $\Gamma_1 = \overline{AB}$, $\Gamma_2 = \overline{BC}\cup\overline{CD}\cup\overline{GA}$, and $\Gamma_3 = \overline{DE}\cup\overline{EF}\cup\overline{FG}$. Using the additive property of the integral, we can split the line integral as

$$\int_\Gamma = \int_{\Gamma_1} + \int_{\Gamma_2} + \int_{\Gamma_3}$$

Moreover, if we recall that w must vanish over Γ_1, and that the heat flux is zero in Γ_2, the line integral over the portions Γ_1 and Γ_2 of the boundary must vanish. Hence we have

$$\int_\Gamma w\left(-k\frac{\partial T}{\partial n}\right)d\Gamma = \int_{\Gamma_3} w\left(-k\frac{\partial T}{\partial n}\right)d\Gamma$$

Using Eqs. (3.4) and (3.5) we see that

$$\int_{\Gamma_3} w\left(-k\frac{\partial T}{\partial n}\right)d\Gamma = \underset{DEFG}{\int} w\,h_c(T-T_a)d\Gamma \qquad\blacksquare$$

The weak form of Eqs. (3.1) – (3.5) is then as follows.

Find the function T in $H^1(\Omega)$ such that $T = T_B$ in $\Gamma_1 = \overline{AB}$ and such that

$$\int_\Omega \left[k\frac{\partial w}{\partial x}\frac{\partial T}{\partial x} + k\frac{\partial w}{\partial y}\frac{\partial T}{\partial y} - wf\right]d\Omega + \int_{\Gamma_3} w\,h_c(T-T_a)d\Gamma = 0 \quad (3.36)$$

for all functions w in $H_0^1(\Omega)$, where Γ_3 is the polygonal line $\overline{DEFG}$ in Fig. 3.2.

REMARKS

Only the Dirichlet boundary condition $T = T_a$ on Γ_1 was required to be satisfied by T, hence, the name *essential* boundary condition. The boundary conditions given by Eqs. (3.3) – (3.5) were imposed in the integral representation of the differential equation by replacing the prescribed values of the boundary fluxes in the line integral of the weak form. These conditions will therefore be satisfied in the integral or weak sense and are called *natural* boundary conditions because they can be naturally inserted in Eq. (3.35). As a consequence, it will be very easy to deal with any boundary condition involving a flux at the boundary.

A well-posed boundary value problem requires that over every portion of the boundary either the unknown function or its normal derivative must be prescribed. When the former is the case and a Dirichlet boundary condition is given, such as $T(0) = T_0$ in our one-dimensional example, the flux normal to the boundary can be calculated after the solution T has been obtained. However, because flux boundary conditions are natural, they are not exactly satisfied by the approximation to the solution, and direct differentiation of the numerical solutions can yield very poor approximations to the fluxes. On the other hand, natural boundary conditions are introduced directly into the weak formulation and are therefore satisfied in the weak sense. Thus it is more natural to recover the fluxes from the boundary residuals. In our one-dimensional example, we can write

$$-k\frac{dT}{dx}\bigg|_{x=0} = \frac{1}{w(0)}\left[w(L)q - \int_0^L \left(k\frac{dw}{dx}\frac{dT}{dx} - wf\right)dx\right] \quad (3.37)$$

However, for the expression to be meaningful, we must have $w(0) \neq 0$. This is not a problem though because everything on the right hand side of Eq. (3.37), with the exception of $w(0)$, is known and T has already been properly calculated. This is therefore an independent a posteriori calculation. We will see in sections 3.6.4 and 3.7.2 that by choosing w in Eq. (3.37) in such a way that $w(0) \neq 0$, and the functions $w(x)$ are a consistent extension of those used to obtain T in the weak formulation, we can develop a practical way to obtain accurate approximations to the heat fluxes normal to the boundaries.

Weak weighted residual formulations for any second–order linear differential operator can be obtained in the manner described above. Nonlinear problems must be treated on a case by case basis, but we can always generate a weak form. We will deal with a variety of different situations throughout this book, particularly when we reach the applications.

3.5 GALERKIN METHOD

Throughout this book, we will construct finite element algorithms based on the Galerkin method (Galerkin, 1915) and its generalization, the Petrov-Galerkin method (Andersen and Mitchell, 1979). We will consider only those aspects of the Galerkin method necessary for the development of our finite element algorithms. For a more comprehensive discussion of Galerkin methods, refer to Fletcher (1984).

The basic difficulty with the solution of the weak form of the problems stated in the previous section is that the spaces $H_0^1(\Omega)$ and $H^1(\Omega)$ are infinite dimensional. The idea of the Galerkin method involves two steps:

1. Replace the Sobolev Space $H^1(\Omega)$ by a finite dimensional subspace $H^1(n,\Omega)$ of dimension n, and $H_0^1(\Omega)$ by the corresponding subspace $H_0^1(n,\Omega)$.

2. Choose a set of functions $\{\phi_i\}_{i=1}^n$ in $H_0^1(\Omega)$ that constitute a basis for $H^1(n,\Omega)$, i.e., if u is in $H^1(n,\Omega)$ then u can be uniquely expressed as a linear combination of the functions ϕ_i in the form

$$u = c_1\phi_1 + c_2\phi_2 + \ldots + c_n\phi_n \tag{3.38}$$

and set the weighting functions w equal to the functions ϕ_i.

The unknown coefficients c_i, i = 1, . . ., n, are obtained from the weighted residuals formulation

$$\int_\Omega w_i R(c_j, \phi_j) d\Omega = 0 \tag{3.39}$$

where $w_i = \phi_i$ for i = 1, . . , n. Clearly, Eq. (3.39) gives a system of n linear algebraic equations in the n unknown parameters c_i, i = 1, . ., n. Solution of this system of equations yields the Galerkin approximation to the solution of our problem.

In Eq. (3.39) the functions w_i are called *test* functions, the functions u are called *trial* functions, and the functions ϕ_i are called *shape* functions. A Galerkin method is one in which the test functions are equal to the trial functions. It is obvious also that given a problem of the form of Eqs. (3.9) – (3.12), there are many ways to obtain approximating spaces $H^1(n,\Omega)$, as we will see soon. However, before going any further, let us first ask ourselves the following question: The solution of Eq. (3.39) for a finite n can, in general, yield only an approximation to the exact solution to the problem. How good is our approximation and how can we guarantee that a sequence of solutions will converge to the exact solution as we increase n, i.e., the dimension of the approximating space? The answer to the second part of the question is relatively simple: convergence will occur as long as the set of basic functions $\{\phi_i\}_{i=1}^n$ of $H^1(n,\Omega)$ is complete in $H^1(\Omega)$ as $n \to \infty$. This is the same as saying that any function u in $H^1(\Omega)$ can be approximated to an arbitrary degree of closeness by a linear combination of the basis functions ϕ_i. The first part of the question is related to the analysis of the error in the approximation and the answer depends on the particular choice of the functions ϕ_i. We will state some results later on. Those readers interested in the mathematical aspects of the error bounds should consult Strang and Fix (1973), Mitchell and Wait (1977), or Carey and Oden (1983).

3.5.1 Galerkin Method in One Dimension

We will illustrate the use of Galerkin's method to solve two-point boundary value problems through a simple example. Consider the equation

$$\frac{d^2u}{dx^2} = 1 \qquad \Omega = \{x / 0 < x < 1\} \qquad (3.40)$$

with

$$u(0) = 0 \qquad (3.41)$$

and

$$\left.\frac{du}{dx}\right|_{x=1} = 1 \qquad (3.42)$$

The problem admits a simple analytical solution given by

$$u^*(x) = \frac{1}{2}x^2 \qquad (3.43)$$

The weak form of Eqs. (3.40) – (3.42) is

$$\int_0^1 \left(\frac{dw}{dx} \frac{du}{dx} + w \right) dx - w(1) = 0 \qquad (3.44)$$

Notice also that because the Dirichlet condition at $x = 0$ is homogeneous, the solution to Eq. (3.44) must be in $H_0^1(\Omega)$, i.e., in this case the set of trial functions is equal to the space of test functions.

We will solve Eq. (3.44) using several different Galerkin approximations.

3.5.1.1 Approximation 1.

Choose the subspace $H_0^1(\Omega)$ spanned by the functions

$$\phi_i(x) = \sin i \pi x \quad i = 1, . ., n \qquad (3.45)$$

The functions ϕ_i are in $H^1([0,1])$ (Exercise 3.13). Moreover, they satisfy the homogeneous Dirichlet condition $\phi_i(0) = 0$. Hence these are admissible functions. We also know from the theory of generalized Fourier series that they constitute a complete set of $L^2([0,1])$, hence we should expect convergence in L^2.

Let us first take $n = 1$, i.e., we look for a solution of the form $u_1(x) = a_1 \sin \pi x$. The Galerkin formulation requires that $w_1(x) = \phi_1(x) = \sin \pi x$. Replacing u and w in Eq. (3.44), we obtain

$$\int_0^1 \left[(\pi \cos \pi x)(\pi a_1 \cos \pi x) + \sin \pi x \right] dx - \sin \pi = 0$$

The integration yields

$$\frac{\pi^2}{2} a_1 - \frac{2}{\pi} = 0$$

Hence

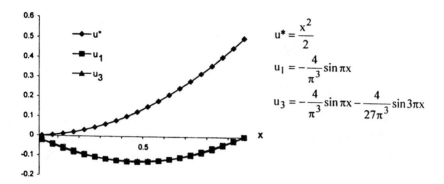

Figure 3.6 Exact solution and Galerkin approximation to Eqs. (3.40) - (3.42) using trial function in Eq. (3.45).

$$u_1(x) = -\frac{4}{\pi^3}\, \sin \pi\, x$$

The exact solution and $u_1(x)$ are depicted in Fig. 3.6. Clearly, $u_1(y)$ is not a good approximation to u^*. If we increase the number of approximation functions to $n = 2$, we have

$$u_2(x) = a_1 \sin \pi\, x - a_2 \sin 2\pi\, x$$

and $w_1(x) = \sin\pi x$, $w_2(x) = \sin 2\pi x$. Substituting into Eq. (3.44), we obtain a system of two equations for a_1, a_2:

$$\int_0^1 \left(\frac{d w_1}{dx}\frac{du}{dx} + w_1 \right) dx - w_1(1)$$

$$= \int_0^1 \left[(\pi \cos \pi x)(\pi a_1 \cos \pi x + 2\pi a_2 \cos 2\pi x) + \sin \pi x \right] dx = 0$$

$$\int_0^1 \left(\frac{d w_2}{dx}\frac{du}{dx} + w_2 \right) dx - w_2(1)$$

$$= \int_0^1 \left[(2\pi \cos 2\pi x)(\pi a_1 \cos \pi x + 2\pi a_2 \cos 2\pi x) + \sin 2 \pi x \right] dx = 0$$

The integration produces the system of equations

$$\frac{\pi^2}{2} a_1 + \frac{2}{\pi} = 0$$

$$\pi^2 a_2 = 0$$

with solution $a_1 = -4/\pi^3$ and $a_2 = 0$. Hence no improvement is obtained using $n = 2$. Further increasing the dimension of the approximation space to $n = 3$ yields (Exercise 3.14) $a_1 = -4/\pi^3$, $a_2 = 0$, $a_3 = -4/27\pi^3$. Hence

$$u_3(x) = -\frac{4}{\pi^3} \sin \pi x - \frac{4}{27 \pi^3} \sin 3\pi x$$

The function $u_3(x)$ is also shown in Fig. 3.6. We see that the improvement over u_1 is marginal. Even though the set of functions in Eq. (3.45) is an admissible set, it does not appear to be a very fortunate choice. On the other hand, if instead of Eq. (3.45) we choose the functions

$$\phi_i(x) = \sin \frac{(2i-1)}{2} \pi x \tag{3.46}$$

which are also admissible functions, for $n = 1$ we obtain

$$u_1(x) = \frac{8}{\pi^2} \left(1 - \frac{2}{\pi} \right) \sin \frac{\pi}{2} x \tag{3.47}$$

which is already a much better approximation than the ones obtained before (Exercise 3.15). Furthermore, if instead of the problem defined by Eqs. (3.40) – (3.41), we apply the functions in Eq. (3.45) to the following two-point boundary value problem

$$\frac{d^2 u}{dx^2} = 1 \qquad 0 < x < 1$$

with

$$u(0) = u(1) = 0$$

the Galerkin solution to this problem using the functions ϕ_i given by Eq. (3.45) will converge rapidly to the exact solution (Exercise 3.16).

REMARKS

1. The difficulty with solving Eqs. (3.40) - (3.42) using the function given by Eq. (3.45) is that even though they are admissible functions, they all vanish at x = 1. The information about the boundary condition at x = 1 was lost in Eq. (3.44) and the Galerkin method could not converge to the correct solution. Therefore *the weighting functions must be nonzero over the portions* Γ_2 *and* Γ_3 *of the boundary* Γ.

2. A second difficulty is the fact that the trial functions are nonzero over the whole interval 0 < x < 1. The effect of the wrong behavior at the point x = 1 propagates throughout the interval. These functions are called *global* because each function influences the whole domain Ω.

3. The fact that a set of functions is admissible does not mean that it will be appropriate. If global functions are used, we are faced with the problem of determining whether they are indeed appropriate for the problem at hand.

4. The functions we used here are *orthogonal*, that is,

$$\int_0^1 \phi_i(x)\phi_j(x)dx = 0$$

if i ≠ j. This results in an uncoupled system of equations. Each time we added a new function, the coefficients obtained with the previous solutions remained the same. This is a very helpful property because it makes the solution of the linear system of equations almost trivial. Unfortunately, the functions used in the FEM will not be orthogonal, and solutions of large systems of linear equations will be required.

3.5.1.2 Approximation 2.

Let us return to Eqs. (3.40) – (3.42), but this time we will use polynomial trial functions. The functions $\phi_i(x) = x^i$, i = 1,...,n, are admissible functions, since they are clearly in $H^1(\{0,1\})$ and they vanish at $x = 0$. For $n = 1$, we approximate $u(x)$ by $u_1(x) = a_1x$, and we set $w_1(x) = \phi_1(x)$. Using Eq. (3.44) and solving for a_1, we obtain

$$\int_0^1 (a_1 + x)dx - 1 = 0$$

which yields $a_1 = 1/2$. Therefore the Galerkin approximation for $n = 1$ is $u_1(x) = x/2$. Notice that the approximate solution $u_1(x)$ does not satisfy the boundary condition at $x = 1$, $du_1/dx|_{x=1} = 1/2$, which is only an approximation to the exact value.

Let us now increase the order of the approximation to $n = 2$, i.e., we look for an approximate solution of the form $u_2(x) = a_1x + a_2x^2$. We also have $w_1 = x$, $w_2 = x^2$. Substituting into Eq. (3.44), we obtain

$$\int_0^1 \left[(2a_2x + a_1) + x \right]dx - 1 = 0$$

$$\int_0^1 2x \left[(2a_2x + a_1) + x^2 \right]dx - 1 = 0$$

This yields the system of equations

$$4a_1 + 4a_2 = 2$$

$$3a_1 + 4a_2 = 2$$

with solution $a_1 = 0$, $a_2 = 1/2$. Hence $u_2(x) = x^2/2$, *which is the exact solution!* This is, of course, fortuitous. It occurred because the exact solution happens to be in the approximation space $H^1(n,\Omega)$. This usually occurs only in simple cases. However, if the solution is indeed in the approximation space, the Galerkin method must yield the solution. Furthermore, if we continue to increase the dimension of $H^1(n,\Omega)$, we should still get the exact solution, e.g., for $n = 3, 4, \ldots n$, we must still find $a_2 = 1/2$ and $a_1 = a_3 = a_4, \ldots = 0$ (Exercise 3.17).

REMARKS

1. The polynomial functions show some immediate advantages over the trigonometric functions of the previous section. For one thing, they are easier to differentiate and integrate. However, these are still *global* functions. Hence it may get cumbersome to satisfy more general boundary conditions, e.g., if we have u(0) = p and u(1) = q, it is not clear how to construct the trial functions.

2. In one dimension most of the above problems can be resolved (Exercise 3.18). However, in two and three dimensions the problems quickly become insurmountable. Just consider the simple problem of flow in a 90° duct as shown in Fig. 3.7. It is practically impossible to find trigonometric functions or polynomials that will vanish along the tube boundaries and match the inflow and/or outflow profiles. Therefore, *global* functions must be abandoned.

3. Flux boundary conditions will, in general, not be satisfied by the approximate solution. It will be shown later that the Galerkin solution satisfies flux boundary conditions approximately and that the value of the normal derivative converges to the exact values as $n \to \infty$.

3.5.2. Galerkin Method in Two Dimensions

All of the properties of the Galerkin method discussed above in the context of one-dimensional problems carry over to two- and three-dimensional problems. Plus we have the added complication of having to deal with geometry. In two and three dimensions, we encounter difficulties when trying to use *global* functions. Geometric irregularities soon make it impossible to find admissible functions. This leads us to the concept of *local* functions, or functions of *local support*, which are functions that vanish outside a small subdomain of Ω. In fact, the introduction of local functions leads to the beginning of finite elements. The most practical functions are polynomial functions defined in a piecewise fashion, which we will introduce in the next section.

Galerkin methods in two and three dimensions using basis of transcendental functions (trigonometric, exponential, etc.) have been extensively used in the literature when the geometry is simple (Fletcher, 1984; Finlayson, 1972), namely, rectangular in the coordinate system under consideration (Exercise 3.19). Furthermore, when linked to Fourier series, we obtain the Galerkin spectral method (Gottlieb and

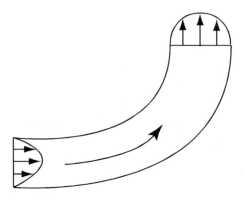

Figure 3.7 Two-dimensional flow in a duct.

Orzsag, 1977). Another family of methods arises when the Galerkin method is applied only in one of the coordinate directions, and the other is left to be resolved by other means, such as the *finite strip* method (Cheung, 1976). In this book, however, we will deal exclusively with the Galerkin method as applied in finite element approximations.

3.6 FINITE ELEMENT METHOD IN ONE DIMENSION

We will now view the finite element method basically as a Galerkin method in which the trial functions are normally chosen to be piecewise polynomial functions and are therefore defined locally. The finite elements arise when we start choosing certain types of approximation spaces $H^1(n,\Omega)$.

3.6.1 Basic Piecewise Linear Spaces

Let us go back to our one-dimensional problem defined by Eqs. (3.40) - (3.42), only this time we will consider a piecewise polynomial approximation to the solution. Divide the domain $0 < x < 1$ into two equal intervals, and seek a solution that is linear over each of the subintervals, that is,

$$u(x) = \begin{cases} a x & 0 \le x \le \dfrac{1}{2} \\ b + cx & \dfrac{1}{2} \le x \le 1 \end{cases} \qquad (3.48)$$

Because of continuity at $x = 1/2$, we must have $a = 2b + c$. Hence we can write

$$u(x) = \begin{cases} (2b + c)x & 0 \le x \le \dfrac{1}{2} \\ b + cx & \dfrac{1}{2} \le x \le 1 \end{cases} \qquad (3.49)$$

which contains two unknown parameters. This reflects the fact that our problem reduces to one of looking for values of $u(x)$ at two points, $x = 1/2$ and $x = 1$. Figure 3.8 shows the exact solution together with a possible piecewise linear approximation of the form of Eq. (3.49). It is not difficult to prove that the functions defined by Eq. (3.49) are in $H_0^1(3,[0,1])$ (Exercise 3.20). We still have some difficulty in the sense that it does not become clear from Eq. (3.49) which functions should be used as the weighting functions. To resolve this, we look for a set of piecewise linear functions that will constitute a convenient basis for the space of piecewise linear functions with nodes at 0, 1/2, and 1. A linear function u between two nodal points x_i and x_{i+1} can be written as

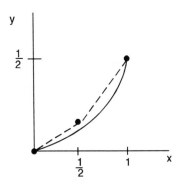

Figure 3.8 Piecewise linear approximation using two elements.

$$u(x) = \left[\frac{x_{i+1} - x}{x_{i+1} - x_i}\right] u_i + \left[\frac{x - x_i}{x_{i+1} - x_i}\right] u_{i+1} \qquad (3.50)$$

Hence we can write

$$u(x) = N_1 u_1 + N_2 u_2 + N_3 u_3 \qquad (3.51)$$

where

$$N_1(x) = \begin{cases} 1 - 2x & 0 \le x \le \dfrac{1}{2} \\[2em] 0 & \text{otherwise} \end{cases} \qquad (3.52a)$$

$$N_2(x) = \begin{cases} 2x & 0 \le x \dfrac{1}{2} \\[1em] 2 - 2x & \dfrac{1}{2} \le x \le 1 \\[1em] 0 & \text{otherwise} \end{cases} \qquad (3.52b)$$

$$N_3(x) = \begin{cases} 2x - 1 & \dfrac{1}{2} \le x \le 1 \\[2em] 0 & \text{otherwise} \end{cases} \qquad (3.52c)$$

The functions $N_i(x)$, i = 1, 2, 3, are shown in Fig. 3.9 and are called *shape* functions, *trial* functions, or *basis* functions. It is not difficult to

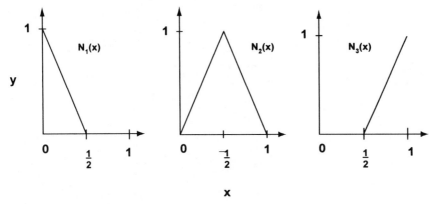

Figure 3.9 Linear 1-D shape functions for a two element approximation.

show that every piecewise linear function between $0 \leq x \leq 1/2$ and $1/2 \leq x \leq 1$ can be written in the form of Eq. (3.51) (Exercise 3.21), where $u_i \equiv u(x_i)$ is the value of the function at the nodal point i.

The most important difference between the shape functions and the functions used in Section 3.5, is that the functions have *local support*, i.e., they vanish outside a maximum of two elements. The implementation of Dirichlet conditions is thus trivial, e.g., to impose $u(0) = 0$ in Eq. (3.51), we simply set $u_1 = u_1(x_1) = u(0) = 0$. The approximation spaces are then

$$H^1(3,[0,1]) = \left\{ u(x) \, / \, u(x) = N_1(x)u_1 + N_2(x)u_2 + N_3(x)u_3 \right\}$$

and

$$H_0^1(3,[0,1]) = \left\{ u(x) \, / \, u(x) = N_2(x)u_2 + N_3(x)u_3 \right\}$$

Notice that in this case $H^1(n,\Omega)$ has dimension 3, while $H_0^1(n,\Omega)$ has dimension 2. In general, $\dim H^1(n,\Omega) \geq \dim H_0^1(n,\Omega)$.

In a Galerkin formulation, we set $w_i(x) = N_i$, and replacing into Eq. (3.44) we obtain

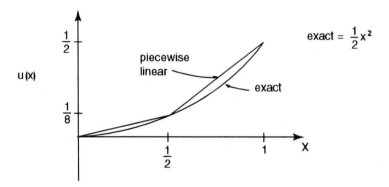

Figure 3.10 Exact solution and two linear element approximation to Eqs. (3.40) - (3.42).

$$\int_0^1 \left\{ \frac{dN_2}{dx} \left[\frac{dN_2}{dx} u_2 + \frac{dN_3}{dx} u_3 \right] + N_2 \right\} dx - N_2(1) = 0 \qquad (3.53)$$

$$\int_0^1 \left\{ \frac{dN_3}{dx} \left[\frac{dN_2}{dx} u_2 + \frac{dN_3}{dx} u_3 \right] + N_3 \right\} dx - N_3(1) = 0 \qquad (3.54)$$

After integration, Eqs. (3.53) – (3.54) yield a system of two algebraic equations in the unknown u_2 and u_3 with solution (Exercise 3.22)

$$u_2 = \frac{1}{8} \quad u_3 = \frac{1}{2}$$

The approximate solution can be written in piecewise polynomial form as

$$u(x) = \begin{cases} \dfrac{1}{4}x & 0 \le x \le \dfrac{1}{2} \\[3mm] \dfrac{1}{4}(3x-1) & \dfrac{1}{2} \le x \le 1 \end{cases} \qquad (3.55)$$

and is shown in Fig. 3.10.

REMARKS

1. Notice that the approximate solution is *exact* at the nodal points. This phenomenon is called *superconvergence,* and it happens if in Eq. (3.20) L has constant coefficients and the right-hand side $f(x)$ is in $H^1(n,\Omega)$. In our example, the right-hand side is $f(x) \equiv 1$, which is a special case of a piecewise linear function. A proof of this result can be found in the work by Hughes (1987). Of course, this is a fortunate situation and only occurs in rather simple cases. In more complex problems we cannot expect to be so lucky. In general, we will obtain finite rates of convergence that will be discussed in Section 3.6. However, we will also see how sometimes an exact solution to a simple problem can provide us improved accuracy in very complex cases when we look at convective transport and the Petrov-Galerkin methods.

2. In Eqs. (3.54) the natural boundary condition $u'(1) = 1$ only affects the second equation because $N_2(1) = 0$. Hence the fact that the shape functions have a local support also has the effect of localizing flux boundary conditions. It can be easily seen that no matter how many intervals we can subdivide the domain $\Omega = (0,1)$, the boundary condition at the right-hand side will always contribute only to the last equation.

3. We now define our first finite element space. Consider a partition of the interval $0 \le x \le L$ into n subintervals, which we will call *elements*:

$$e_i = \{ x \, / \, x_i \le x \le x_{i+1} \} \qquad (3.56)$$

such that

(i) $e_i \cap e_j = \phi$ if e_i and e_j are not adjacent elements, and $e_i \cap e_{i+1} = \{ x_{i+1} \}$ if they are adjacent. Hence the elements do not overlap and must have, at most, an end point in common.

(ii) The union $\bigcup\limits_{i=1}^{n} e_i = [0,1]$. Therefore the whole domain Ω plus its boundary is covered by the partition.

The element's end points x_i are called *nodes*, and we denote the length of each element by

$$h_i = x_{i+1} - x_i \qquad (3.57)$$

The subspace $H^1(n+1,\Omega)$ of $H^1(\Omega)$, where $\Omega = \{x / 0 < x < L\}$ of functions that are piecewise linear over each element, is defined by

$$H^1(n+1,\Omega) = \left\{ u / u(x) = \sum_{i=1}^{n+1} N_i(x)u_i \right\} \qquad (3.58)$$

where the shape functions $N_i(x)$ are given by

$$N_1(x) = \begin{cases} \dfrac{x_2 - x}{h_1} & x_1 = 0 \le x \le x_2 \\[4mm] 0 & \text{otherwise} \end{cases} \qquad (3.59a)$$

$$N_i(x) = \left. \begin{cases} \dfrac{x - x_{i-1}}{h_{i-1}} & x_{i-1} \le x \le x_i \\[4mm] \dfrac{x_{i+1} - x}{h_i} & x_i \le x \le x_{i+1} \\[4mm] 0 & \text{otherwise} \end{cases} \right\} \; i = 2, \ldots, n \qquad (3.59b)$$

$$N_{n+1}(x) = \begin{cases} \dfrac{x - x_n}{h_n} & x_n \le x \le x_{n+1} = L \\[4mm] 0 & \text{otherwise} \end{cases} \qquad (3.59c)$$

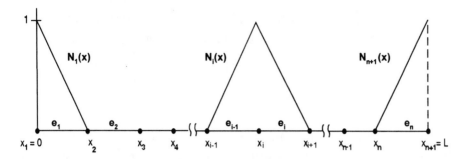

Figure 3.11 General piecewise linear shape functions.

and are shown in Fig. 3.11. Note that the elements are of arbitrary size and the shape functions $N_i(x)$ are equal to 1 at node i and zero at all other nodes, i.e., they satisfy

$$N_i(x_j) = \delta_{ij} \qquad 3.60)$$

Furthermore, if a Dirichlet condition is given at one or both ends, they are readily imposed. For example, if the boundary conditions are $u(0) = a$ and $u(L) = b$, the set of admissible functions takes the form

$$u(x) = N_1(x) \cdot a + \sum_{i=2}^{n} N_i(x) u_i + N_{n+1}(x) \cdot b \qquad (3.61)$$

The above expression satisfies $u(0) = a$, $u(L) = b$, and $u(x_i) = u_i$ at each node, a direct consequence of Eq. (3.60).

    4. When we derived Eq. (3.53) and (3.54), we only considered $w_2 = N_2$ and $w_3 = N_3$, but we ignored $w_1 = N_1$ at node 1, where the Dirichlet boundary condition $u(0) = 0$ is given. As we showed in Section 3.4, the weighting functions w must vanish at points where a Dirichlet boundary condition is prescribed.

    To better understand this in the general case, let us write the weighting functions in $H^1(n+1, \Omega)$ as

$$w(x) = N_1(x) w_1 + N_2(x) w_2 + \ldots + N_{n+1}(x) w_n$$

as we did above for u(x) in Eq. (3.58). The requirement that w(x) belongs to $H_0^1(n+1, \Omega)$ can now be imposed by setting $w_1 = w_{n+1} = 0$, i.e.,

$$w(x) = N_2(x)w_2 + N_3(x)w_3 + \ldots + N_n(x)w_n$$

Notice that this expression contains exactly the same number of degrees of freedom as in Eq. (3.61) after applying the Dirichlet boundary conditions on u(x). Therefore, only the weighting functions $w_i = N_i$ that constitute a basis for the space $H_0^1(n+1, \Omega)$ need to be considered to generate the system of linear algebraic equations.

3.6.2 Heat Conduction in One Dimension

Consider the solution of Eqs. (3.6) - (3.8). The weak form is given by

$$\int_0^L \left(k \frac{dw}{dx} \frac{dT}{dx} - wf \right) dx - w(L)q = 0 \qquad (3.62)$$

Let us use a finite element discretization consisting of n elements not necessarily of equal size. The admissible approximation trial functions T(x) must satisfy Eq. (3.7) and have the form

$$T(x) = N_1(x)T_o + \sum_{i=2}^{n+1} N_i(x)T_i \qquad (3.63)$$

The weighting functions w(x) must be in the space $H_0^1(n+1, \Omega)$. Hence they are of the form

$$w(x) = N_1(x) \cdot 0 + \sum_{i=2}^{n+1} N_i(x)w_i = \sum_{i=2}^{n+1} N_i(x)w_i \qquad (3.64)$$

All the admissible weighting functions can be expressed as a linear combination of $N_i(x)$ with $i = 2, \ldots, n + 1$. In particular, we can choose $w_i(x) = N_i(x)$, $i = 2, \ldots, n + 1$. Therefore substituting the weighting and trial functions into Eq. (3.61) will give rise to a system of n equations in as many unknowns to be solved for our approximate solution. However, before getting to that, let us use the additivity of the integral to rewrite Eq. (3.61) in the form

$$\int_0^L \left(k \frac{dw}{dx} \frac{dT}{dx} - wf \right) dx - w(L)q = \sum_{i=1}^{n} \int_{x_i}^{x_{i+1}} \left(k \frac{dw}{dx} \frac{dT}{dx} - wf \right) dx - w(L)q$$

$$(3.65)$$

We need only concern ourselves with the integration over one element at a time. Thus we can turn our attention to the evaluation of the expression

$$\int_{e_i} \left(k \frac{dw}{dx} \frac{dT}{dx} - wf \right) dx = \int_{x_i}^{x_{i+1}} \left(k \frac{dw}{dx} \frac{dT}{dx} - wf \right) dx \quad (3.66)$$

Because of the local character of the shape functions, over any one element, only two of the shape functions need to be considered, as shown in Fig. 3.12. Replacing u(x) and w(x) into Eq. (3.66), we obtain

$$\int_{x_i}^{x_{i+1}} \left[k \frac{dN_i}{dx} \left(\frac{dN_i}{dx} T_i + \frac{dN_{i+1}}{dx} T_{i+1} \right) - N_i f \right] dx$$

and

$$\int_{x_i}^{x_{i+1}} \left[k \frac{dN_{i+1}}{dx} \left(\frac{dN_i}{dx} T_i + \frac{dN_{i+1}}{dx} T_{i+1} \right) - N_{i+1} f \right] dx$$

It is customary in finite element analysis to interpolate all data functions using the shape functions, unless the data have discontinuities, in which case it may be more convenient to approximate it as piecewise constant over the element. Furthermore, it has been proved (Strang and Fix, 1973) that the error incurred in approximating the data using the shape functions is smaller than the rest of the errors in the approximate solution process. We will deal with those issues in more detail later on.

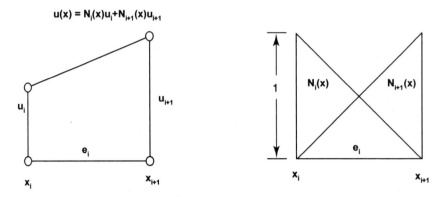

Figure 3.12 Linear element interpolation and element shape functions.

For the time being, let us assume that $k(x)$ is approximated by a piecewise constant value k_i over each element, and $f(x)$ is interpolated using the shape functions, i.e.,

$$f(x) \cong N_i(x)\, f_i + N_{i+1}(x)\, f_{i+1} \qquad x_i \leq x \leq x_{i+1} \qquad (3.67)$$

The integrals above can then be calculated and written in the form

$$\frac{k_i}{h_i}(T_i - T_{i+1}) = h_i\left(\frac{1}{3} f_i + \frac{1}{6} f_{i+1}\right)$$

$$(3.68)$$

$$\frac{k_i}{h_i}(-T_i + T_{i+1}) = h_i\left(\frac{1}{6} f_i + \frac{1}{3} f_{i+1}\right)$$

Of course, the equal signs do not hold; this is just the customary way to express the *element equations* in matrix form:

$$\frac{k_i}{h_i}\begin{bmatrix} 1 & -1 \\ -1 & 1 \end{bmatrix}\begin{bmatrix} T_i \\ T_{i+1} \end{bmatrix} = \frac{h_i}{6}\begin{bmatrix} 2f_i + f_{i+1} \\ f_i + 2f_{i+1} \end{bmatrix} \qquad (3.69)$$

It is also clear that every element will generate a relation of the form of Eq. (3.69), except for the first and last ones, in which the contribution of the boundary condition at $x = 0$ and $x = L$ must be included. Element 1 yields

$$\frac{k_1}{h_1} T_2 = \frac{h_1}{6}(f_1 + 2f_2) + \frac{k_1}{h_1} T_0 \qquad (3.70)$$

and element n gives

$$\frac{k_n}{h_n}\begin{bmatrix} 1 & -1 \\ -1 & 1 \end{bmatrix}\begin{bmatrix} T_n \\ T_{n+1} \end{bmatrix} = \frac{h_n}{6}\begin{bmatrix} 2f_n + f_{n+1} \\ f_n + 2f_{n+1} \end{bmatrix} + \begin{bmatrix} 0 \\ q \end{bmatrix} \qquad (3.71)$$

The assembly of all element contributions* yields the finite element equations

$$\begin{bmatrix} \frac{k_1}{h_1}+\frac{k_2}{h_2} & -\frac{k_2}{h_2} & 0 & 0 & 0 & 0 & 0 \\ -\frac{k_2}{h_2} & \frac{k_2}{h_2}+\frac{k_3}{h_3} & -\frac{k_3}{h_3} & 0 & \cdots & 0 & 0 & 0 \\ & -\frac{k_3}{h_3} & \frac{k_3}{h_3}+\frac{k_4}{h_4} & -\frac{k_4}{h_4} & 0 & 0 & 0 \\ & & \vdots & & \ddots & & \vdots \\ 0 & 0 & 0 & 0 & \cdots & -\frac{k_{n-1}}{h_{n-1}} & \frac{k_{n-1}}{h_{n-1}}+\frac{k_n}{h_n} & -\frac{k_n}{h_n} \\ 0 & 0 & 0 & 0 & \cdots & 0 & -\frac{k_n}{h_n} & \frac{k_n}{h_n} \end{bmatrix}\begin{Bmatrix} T_2 \\ T_3 \\ T_4 \\ \vdots \\ T_n \\ T_{n+1} \end{Bmatrix}$$

---

*For an elementary treatment of the basic concept of assemblage of the equations, see Pepper and Heinrich (1992), or Allair (1985).

$$
= \begin{bmatrix}
\dfrac{h_1}{6}(f_1 + 2f_2) + \dfrac{h_2}{6}(2f_2 + f_3) + \dfrac{k_1}{h_1}T_o \\[2mm]
\hline
\dfrac{h_2}{6}(f_2 + 2f_3) + \dfrac{h_3}{6}(2f_3 + f_4) \\[2mm]
\hline
\dfrac{h_3}{6}(f_3 + 2f_4) + \dfrac{h_4}{6}(2f_4 + f_5) \\[2mm]
\hline
\vdots \\[2mm]
\hline
\dfrac{h_{n-1}}{6}(f_{n-1} + 2f_n) + \dfrac{h_n}{6}(2f_n + f_{n+1}) \\[2mm]
\hline
\dfrac{h_n}{6}(f_n + 2f_{n+1}) + q
\end{bmatrix}
\tag{3.72}
$$

or

$$
\mathbf{K}\,\mathbf{T} \;=\; \mathbf{F} \tag{3.73}
$$

we refer to $\mathbf{K}$ as the *global stiffness* matrix, $\mathbf{T}$ is the degrees of freedom vector, and $\mathbf{F}$ the global right-hand side. The matrix $\mathbf{K}$ is also called the *conductivity* matrix; however, the name stiffness matrix that originated in structural analysis is now often used to denote the coefficient matrices that arise in all kinds of finite element discretizations.

Notice that the matrix K is *symmetric* and *tridiagonal*, the latter property deriving from the way the nodes are numbered. We will discuss the structure of the stiffness matrix in more detail later, when we consider the problem of mesh generation in two dimensions.

It is not difficult to prove that the matrix $\mathbf{K}$ is positive definite and diagonally dominant (see Appendix A). Either one of these properties is sufficient to guarantee that $\mathbf{K}$ is nonsingular. Hence the solution to Eq. (3.73) can be written as

$$
\mathbf{T} \;=\; \mathbf{K}^{-1}\mathbf{F} \tag{3.74}
$$

We assume that the reader is already familiar with the most common techniques utilized in the solution of the system of linear algebraic equations. Let us now examine how good our approximation will be.

3.6.3 Error in the Finite Element Approximation

The mathematical properties of finite element approximation have been thoroughly studied, and there are many excellent books that contain detailed information. Some of those already referenced (Hughes, 1987; Carey and Oden, 1983; Oden and Carey, 1983; Mitchell and Wait, 1977; Strang and Fix, 1973; Hall and Porshing, 1990) are a good start for the mathematically inclined reader. One soon finds out that the amount of work that has been done in the area of stability and error analysis is exhaustive.

We will only state the main results pertaining to FEM using piecewise linear shape functions for the solution of a problem slightly more general than that of Eqs. (3.6) – (3.8). Consider a second order linear differential equation of the form

$$-\frac{d}{dx}\left(p(x)\frac{du}{dx}\right) + q(x)u = f(x) \qquad 0 < x < L \qquad (3.75)$$

Let us denote the exact solution by u^* and the finite element solution by u^h. Define the mesh parameter h as

$$h = \max_i h_i \qquad (3.76)$$

and we will simply assume there are two boundary conditions applied at $x = 0$ and $x = L$, i.e., we have a well-posed problem. However, the results are independent of the particular boundary conditions. We state our main result in the following way.

If $p(x)$ and $q(x)$ are piecewise continuous and bounded in $0 \leq x \leq L$ with $p(x) \geq p_o > 0$, $q(x) \geq 0$, then the Galerkin finite element approximation using piecewise linear trial functions satisfies (Strang and Fix, 1973)

$$\left\|u^* - u^h\right\|_o \leq c_1 h^2 \left\|u''\right\|_o \leq c_2 h^2 \left\|f\right\|_o \qquad (3.77)$$

where $c_1 = p_{max} / \pi$, $c_2 = p_{max} / \left(\pi^2 p_{min}\right)$ and $\left\|\ \right\|_o$ is the norm in the L^2 space, i.e.,

$$\| u \|_{0} = \left(\int_{0}^{L} u^2 dx \right)^{\frac{1}{2}} \tag{3.78}$$

The proofs are rather lengthy and can be found in the works by Strang and Fix (1973) and Carey and Oden (1983). It also follows from the proofs that the derivative $(u*)'$ is approximated to first order accuracy. It is interesting to observe that the error can be estimated in terms of the problem data.

In the above estimate the error is measured in the L^2 norm and is therefore a global error. Error estimates in the maximum norm, defined as

$$\| u \|_{\infty} = \max_{0 \le x \le L} |u(x)|$$

can also be obtained (Douglas et al., 1975), but as pointed out by Nitsche (1979), these can never be optimal with respect to the power of h. For all practical purposes, however, the error estimates obtained in terms of the L^2 norms will hold when applied pointwise. Therefore we will be satisfied with the results in terms of the mean square norm.

Example 3.4

To illustrate the above estimates, let us look into the solution to the equation

$$-\frac{d^2u}{dx^2} + qu = q \qquad 0 < x < 1 \tag{3.79}$$

This equation arises in the analysis of heat exchanger fins (Kreith, 1958), where q is the ratio of convective heat transfer to conductive heat transfer. For the purposes of our example, we can choose q = 1 and the boundary conditions u(0) = 0, u(1) = 1.

The error in the solution of the above equation at the point x = 0.5 using 4, 8, 16, and 32 linear elements of equal size is shown in Fig. 3.13. The plot shows ln e versus ln h, where

$$e = \left| u^*(.5) - u^h(.5) \right|$$

$$u^* = \frac{1}{e^2 - 1}\left(e^x - e^{2-x}\right) + 1$$

A second line shows the error in the derivative at $x = 0.5$. Because of the logarithmic nature of the plots, the rate of convergence is given by the slope of the lines (Exercise 3.23). The second-order convergence of u and first-order convergence of its derivative are almost exact in this example. ∎

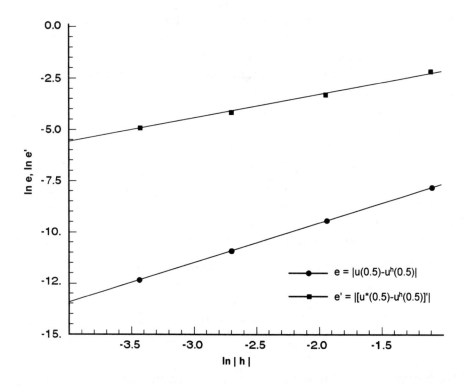

Figure 3.13 Convergence rates in the approximation of second order self adjoint equations using linear elements.

The question now arises, what happens if the second derivative of u is not in $L^2(\Omega)$? This is certainly the case in Eq. (3.23), where f(x) is the Dirac delta function. The answer is not difficult to find once the previous analysis has been performed. The result is that the rate of convergence drops to first order in general, unless there is a mesh point located at points where u is discontinuous. Finally, if all we can say is that u is in $H^1(\Omega)$, then nothing can be said of the rate of convergence, which may be arbitrarily slow as h goes to zero. However, convergence still occurs (Strang and Fix, 1973).

3.6.4 Boundary Fluxes

There are many practical situations, particularly in heat transfer and fluid flow, when boundary fluxes or stresses are the quantities of interest in the calculation. However, these are not given directly by the finite element solution and must be obtained a posteriori once the unknown dependent variables have been found. Fluxes may be of interest at points in the interior of the domain as well as on the boundary. We will address the calculation of fluxes in the interior later on. Here we will consider only boundary fluxes.

Example 3.5

To establish the ideas, let us return to the one-dimensional problem defined by Eqs. (3.40) – (3.42). We will look into estimating the boundary flux at x = 0 from the finite element solution using two linear elements, which is given by Eq. (3.55).

One obvious way to estimate the value of the derivatives is by differentiation of the solution $u^h(x)$. In this case we obtain

$$\left.\frac{du^h}{dx}\right|_{x=0} = \left.\frac{d}{dx}\left(\frac{x}{4}\right)\right|_{x=0} = \frac{1}{4}$$

If we compare this value with that of the exact solution, which is u'(0) = 0, we see that it is subject to a significant error, even though the function is obtained exactly at the nodal points. Furthermore, if we evaluate the derivative at x = 1, we obtain

$$\left.\frac{d\,u^h}{d\,x}\right|_{x=1} = \left.\frac{d}{d\,x}\left(\frac{3x}{4} - \frac{1}{4}\right)\right|_{x=1} = \frac{3}{4} \neq 1 \qquad \text{!!!}$$

Hence the solution u^h does *not even satisfy the boundary condition* at the right-hand side. As was mentioned in the previous section, the derivatives are approximated to one order of accuracy less than the function itself. In the case of linear elements that means $O(h)$. In this case, we can expect that the error in the flux at $x = 0$, as given by the derivative of u^h, will behave as

$$e' \cong 0.3\,h$$

when $h \rightarrow 0$ (Exercise 3.24), and in order to attain an accuracy of m significant digits, we would need to use a mesh with elements of size $h < 1/0.3 \times 10^{-m}$. For $m = 4$, i.e., for the error to be less than 0.0001, a regular mesh would require 33,333 elements. ∎

A more accurate calculation of the boundary fluxes, albeit requiring more work, consists in using the boundary residual equations (Marshall et al., 1978) derived from the finite element formulation. The weighted residual expression of Eqs. (3.40) - (3.42) for arbitrary weighting functions w is

$$\int_0^1 \left(\frac{dw}{dx}\frac{du}{dx} + w\right)dx - \left.w\frac{du}{dx}\right|_0^1 = 0 \qquad (3.80)$$

Notice that we have not imposed the requirement that $w(0) = 0$ in Eq. (3.80). In fact, if we choose w such that $w(0) = 1$ in Eq. (3.80), we obtain an expression for the derivative at $x = 0$, i.e.,

$$\left.\frac{du}{dx}\right|_{x=0} = \left.w\frac{du}{dx}\right|_{x=1} - \int_0^1 \left(\frac{dw}{dx}\frac{du}{dx} + w\right)dx$$

which allows us to represent boundary fluxes in terms of the weighted residuals.

Example 3.6

Let us now perform a finite element discretization using two elements and letting $w_1(x) = N_1(x)$, so that $w_1(0) = 1$. The resulting element equations are

Element e_1

$$\begin{bmatrix} 2 & -2 \\ -2 & 2 \end{bmatrix} \begin{bmatrix} u_1 \\ u_2 \end{bmatrix} = \begin{bmatrix} -\dfrac{1}{4} + \dfrac{du}{dx}\Big|_{x=0} \\ -\dfrac{1}{4} \end{bmatrix}$$

Element e_2

$$\begin{bmatrix} 2 & -2 \\ -2 & 2 \end{bmatrix} \begin{bmatrix} u_2 \\ u_3 \end{bmatrix} = \begin{bmatrix} -\dfrac{1}{4} \\ \dfrac{3}{4} \end{bmatrix}$$

where $-du/dx\big|_{x=0}$, which is unknown, has been retained in the first equation. Upon assembly, the final system of linear equations is

$$\begin{bmatrix} 2 & -2 & 0 \\ \hline -2 & 4 & -2 \\ 0 & -2 & 2 \end{bmatrix} \begin{bmatrix} u_1 = 0 \\ u_2 \\ u_3 \end{bmatrix} = \begin{bmatrix} -\dfrac{1}{4} \\ -\dfrac{1}{2} \\ \dfrac{3}{4} \end{bmatrix} + \begin{bmatrix} \dfrac{du}{dx}\Big|_{x=0} \\ 0 \\ 0 \end{bmatrix}$$

The system above has been partitioned in such a way that the lower 2 x 2 system of equations is the same as that obtained from Eq. (3.53), where we assumed that $w_1(x) \equiv 0$. Furthermore, notice that the first equation is not needed to solve the 2 x 2 subsystem for u_2 and u_3; hence the first equation can be viewed as an equation for $-du/dx\big|_{x=0}$ and can be evaluated explicitly after u_2 and u_3 have been calculated. The weighted residuals expression is

$$\frac{du}{dx}\bigg|_{x=0} = w_1(1)\frac{du}{dx}\bigg|_{x=1} - \int_0^1\left(\frac{dw_1}{dx}\frac{du}{dx} + w_1\right)dx$$

and yields

$$\frac{du}{dx}\bigg|_{x=0} = 0 - 2\,u_2 + \frac{1}{4} = -2\frac{1}{8} + \frac{1}{4} = 0$$

which is *exact*! The fact is that the approximation to the boundary fluxes obtained this way is *superconvergent*.

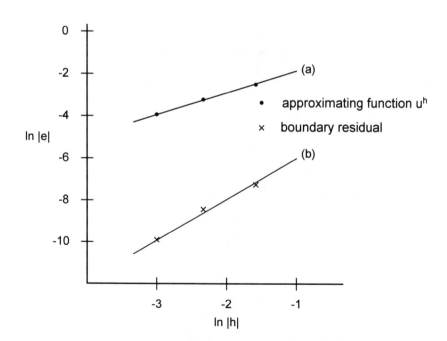

Figure 3.14 Behavior of the error in boundary fluxes using (a) differentiation of the approximating polynomial and (b) from the boundary residual.

For linear elements, Carey (1982) has shown that the boundary fluxes are second-order accurate, or $O(h^2)$ when calculated using the boundary residual equations. This behavior is illustrated in Fig. 3.14, for Example 3.4, where the error in the flux at $x = 0$ is calculated by direct differentiation of the derivative and by the boundary residuals method. The first- and second-order convergence rates are evident . ■

Those familiar with finite element structural analysis will recognize a direct analogy with the structural system. Displacements are calculated for those boundary nodes where the applied loads are known and reactions where the displacements are prescribed.

3.6.5 Interelement Conditions

One of the most powerful aspects of finite element analysis is the ability of the method to deal with all kinds of different conditions in a simple and straightforward way. To illustrate to what extent we can incorporate other types of conditions, let us consider the following boundary value problem

$$-\frac{d}{dx}\left(k\frac{du}{dx}\right) = f \qquad 0 < x < 1 \qquad (3.81)$$

where

$$u(0) = 0 \qquad (3.82)$$

with the convective heat transfer boundary condition

$$-k\frac{du}{dx}\bigg|_{x=1} = h(u - u_\infty)\bigg|_{x=1} \qquad (3.83)$$

In addition, we will introduce an interface condition of the form

$$\left(-k_2\frac{du}{dx}\right)\bigg|_{x=\frac{1}{3}} - \left(-k_1\frac{du}{dx}\right)\bigg|_{x=\frac{1}{3}} = c$$

This kind of interface condition arises, for example, if we have a change of phase interface at $x = 1/3$. In this case, c would be the latent heat times the velocity of the interface. Furthermore, the functions $k(x)$ and $f(x)$ will be discontinuous, and are given in Fig. 3.15.

We introduce the *jump operator* $\left[\left|\cdot\right|\right]$ defined as

$$\left[\left|f(a)\right|\right] \equiv \lim_{x \to a^+} f(x) - \lim_{x \to a^-} f(x) \equiv f\left(a^+\right) - f\left(a^-\right) \qquad (3.84)$$

In Eq. (3.84), a^+ requires $x = a + |\Delta x|$ and a^- requires $x = a - |\Delta x|$, i.e., a^+ means we approximate a from the right-hand side and a^- we approximate from the left-hand side. $[|f(a)|]$ is a measure of the *jump* of f(x) at $x = a$, as shown in Fig. 3.15(b) at $x = 2/3$. The interface condition can then be written in the form

$$\left[\left|-k\frac{du}{dx}\right|\right] = c \qquad \text{at } x = \frac{1}{3}$$

In general, the argument of the function under consideration will be omitted if it is clear from the context.

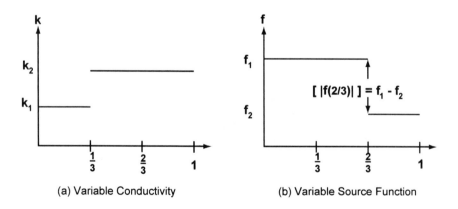

(a) Variable Conductivity   (b) Variable Source Function

Figure 3.15 Data for problem described by Eqs. (3.81) – (3.84).

Let us now consider the weighted residuals formulation of Eq. (3.81), but before we integrate by parts, we will split the integral using its additivity property, i.e.,

$$\int_0^1 w\left(-\frac{d}{dx}k\frac{du}{dx}-f\right)dx = \int_0^{\frac{1}{3}} w\left(-\frac{d}{dx}k\frac{du}{dx}-f\right)dx \qquad (3.85)$$

$$+\int_{\frac{1}{3}}^{\frac{2}{3}} w\left(-\frac{d}{dx}k\frac{du}{dx}-f\right)dx + \int_{\frac{2}{3}}^1 w\left(-\frac{d}{dx}k\frac{du}{dx}-f\right)dx = 0$$

Now integrate each of the subintegrals by parts

$$\int_0^{\frac{1}{3}} w\left(-\frac{d}{dx}k\frac{du}{dx}-f\right)dx = \int_0^{\frac{1}{3}}\left(k_1\frac{dw}{dx}\frac{du}{dx}-wf_1\right)dx$$

$$+\left[w\left(-k_1\frac{du}{dx}\right)\right]_{x=\left(\frac{1}{3}\right)^-} - \left[w\left(-k_1\frac{du}{dx}\right)\right]_{x=0^+}$$

$$\int_{\frac{1}{3}}^{\frac{2}{3}} w\left(-\frac{d}{dx}k\frac{du}{dx}-f\right)dx = \int_{\frac{1}{3}}^{\frac{2}{3}}\left(k_2\frac{dw}{dx}\frac{du}{dx}-wf_1\right)dx$$

$$+\left[w\left(-k_2\frac{du}{dx}\right)\right]_{x=\left(\frac{2}{3}\right)^-} - \left[w\left(-k_2\frac{du}{dx}\right)\right]_{x=\left(\frac{1}{3}\right)^+}$$

$$\int_{\frac{2}{3}}^{1} w\left(-\frac{d}{dx}k\frac{du}{dx}-f\right)dx = \int_{\frac{2}{3}}^{1}\left(k_2\frac{dw}{dx}\frac{du}{dx}-wf_2\right)dx$$

$$+\left[w\left(-k_2\frac{du}{dx}\right)\right]_{x=1^-} - \left[w\left(-k_2\frac{du}{dx}\right)\right]_{x=\left(\frac{2}{3}\right)^+}$$

Adding them all together and using Eq. (3.84) we obtain

$$\int_0^1 w\left(-\frac{d}{dx}k\frac{du}{dx}-f\right)dx = \int_0^1\left(k\frac{dw}{dx}\frac{du}{dx}-wf\right)dx - \left[w\left(-k_1\frac{du}{dx}\right)\right]_{x=0}$$

$$-w\left(\frac{1}{3}\right)\left[\left|-k\frac{du}{dx}\left(\frac{1}{3}\right)\right|\right]-w\left(\frac{2}{3}\right)\left[\left|-k\frac{du}{dx}\left(\frac{2}{3}\right)\right|\right]+\left[w\left(-k_2\frac{du}{dx}\right)\right]_{x=1}=0$$

where the weighting functions are continuous over the domain in our finite element formulations. If, in addition, we make use of the interface condition at x = 1/3, the continuity of the flux at x = 2/3 and the convective boundary condition at x = 1, we finally arrive at

$$\int_0^1\left(k\frac{dw}{dx}\frac{du}{dx}-wf\right)dx = \left[w\left(-k_1\frac{du}{dx}\right)\right]_{x=0} + w\left(\frac{1}{3}\right)c - w(1)h\big(u(1)-u_\infty\big)$$

$$(3.86)$$

which is the weighted residuals formulation of Eqs. (3.81) – (3.85) incorporating the interface condition given by Eq. (3.84).

Before we go any further in the solution of Eq. (3.86) using finite elements, let us address the question of how to treat the functions k(x) and f(x). If the function k(x) does not vary too much within an element, it will be sufficient to consider its average value as a constant over the element. We must make sure, however, that a node is placed at points of discontinuity of k(x) to preserve the accuracy; however, this is not necessary for the method to converge. In this case, the interface condition at x = 1/3 requires that a node should be placed there anyway.

The function f is customarily approximated using the shape functions, and again it is wise to place a node at the points of discontinuity of f to guarantee maximum accuracy, although this is not necessary for convergence.

Example 3.7

To illustrate the advantages of placing a node at the points of the discontinuity of the data functions, let us use three equal linear elements to discretize the interval [0,1]. It is easy to show that the contribution to the right-hand side vector using linear interpolation on f is identical to the contribution obtained using the exact piecewise constant representation (Exercise 3.25).

If no node is placed at $x = 2/3$, as seen in Fig. 3.16, linear interpolation of f produces the same approximation regardless of the position of the jump within the element. This creates the potential to introduce a significant error.

The Galerkin finite element approximation using three linear elements yields the following element equations:

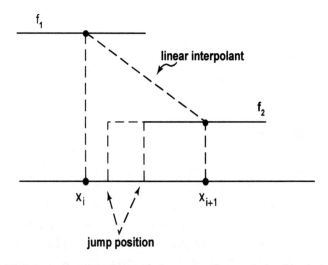

Figure 3.16 Schematic of the interpolation error if no node is placed at a jump discontinuity point.

Element e_1

$$3k_1 \begin{bmatrix} 1 & -1 \\ -1 & 1 \end{bmatrix} \begin{bmatrix} u_1 \\ u_2 \end{bmatrix} = \frac{1}{6} \begin{bmatrix} f_1 \\ f_1 \end{bmatrix} + \begin{bmatrix} -k_1 \dfrac{du}{dx} \Big|_{x=0} \\ c \end{bmatrix}$$

Element e_2

$$3k_2 \begin{bmatrix} 1 & -1 \\ -1 & 1 \end{bmatrix} \begin{bmatrix} u_2 \\ u_3 \end{bmatrix} = \frac{1}{6} \begin{bmatrix} f_1 \\ f_1 \end{bmatrix}$$

The reader may observe a slight ambiguity in the finite element formulation because c does not appear in the first equation of element e_2. The reason is that it was already considered in element e_1 and if it appears again in e_2, we would be applying the flux jump twice. A way around this could be to consider one-half of c in each of the elements.

Element e_3

$$\begin{bmatrix} 3k_2 & -3k_2 \\ -3k_2 & 3k_2 + h \end{bmatrix} \begin{bmatrix} u_3 \\ u_4 \end{bmatrix} = \begin{bmatrix} \dfrac{1}{6} f_2 \\ \dfrac{1}{6} f_2 + h u_\infty \end{bmatrix}$$

which leads to the solution of the system of equations,

$$\begin{bmatrix} 3k_1 + 3k_2 & -3k_2 & 0 \\ -3k_2 & 6k_2 & -3k_2 \\ 0 & -3k_2 & 3k_2 + h \end{bmatrix} \begin{bmatrix} u_2 \\ u_3 \\ u_4 \end{bmatrix} = \begin{bmatrix} \dfrac{1}{3} f_1 + c \\ \dfrac{1}{6} (f_1 + f_2) \\ \dfrac{1}{6} f_2 + h u_\infty \end{bmatrix} \qquad (3.87)$$

where all of the conditions have been incorporated. The heat flux at $x = 0$ is calculated after the solution of Eq. (3.87) has been found from the first equation in element e_1, i.e.,

$$-k_1 \frac{du}{dx}\Big|_{x=0} = -3k_1u_2 - \frac{1}{6}f_1 \tag{3.88}$$

Remember that, in general, the solution $u^h(x)$ will not satisfy the boundary condition at $x = 1$. However, this is not important because the boundary condition is correctly imposed in Eq. (3.86). ■

REMARKS

1. At every internal nodal point, if no interface conditions are imposed, we always satisfy a continuity-of-flux condition of the form

$$\left[\left|-k\frac{du}{dx}(x_i)\right|\right] = 0 \tag{3.89}$$

in the weak sense. This is important because, as shown in Fig. 3.17, the piecewise linear finite element solution u^h does not have continuous derivatives at the nodal points x_i, i.e.,

$$\frac{du^h}{dx}\Big|_{x_i^+} \neq \frac{du^h}{dx}\Big|_{x_i^-}$$

2. If an interface occurs, by placing a node at the interface, we automatically satisfy the continuity-of-flux condition in the weak formulation.

3. Any and all conditions involving fluxes, either at the boundary or in the interior of the computational regions, are dealt with directly using the fluxes normal to the boundaries or the jump operators in the interior of the domain.

4. For reasons of accuracy, it is desirable to place nodes at points of discontinuity in the data functions.

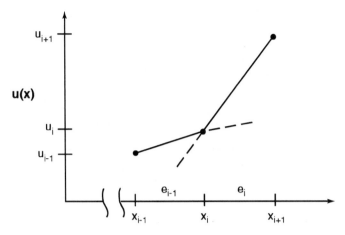

Figure 3.17 Schematic of discontinuity of the derivative at a nodal point.

3.7 FINITE ELEMENT METHOD IN TWO DIMENSIONS

Practically all of the basic concepts introduced in the previous section extend directly to two- and three-dimensional problems. The big difference in two and three dimensions is the need to deal with geometry, which can introduce enormous complications.

3.7.1 Basic Piecewise Bilinear Space

The most elementary geometric figure that defines an area is a triangle. However, throughout this book our finite element approximations will be based on quadrilateral shapes, which are geometrically more complex. The reason for this is that quadrilaterals can offer distinct advantages over elements based on a triangular geometry.

In two dimensions, in addition to approximating the functions, we may have to approximate the domain Ω when it is irregular. In this section we will consider only polygonal regions whose boundaries are always parallel to one of the coordinate axes, such as the domain of Figure 3.2. We can subdivide the domain using rectangular elements as shown in Fig. 3.18, where one such element has been identified in the interior. We will associate four nodes with this element, one at each corner of the rectangle, in a local coordinate system (x',y') with its origin at the lower left corner and using the node numbering convention shown

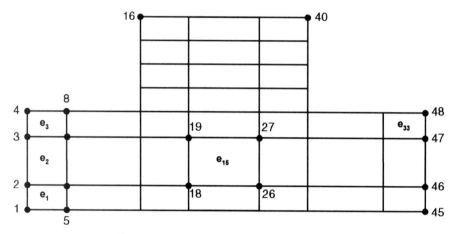

Figure 3.18 A polygonal domain subdivided into rectangular elements.

in Fig. 3.19(a). The two-dimensional shape functions associated with each node are

$$N_1(x',y') = \left(1 - \frac{x'}{\Delta x}\right)\left(1 - \frac{y'}{\Delta y}\right)$$

$$N_2(x',y') = \left(\frac{x'}{\Delta x}\right)\left(1 - \frac{y'}{\Delta y}\right)$$

$$N_3(x',y') = \left(\frac{x'}{\Delta x}\right)\left(\frac{y'}{\Delta y}\right) \tag{3.90}$$

$$N_4(x',y') = \left(1 - \frac{x'}{\Delta x}\right)\left(\frac{y'}{\Delta y}\right)$$

The function $N_1(x',y')$ is shown in Fig. 3.19(b). It is clear from Eq. (3.90) that the two-dimensional functions are obtained from the products of one-dimensional linear shape functions, hence the name *bilinear*. They satisfy the condition that they are equal to 1 at one node and 0 at all others, i.e.,

$$N_i(x_j',y_j') = \delta_{ij} \tag{3.91}$$

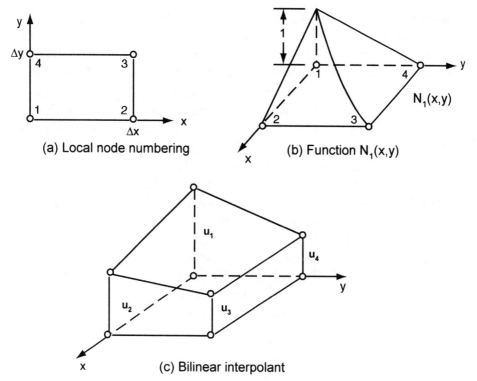

(a) Local node numbering

(b) Function $N_1(x,y)$

(c) Bilinear interpolant

Figure 3.19 Bilinear element.

and they are linear along the boundaries but not linear in the interior of the element. A function $u(x,y)$ is interpolated using the shape functions in the way shown in Fig. 3.18(c); the interpolant is given by

$$u(x',y') \cong \sum_{i=1}^{4} N_i(x',y') u_i \qquad (3.92)$$

where, as before, $u_i \equiv u(x'_i,y'_i)$. Clearly, the shape functions in the global coordinate system that describes the domain Ω are obtained by a simple translation of the axis to node i in the element.

As shown in Fig. 3.18, the nodal numbers of element e_{15} in the global system would be 18, 26, 27, and 19 rather than 1, 2, 3, and 4. This information will be needed in the assembly procedure for each element. Furthermore, if we consider two adjacent elements along the common boundary, both elements must reduce to the same straight line that guarantees the continuity of the function over the whole domain Ω. This property, rather innocent at first sight is called *conformity* and is crucial to guarantee convergence of the method. For second-order operators, it means that continuity of the function should be preserved in order to guarantee that the approximation space is indeed a subspace of $H^1(\Omega)$.

The concept has more profound implications when fourth-order operators are considered, as in the case of the stress analysis of plates, where it gives rise to the concept of *the patch test*. We will stay clear of this aspect of finite element approximations in this book, since we intend to restrict ourselves exclusively to second-order operators. We will discuss the concept of conformity in more detail when we introduce *isoparametric transformations* in Chapter 4.

The local character of bilinear shape functions should be obvious. A shape function is non-zero for (at most) four elements sharing a common node. Fig. 3.20 shows typical bilinear shape functions for

        1. a boundary corner node,
        2. a boundary side node, and
        3. an interior node.

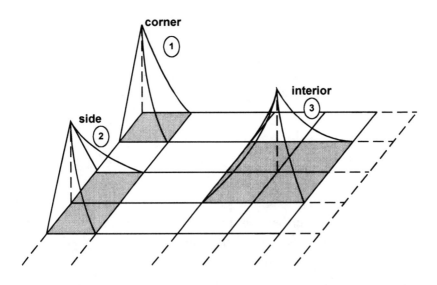

Figure 3.20 Three different types of bilinear shape functions in global coordinates.

Example 3.8

Let us now return to Fig. 3.18, which defines a two-dimensional domain discretized by 33 bilinear elements and 48 nodes. Each element e_i has size (h_i, k_i) and satisfies the condition that the intersection $e_i \cap e_j$ is either empty if e_i and e_j are not adjacent elements or is equal to their common boundary if e_i and e_j are adjacent. The union of all elements covers the entire domain, i.e., $\bigcup_{i=1}^{33} e_i = \Omega$.

The subspace $H^1(n,\Omega)$ associated with this mesh has dimension $n = 33$ and is given by

$$H^1(n,\Omega) = \left\{ u / u(x) = \sum_{i=1}^{n} N_i(x,y) u_i \right\}$$

The global shape functions N_i can be easily obtained from Eq. (3.90) for each individual element (Exercise 3.28). ∎

3.7.2 Heat Conduction in Two Dimensions

Consider the solution of a simple, heat conduction boundary value problem defined by

$$-k\left(\frac{\partial^2 u}{\partial x^2} + \frac{\partial^2 u}{\partial y^2} \right) = f \qquad \text{in } \Omega \qquad (3.93)$$

where k and f are constant and Ω is the rectangular region defined by $0 < x < a$ and $0 < y < b$. We will apply the following boundary conditions:

$$u(0,y) = u_o \qquad\qquad 0 \le y \le b \qquad (3.94a)$$

$$k\frac{\partial u}{\partial y}\bigg|_{y=0} = 0 \qquad\qquad 0 \le x \le a \qquad (3.94b)$$

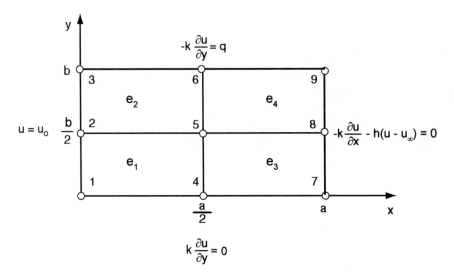

Figure 3.21 Domain, mesh, and boundary conditions for heat conduction problem defined by Eqs. (3.93) – (3.94).

$$\left[-k\frac{\partial u}{\partial x} - h(u - u_\infty)\right]_{x=a} = 0 \qquad 0 \le y \le b \qquad (3.94c)$$

$$-k\frac{\partial u}{\partial y}\bigg|_{y=b} = q \qquad 0 \le x \le a \qquad (3.94d)$$

We will assume that all physical parameters are constant and subdivide Ω into four equal bilinear elements. The situation is depicted in Fig. 3.21. The different parts of the boundary are

$$\Gamma_1 = \left\{(x,y)/x = 0, 0 \le y \le b\right\}$$
$$\Gamma_2 = \left\{(x,y)/y = 0, y = b, 0 \le x \le a\right\}$$
$$\Gamma_3 = \left\{(x,y)/x = a, 0 \le y \le b\right\}$$

The weighted residuals form of Eq. (3.93) gives (Exercise 3.27)

$$\int_{\Omega}\left[k\left(\frac{\partial w}{\partial x}\frac{\partial u}{\partial x}+\frac{\partial w}{\partial y}\frac{\partial u}{\partial y}\right)-wf\right]d\Omega$$

$$(3.95)$$

$$+\int_{\Gamma_2}wqd\Gamma+\int_{\Gamma_3}wh(u-u_\infty)d\Gamma=0$$

where only the segment of Γ_2 corresponding to $y = b$ must be considered, since $q = 0$ along $y = 0$.

The Galerkin finite element formulation is obtained upon approximating the dependent variable u by

$$u(x,y)=\sum_{i=1}^{9}N_i(x,y)u_i$$

and setting the weighting functions w_i equal to N_i. We will *not require* the weighting functions to vanish over Γ_1 in order to construct the boundary residual equations for calculating the heat flux along Γ_1. The final form of Eq. (3.95) is

$$\int_0^a\int_0^b k\left[\frac{\partial N_i}{\partial x}\left(\sum_{j=1}^9\frac{\partial N_j}{\partial x}u_j\right)+\frac{\partial N_i}{\partial y}\left(\sum_{j=1}^9\frac{\partial N_j}{\partial y}u_j\right)\right]dx\,dy$$

$$+\int_0^b N_i h\left(\sum_{j=1}^9 N_j u_j\right)dy\bigg|_{x=a}=$$

$$\int_0^a\int_0^b N_i f\,dx\,dy-\left[\int_0^a N_i q\,dx\right]_{y=b}+\left[\int_0^b N_i h u_\infty\,dy\right]_{x=a}\quad(3.96)$$

$$-\left[\int_0^b N_i\left(-k\frac{du}{dx}\right)dy\right]_{x=0}$$

$$i=1,2,\ldots9$$

where the last term includes the heat flux along the boundary $x = 0$.

It should be perfectly clear at this point that Eq. (3.96) represents a system of nine equations for the six unknown temperatures at nodes 4 through 9. These are to be obtained through the solution of the resulting 6 x 6 subsystem of linear equations that does not include u_1, u_2, u_3 as unknowns. Three values of the heat flux in the x direction at nodes 1 through 3 will be obtained from the solution of a 3 x 3 subsystem of linear equations after the temperatures have been calculated.

The element equations can be readily calculated, e.g., for element e_4 we have

$$
\int_{\frac{a}{2}}^{a} \int_{\frac{b}{2}}^{b} k \left[\frac{\partial N_i}{\partial x} \left(\frac{\partial N_5}{\partial x} u_5 + \frac{\partial N_8}{\partial x} u_8 + \frac{\partial N_9}{\partial x} u_9 + \frac{\partial N_6}{\partial x} u_6 \right) \right.
$$

$$
+ \frac{\partial N_i}{\partial y} \left(\frac{\partial N_5}{\partial y} u_5 + \frac{\partial N_8}{\partial y} u_8 + \frac{\partial N_9}{\partial y} u_9 + \frac{\partial N_6}{\partial y} u_6 \right) \Bigg] dx dy
$$

$$
+ \left[\int_{\frac{b}{2}}^{b} N_i h (N_5 u_5 + N_8 u_8 + N_9 u_9 + N_6 u_6) dy \right]_{x=a}
$$

$$
= \int_{\frac{a}{2}}^{a} \int_{\frac{b}{2}}^{b} N_i f dx dy + \left[\int_{a}^{\frac{a}{2}} N_i q dy \right]_{y=b} + \left[\int_{\frac{b}{2}}^{b} N_i h u_\infty dy \right]_{x=a}
$$

$$
\tag{3.97}
$$

$$
i = 5, 8, 9, 6
$$

The shape functions are easily calculated from Eq. (3.90) using

$$
x' = x - a/2
$$
$$
y' = y - b/2
$$

Substituting the shape functions into Eq. (3.97) yields

$$\left\{ \frac{kb}{6a} \begin{bmatrix} 2 & -2 & -1 & 1 \\ -2 & 2 & 1 & -1 \\ -1 & 1 & 2 & -2 \\ 1 & -1 & -2 & 2 \end{bmatrix} + \frac{ka}{6b} \begin{bmatrix} 2 & 1 & -1 & -2 \\ 1 & 2 & -2 & -1 \\ -1 & -2 & 2 & 1 \\ -2 & -1 & 1 & 2 \end{bmatrix} + \frac{hb}{12} \begin{bmatrix} 0 & 0 & 0 & 0 \\ 0 & 2 & 1 & 0 \\ 0 & 1 & 2 & 0 \\ 0 & 0 & 0 & 0 \end{bmatrix} \right] \left\{ \begin{matrix} u_5 \\ u_8 \\ u_9 \\ u_6 \end{matrix} \right\}$$

$$= \frac{abf}{16} \begin{bmatrix} 1 \\ 1 \\ 1 \\ 1 \end{bmatrix} - \frac{aq}{4} \begin{bmatrix} 0 \\ 0 \\ 1 \\ 1 \end{bmatrix} + \frac{hbu_\infty}{4} \begin{bmatrix} 0 \\ 1 \\ 1 \\ 0 \end{bmatrix}$$

$$(3.98)$$

where the contribution of each term in Eq. (3.97) is shown. The calculation of the rest of the element relations and the assembly are left to Exercise 3.29. The final system of equations is

$$\left\{ \frac{kb}{6a} \begin{bmatrix} 2 & 1 & 0 & -2 & -1 & 0 & 0 & 0 & 0 \\ 1 & 4 & 1 & -1 & -4 & -1 & 0 & 0 & 0 \\ 0 & 1 & 2 & 0 & -1 & -2 & 0 & 0 & 0 \\ -2 & -1 & 0 & 4 & 2 & 0 & -2 & -1 & 0 \\ -1 & -4 & -1 & 2 & 8 & 2 & -1 & -4 & -1 \\ 0 & -1 & -2 & 0 & 2 & 4 & 0 & -1 & -2 \\ 0 & 0 & 0 & -2 & -1 & 0 & 2 & 1 & 0 \\ 0 & 0 & 0 & -1 & -4 & -1 & 1 & 4 & 1 \\ 0 & 0 & 0 & 0 & -1 & -2 & 0 & -1 & 2 \end{bmatrix} \right.$$

$$
+ \frac{ka}{6b}
\begin{bmatrix}
2 & -2 & 0 & 1 & -1 & 0 & 0 & 0 & 0 \\
-2 & 4 & -2 & -1 & 2 & -1 & 0 & 0 & 0 \\
0 & -2 & 2 & 0 & -1 & 1 & 0 & 0 & 0 \\
1 & -1 & 0 & 4 & -4 & 0 & 1 & -1 & 0 \\
-1 & 2 & -1 & -4 & 8 & -4 & -1 & 2 & -1 \\
0 & -1 & 1 & 0 & -4 & 4 & 0 & -1 & 1 \\
0 & 0 & 0 & 1 & -1 & 0 & 2 & -2 & 0 \\
0 & 0 & 0 & -1 & 2 & -1 & -2 & 4 & -2 \\
0 & 0 & 0 & 0 & -1 & 1 & 0 & -2 & 2
\end{bmatrix}
$$

$$
+ \frac{hb}{12}
\begin{bmatrix}
0 & 0 & 0 & 0 & 0 & 0 & 0 & 0 & 0 \\
0 & 0 & 0 & 0 & 0 & 0 & 0 & 0 & 0 \\
0 & 0 & 0 & 0 & 0 & 0 & 0 & 0 & 0 \\
0 & 0 & 0 & 0 & 0 & 0 & 0 & 0 & 0 \\
0 & 0 & 0 & 0 & 0 & 0 & 0 & 0 & 0 \\
0 & 0 & 0 & 0 & 0 & 0 & 0 & 0 & 0 \\
0 & 0 & 0 & 0 & 0 & 0 & 2 & 1 & 0 \\
0 & 0 & 0 & 0 & 0 & 0 & 1 & 4 & 1 \\
0 & 0 & 0 & 0 & 0 & 0 & 0 & 1 & 2
\end{bmatrix}
\left\}
\begin{bmatrix}
u_1 \\ u_2 \\ u_3 \\ u_4 \\ u_5 \\ u_6 \\ u_7 \\ u_8 \\ u_9
\end{bmatrix}
= \frac{abf}{16}
\begin{bmatrix}
1 \\ 2 \\ 1 \\ 2 \\ 4 \\ 2 \\ 1 \\ 2 \\ 1
\end{bmatrix}
- \frac{aq}{4}
\begin{bmatrix}
0 \\ 0 \\ 1 \\ 0 \\ 0 \\ 2 \\ 0 \\ 0 \\ 1
\end{bmatrix}
$$

$$
+ \frac{hbu_\infty}{4}
\begin{bmatrix}
0 \\ 0 \\ 0 \\ 0 \\ 0 \\ 0 \\ 1 \\ 2 \\ 1
\end{bmatrix}
- \frac{b}{12}
\begin{bmatrix}
2\bar{q}_1 + \bar{q}_2 \\
\bar{q}_1 + 4\bar{q}_2 + \bar{q}_3 \\
\bar{q}_2 + 2\bar{q}_3 \\
0 \\
0 \\
0 \\
0 \\
0 \\
0
\end{bmatrix}
\tag{3.99}
$$

To obtain the last term, we assume that along the boundary $x = 0$, the heat flux can be written in the form

$$-k\frac{\partial u}{\partial n} = k\frac{\partial u}{\partial x} = \sum_{i=1}^{3} N_i(0,y)\bar{q}_i \qquad (3.100)$$

and replace it into the last integral in Eq. (3.96). Here $\bar{q}$ represents the local heat flux at node i, and Eq. (3.100) gives the local heat flux distribution along the boundary $x = 0$.

To simplify the presentation of the rest of the procedure, consider the following specific case.

EXAMPLE 3.9

Let us set $a = b = f = h = k = q = u_\infty = 1$ and $u_0 = 10$. Then Eq. (3.99) reduces to

$$\begin{bmatrix} 4 & -1 & 0 & -1 & -2 & 0 & 0 & 0 & 0 \\ -1 & 8 & -1 & -2 & -2 & -2 & 0 & 0 & 0 \\ 0 & -1 & 4 & 0 & -2 & -1 & 0 & 0 & 0 \\ -1 & -2 & 0 & 8 & -2 & 0 & -1 & -2 & 0 \\ -2 & -2 & -2 & -2 & 16 & -2 & -2 & -2 & -2 \\ 0 & -2 & -1 & 0 & -2 & 8 & 0 & -2 & -1 \\ 0 & 0 & 0 & -1 & -2 & 0 & 5 & -\frac{1}{2} & 0 \\ 0 & 0 & 0 & -2 & -2 & -2 & -\frac{1}{2} & 10 & -\frac{1}{2} \\ 0 & 0 & 0 & 0 & -2 & -1 & 0 & -\frac{1}{2} & 5 \end{bmatrix} \begin{bmatrix} u_1 \\ u_2 \\ u_3 \\ u_4 \\ u_5 \\ u_6 \\ u_7 \\ u_8 \\ u_9 \end{bmatrix}$$

$$= \frac{3}{8} \begin{bmatrix} 1 \\ 2 \\ -3 \\ 2 \\ 4 \\ -6 \\ 5 \\ 10 \\ 1 \end{bmatrix} - \frac{1}{2} \begin{bmatrix} 2\bar{q}_1 + \bar{q}_2 \\ \bar{q}_1 + 4\bar{q}_2 + \bar{q}_3 \\ \bar{q}_2 + 2\bar{q}_3 \\ 0 \\ 0 \\ 0 \\ 0 \\ 0 \\ 0 \end{bmatrix} \tag{3.101}$$

Because of the Dirichlet boundary condition on the left-hand side we know that $u_1 = u_2 = u_3 = 10$. Hence the first three equations can be written as

$$\begin{bmatrix} 2 & 1 & 0 \\ 1 & 4 & 1 \\ 0 & 1 & 2 \end{bmatrix} \begin{bmatrix} \bar{q}_1 \\ \bar{q}_2 \\ \bar{q}_3 \end{bmatrix} = \frac{3}{4} \begin{bmatrix} 1 \\ 2 \\ -3 \end{bmatrix} - \begin{bmatrix} 60 - 2u_4 - 4u_5 \\ 120 - 4u_4 - 4u_5 - 4u_6 \\ 60 - 4u_5 - 2u_6 \end{bmatrix} \tag{3.102}$$

and the remaining equations are

$$
\begin{bmatrix}
8 & -2 & 0 & -1 & -2 & 0 \\
-2 & 16 & -2 & -2 & -2 & -2 \\
0 & -2 & 8 & 0 & -2 & -1 \\
-1 & -2 & 0 & 5 & -\frac{1}{2} & 0 \\
-2 & -2 & -2 & -\frac{1}{2} & 10 & -\frac{1}{2} \\
0 & -2 & -1 & 0 & -\frac{1}{2} & 5
\end{bmatrix}
\begin{bmatrix}
u_4 \\ u_5 \\ u_6 \\ u_7 \\ u_8 \\ u_9
\end{bmatrix}
= \frac{3}{8}
\begin{bmatrix}
2 \\ 4 \\ -6 \\ 5 \\ 10 \\ 1
\end{bmatrix}
+
\begin{bmatrix}
30 \\ 60 \\ 30 \\ 0 \\ 0 \\ 0
\end{bmatrix}
=
\begin{bmatrix}
30.75 \\ 61.5 \\ 27.75 \\ 1.875 \\ 3.75 \\ 0.375
\end{bmatrix}
$$

$$(3.103)$$

Both systems of equations are diagonally dominant, and hence unique solutions exist. Furthermore, Eq. (3.103) is clearly independent of Eq. (3.102) and can be solved immediately. The values of u_4, u_5, and u_6 are then replaced into Eq. (3.102) to find the nodal heat fluxes $\bar{q}_i$. Of course, if these are not of interest, Eq. (3.102) can simply be ignored.

The solution to Eq. (3.103) is given in Table 3.1, together with the values at nodes 4 through 9 obtained using finer meshes of 10 x 10, 20 x 20, and 40 x 40 elements, and depicted in Fig. 3.22.

Table 3.1 Solution of Eq. (3.103)

Node	Mesh			
	2 x 2	10 x 10	20 x 20	40 x 40
4	7.8921	7.8672	7.8665	7.8663
5	7.8194	7.7902	7.7895	7.7892
6	7.4690	7.4724	7.4723	7.4722
7	5.6368	5.6111	5.6105	5.6102
8	5.5556	5.5361	5.5355	5.5353
9	5.2521	5.2410	5.2406	5.2405

We now substitute the values of u_i into Eq. (3.102) to get

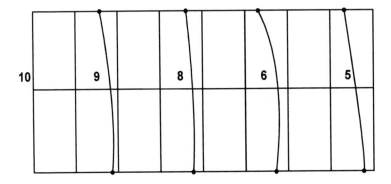

Figure 3.22 Graphical representation of the solution of Eq. (3.102) showing isotherm contours.

$$\begin{bmatrix} 2 & 1 & 0 \\ 1 & 4 & 1 \\ 0 & 1 & 2 \end{bmatrix} \begin{bmatrix} \bar{q}_1 \\ \bar{q}_2 \\ \bar{q}_3 \end{bmatrix} = \begin{bmatrix} -12.188 \\ -25.778 \\ -16.034 \end{bmatrix} \tag{3.104}$$

with solution

$$\bar{q}_1 = -4.1496$$

$$\bar{q}_2 = -3.8889$$

$$\bar{q}_3 = -6.0726$$

It is interesting to compare these numbers with the ones obtained taking the derivatives of the interpolating function at the boundary, which yields $\bar{q}_1 = -4.2158$, $\bar{q}_2 = -4.3612$, and $\bar{q}_3 = -5.062$.

REMARKS

1. The last two columns of Table 3.1 show a maximum difference of less than 0.006%. Therefore the solution obtained with the 40 x 40 element mesh can be considered fully converged, and we call it the "exact" solution. Comparing the solution obtained with the 2 x 2 element mesh with the exact solution, we find a maximum error at the

nodes of less than 0.5%, which is remarkably accurate for such a coarse discretization.

2. The heat fluxes at nodes 1, 2 and 3, calculated using the boundary residuals for the four meshes, are given in Table 3.2, which shows that the fluxes calculated in the coarse mesh are very inaccurate.

Table 3.2 Heat fluxes at nodes 1, 2, and 3.

Node	Mesh			
	2 x 2	10 x 10	20 x 20	40 x 40
1	-4.1496	-4.0702	-4.0722	-4.0729
2	-3.8889	-4.1945	-4.2789	-4.2792
3	-6.0726	-7.1009	-7.5424	-7.9840

There is also a slight oscillation in the results; this is due to the coarseness of the mesh, and it disappears very quickly with refinement. Notice also that between the solutions obtained in the 20 x 20 mesh and the 40 x 40 mesh, there is still a 5% change in the flux at node 3.

In general, to calculate the fluxes (or stresses) is much more difficult than calculating the function itself, as illustrated in this example.

If we approximate the flux at node 3, using the interpolating polynomial, the 40 x 40 mesh yields the value -6.772, which is over 15% in error compared with the solution obtained using the boundary residuals.

3.7.3 Error in Two-Dimensional Approximations

It is relatively easy to prove convergence of the finite element approximation to the exact solution as the computational mesh size goes uniformly to zero over the whole domain (Strang and Fix, 1973). To obtain useful error estimates in the L^2 norm is considerably more difficult, however. A result analogous to that of Eq. (3.77) can be obtained (Strang and Fix, 1973; Hall and Porshing, 1990). If u* denotes the exact solution and u^h the finite element solution, in our case we have

$$\left\| u^* - u^h \right\|_0 \le C_1 h^2 \left\| u^* \right\|_2 \qquad (3.105)$$

where h, defined as $h = \max\,(|x_i - x_j|, |y_i - y_j|)$, is the maximum mesh size in either the x- or y-direction, and the norm $\|\bullet\|_2$, defined as

$$\|u\|_2 = \left(\int_{\Omega}\left[u^2 + \left(\frac{\partial u}{\partial x}\right)^2 + \left(\frac{\partial u}{\partial y}\right)^2 + \left(\frac{\partial^2 u}{\partial x^2}\right)^2 + \left(\frac{\partial^2 u}{\partial x \partial y}\right)^2 + \left(\frac{\partial^2 u}{\partial y^2}\right)^2\right]d\Omega\right)^{\frac{1}{2}}$$

(3.106)

is the norm in the space $H^2(\Omega)$.

The interested reader is encouraged to verify these estimates in Exercise 3.31. As in the one-dimensional case, FEM in two dimensions (and three dimensions) using bilinear (trilinear) elements is a second-order method under the conditions implied by Eq. (3.105) i.e., that u* is in the space $H^2(\Omega)$ and therefore has square integrable second partial derivatives.

We will discuss the error in the partial derivatives in the next chapter. For the time being, let us just say that it is $O\,(h)$.

3.7.4 Interelement Conditions

As we did before in one dimension, let us consider a very general boundary value problem where all three types of boundary conditions and additional interface conditions are imposed, i.e.,

$$-\nabla \cdot k(x,y)\nabla u + b(x,y)u = f(x,y) \qquad \mathbf{x} \in \Omega \qquad (3.107)$$

$$u = u_0 \qquad \mathbf{x} \in \Gamma_1 \qquad (3.108)$$

$$-k\frac{\partial u}{\partial n} = g \qquad \mathbf{x} \in \Gamma_2 \qquad (3.109)$$

$$-k\frac{\partial u}{\partial n} = h(u - u_\infty) \qquad \mathbf{x} \in \Gamma_3 \qquad (3.110)$$

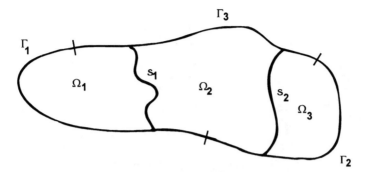

Figure 3.23 Two-dimensional domain with interfaces.

over a domain Ω with boundary Γ as depicted in Fig. 3.23; we show two interfaces S_1 and S_2, over each of which we have a jump condition of the form

$$\left[\!\left[-k\frac{\partial u}{\partial n}\right]\!\right] = \alpha \qquad\qquad \mathbf{x} \in S_1 \qquad\qquad (3.111)$$

$$\left[\!\left[-k\frac{\partial u}{\partial n}\right]\!\right] = \beta(u-\gamma) \qquad \mathbf{x} \in S_2 \qquad\qquad (3.112)$$

A condition of the form of Eq. (3.112) arises if we have a reaction occurring along S_2. This is similar to modeling biological activity at the interface in a population model.

In Eqs. (3.111) and (3.112) the *jump* is defined as

$$\left[\!\left[-k\frac{\partial u}{\partial n}\right]\!\right]_{a\,\in S} = \left(-k\frac{\partial u}{\partial n}\right)^{+}_{a} - \left(-k\frac{\partial u}{\partial n}\right)^{-}_{a}$$

$$= \left[(-k\nabla u)_{x=a^{+}} - (-k\nabla u)_{x=a^{-}}\right]\cdot\hat{n} \qquad (3.113)$$

where the plus and minus limits refer to the unit normal vector to the interface, as shown in Fig. 3.24.

Clearly, Eq. (3.113) is independent of the direction in which the normal is defined. Suppose the normal points in the opposite direction in

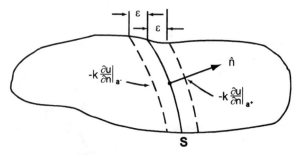

Figure 3.24 Notation for the definition of the jump operator.

Fig. 3.24. Then we must replace **n** by −**n** in Eq. (3.113), but at the same time the plus and minus derivatives must be exchanged because they refer to the direction of **n**.

The weighted residuals expression for Eq. (3.107), with $f = 0$ and

$$R(x,y) = -\nabla \cdot k\nabla u + bu$$

can be written as the sum of three integrals over each of the regions Ω_i, i.e.,

$$\int_\Omega wR\,d\Omega = \int_{\Omega_1} wR\,d\Omega + \int_{\Omega_2} wR\,d\Omega + \int_{\Omega_3} wR\,d\Omega = 0$$

Applying Eq. (3.35) in each of the three integrals above yields

$$\int_{\Omega_1} wR\,d\Omega = \int_{\Omega_1}\left[k\left(\frac{\partial w}{\partial x}\frac{\partial u}{\partial x} + \frac{\partial w}{\partial y}\frac{\partial u}{\partial y}\right) + wbu\right]d\Omega$$

$$+ \int_{\Gamma_3} w\left[-k\frac{\partial u}{\partial n}\right]d\Gamma + \int_{S_1} w\left[-k\frac{\partial u}{\partial n}\right]^- dS_1$$

$$\tag{3.114}$$

$$\int_{\Omega_2} wR\,d\Omega = \int_{\Omega_2}\left[k\left(\frac{\partial w}{\partial x}\frac{\partial u}{\partial x} + \frac{\partial w}{\partial y}\frac{\partial u}{\partial y}\right) + wbu\right]d\Omega + \int_{\Gamma_2} w\left[-k\frac{\partial u}{\partial n}\right]d\Gamma$$

$$+ \int_{\Gamma_3} w\left[-k\frac{\partial u}{\partial n}\right]d\Gamma + \int_{S_1} w\left[-k\frac{\partial u}{\partial n}\right]^+ dS_1 + \int_{S_2} w\left[-k\frac{\partial u}{\partial n}\right]^- dS_2$$

$$\tag{3.115}$$

$$\int_{\Omega_3} wR\,d\Omega = \int_{\Omega_3}\left[k\left(\frac{\partial w}{\partial x}\frac{\partial u}{\partial x}+\frac{\partial w}{\partial y}\frac{\partial u}{\partial y}\right)+wbu\right]d\Omega + \int_{\Gamma_2} w\left[-k\frac{\partial u}{\partial n}\right]d\Gamma_2$$

$$+\int_{\Gamma_3} w\left[-k\frac{\partial u}{\partial n}\right]d\Gamma_3 + \int_{S_2} w\left[-k\frac{\partial u}{\partial n}\right]^{+}dS_2$$

(3.116)

where the line integrals over Γ_2 and Γ_3 in Eqs. (3.114) – (3.116) are taken only over those portions that lie in the corresponding subregion Ω_i.

Combining Eqs. (3.114) - (3.116), we obtain

$$\int_{\Omega}\left[k\left(\frac{\partial w}{\partial x}\frac{\partial u}{\partial x}+\frac{\partial w}{\partial y}\frac{\partial u}{\partial y}\right)+wbu\right]d\Omega = -\int_{\Gamma_2} w\left[-k\frac{\partial u}{\partial n}\right]d\Gamma_2$$

$$-\int_{\Gamma_3} w\left[-k\frac{\partial u}{\partial n}\right]d\Gamma_3 + \int_{S_1}\left[\left(-wk\frac{\partial u}{\partial n}\right)^{+}-\left(-wk\frac{\partial u}{\partial n}\right)^{-}\right]dS_1$$

$$+\int_{S_2}\left[\left(-wk\frac{\partial u}{\partial n}\right)^{+}-\left(-wk\frac{\partial u}{\partial n}\right)^{-}\right]dS_2$$

(3.117)

Notice that in the above equation, the signs of $\left[-wk(\partial u/\partial n)\right]^{+}$ have been changed because the integration over S_1 and S_2 in Eq. (3.117) refers to the positive direction of the normal, as given in Fig. 3.24. Furthermore, the integrands in the interface line integrals are $\left[\left|-wk(\partial u/\partial n)\right|\right]$, which are the jumps in the fluxes across the interfaces.

Example 3.10

Assume that a material interface S, parallel to the y axis, lies within a domain Ω, as shown in Fig. 3.25a. The region Ω_1, to the left of S, has

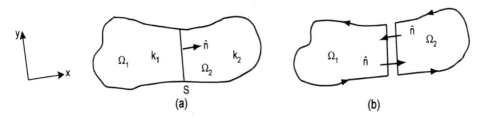

Figure 3.25 Two-material domain for Example 3.9.

thermal conductivity k_1, while Ω_2, to the right, has conductivity k_2. Recall that the heat flux is positive in the positive direction of the outward pointing normal to the region. From Fig. 3.25b we see that

$$\left(-k\frac{\partial u}{\partial n}\right)_S^{(1)} = \left[(-k\nabla u)^{(1)} \cdot n\right]_S = \left[(-k_1\nabla u)^{(1)} \cdot i\right]_S = \left(-k_1\frac{\partial u}{\partial x}\right)_S^{(1)}$$

(3.118)

where $(\partial u/\partial x)^{(1)}$ must be evaluated from the left in Ω_1. On the other hand,

$$\left(-k\frac{\partial u}{\partial n}\right)_S^{(2)} = \left[(-k\nabla u)^{(2)} \cdot n\right] = \left[\left(-k_2\frac{\partial u}{\partial x}\right)^{(2)} \cdot (-i)\right] = -\left(-k_2\frac{\partial u}{\partial x}\right)^{(2)}$$

(3.119)

with $(\partial u/\partial x)^{(2)}$ must be evaluated from the right in Ω_2. Hence

$$\int_S \left[\left(-k\frac{\partial u}{\partial n}\right)^{(2)} + \left(-k\frac{\partial u}{\partial n}\right)^{(1)}\right] dS = \int_S \left[-\left(-k_2\frac{\partial u}{\partial x}\right)^{(2)} + \left(-k_1\frac{\partial u}{\partial x}\right)^{(1)}\right] dS$$

$$= \int_S \left[\left|-k\frac{\partial u}{\partial n}\right|\right] dS$$

(3.120)

Exercise 3.32 shows that if **n** points in the opposite direction in Fig. 3.24, Eq. (3.120) does not change. ∎

Let us now return to Eq. (3.117). Substituting the boundary and interface conditions, we have

$$\int_{\Omega}\left[k\left(\frac{\partial w}{\partial x}\frac{\partial u}{\partial x}+\frac{\partial w}{\partial y}\frac{\partial u}{\partial y}\right)+wbu\right]d\Omega+\int_{\Gamma_3}whud\Gamma-\int_{S_2}w\beta ud S_2$$

$$=-\int_{\Gamma_2}wgd\Gamma+\int_{\Gamma_3}whu_\infty d\Gamma+\int_{S_1}w\alpha d S_1-\int_{S_2}w\beta\gamma d S_2$$

$$(3.121)$$

which is the final form of the weighted residuals formulation of Eqs. (3.107) through (3.112).

Example 3.11

Let us return to the problem of Fig. 3.21, only this time assume an additional interface condition of the form of Eq. (3.111) occurs at $x = a/2$. The rest of the problem remains the same as in Example 3.9.

Clearly, we only need to calculate the additional contribution of the interface to be added in Eqs. (3.95) through (3.98). We have

$$\int_S\left[\left|-wk\frac{\partial u}{\partial n}\right|\right]d S=\int_0^b w\alpha d y$$

The Galerkin form yields element contributions of the form

$$\int_0^{b/2}N_i\alpha d y=\frac{\alpha b}{4}\begin{bmatrix}0\\1\\1\\0\end{bmatrix}$$

and the global contribution to the right-hand side of Eq. (3.99) is

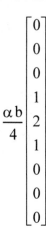

$$\frac{\alpha \, b}{4} \begin{bmatrix} 0 \\ 0 \\ 0 \\ 1 \\ 2 \\ 1 \\ 0 \\ 0 \\ 0 \end{bmatrix}$$

As an exercise (Exercise 3.33), set $\alpha = 10$ and see how this additional term will modify the solution of Eq. (3.103). ■

REMARKS

1. As in the one-dimensional case, two-dimensional finite element discretizations satisfy continuity of fluxes across interelement boundaries in the weak sense. If a jump in the fluxes occurs along any line within the domain, it can be readily incorporated in the finite element model by placing a mesh line along the interface and adding the corresponding contributions to the Galerkin equations.

2. Clearly, these properties extend to three-dimensional domains, with surfaces replacing the interelement boundary lines. The normal to the surface is uniquely defined at every point; hence no conceptual changes are needed. The only additional difficulties are related to geometry and the much larger computational effort involved in three dimensions.

You are now ready to attempt the solution of the problem defined by Fig. 3.2. In Exercise 3.34 a set of quantities is given that constitute representative values for the physical constants and define the problem completely. Notice that you will have to use very small elements to discretize the silicon chip, while you may want to use much larger elements in the ceramic matrix. If you recall Eq. (3.98), the elements of the stiffness matrix are proportional to the ratios of the element sides (a/b and b/a in Eq. (3.98)). Hence you should keep those ratios as close to 1 as possible to avoid ill conditioning of the matrix. In other words, do not use elements that are very long and slender. We will discuss this aspect of discretization in more detail later when we consider mesh generation.

3.8 FINITE ELEMENT METHOD IN THREE DIMENSIONS

The transition to three-dimensional problems does not involve any new concepts. The main difference in the formulation from two-dimensional problems is that area and line integrals are replaced with volume and surface integrals. We will briefly go through the main steps in the formulation.

3.8.1 Trilinear Element

As we did in two dimensions, we will restrict ourselves for the time being to polyfaceted domains in which all boundary surfaces are orthogonal to one of the coordinate axes. Such a volume can then be discretized using right parallelepipeds or "brick" elements. The natural extension of the one-dimensional linear and two-dimensional bilinear elements is the eight-noded trilinear or brick element shown in Fig. 3.26. This is also a Lagrangian element and the shape functions are easily obtained as products of one-dimensional linear functions, as was the case for the bilinear element. Thus, following the notation of Fig. 3.26, define

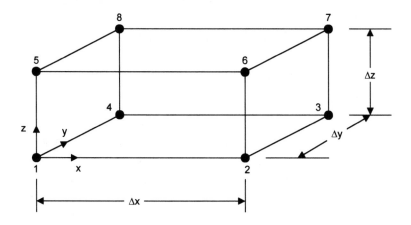

Figure 3.26 Trilinear "brick" hexahedral element.

$$\overline{N}_1(\alpha) = \left(1 - \frac{\alpha}{\Delta\alpha}\right)$$

and

$$\overline{N}_2(\alpha) = \frac{\alpha}{\Delta\alpha}$$

where α is x, y, or z,. The trilinear shape functions become

$$
\begin{aligned}
N_1(x,y,z) &= \overline{N}_1(x)\overline{N}_1(y)\overline{N}_1(z) \\
N_2(x,y,z) &= \overline{N}_2(x)\overline{N}_1(y)\overline{N}_1(z) \\
N_3(x,y,z) &= \overline{N}_2(x)\overline{N}_2(y)\overline{N}_1(z) \\
N_4(x,y,z) &= \overline{N}_1(x)\overline{N}_2(y)\overline{N}_1(z) \\
N_5(x,y,z) &= \overline{N}_1(x)\overline{N}_1(y)\overline{N}_2(z) \\
N_6(x,y,z) &= \overline{N}_2(x)\overline{N}_1(y)\overline{N}_2(z) \\
N_7(x,y,z) &= \overline{N}_2(x)\overline{N}_2(y)\overline{N}_2(z) \\
N_8(x,y,z) &= \overline{N}_1(x)\overline{N}_2(y)\overline{N}_2(z)
\end{aligned}
\tag{3.122}
$$

We can easily visualize that a trilinear shape function may be nonzero over, at most, eight elements. Furthermore, we may have four different types of elements: corner, edge, side, and interior elements.

3.8.2 Heat Conduction in Three Dimensions

The general heat conduction problem in a three-dimensional domain can be written as

$$-k\left(\frac{\partial^2 u}{\partial x^2} + \frac{\partial^2 u}{\partial y^2} + \frac{\partial^2 u}{\partial z^2}\right) = f \quad \text{in } \Omega \tag{3.123}$$

with the boundary conditions

$$u = u_0 \quad \text{in } S_1 \tag{3.124}$$

$$-k\frac{\partial u}{\partial n} = q \quad \text{in } S_2 \tag{3.125}$$

$$-k\frac{\partial u}{\partial n} = h(u - u_\infty) \quad \text{in } S_3 \tag{3.126}$$

where S_1, S_2, and S_3 compose the boundary of Ω and are now surfaces rather than curves.

The weak formulation of Eqs. (3.123) – (3.124) is obtained in a similar way as in two dimensions using Gauss's theorem and consists in finding u in $H^1(\Omega)$ such that u satisfies the Dirichlet boundary conditions over S_1 and

$$\int_\Omega \left\{ k\left[\frac{\partial w}{\partial x}\frac{\partial u}{\partial x} + \frac{\partial w}{\partial y}\frac{\partial u}{\partial y} + \frac{\partial w}{\partial z}\frac{\partial u}{\partial z}\right] - wf \right\} d\Omega$$

$$\tag{3.127}$$

$$+ \int_{S_2} wq\,ds + \int_{S_3} wh(u - u_\infty)\,ds = 0$$

for all weight functions w in $H_0^1(\Omega)$.

The error in the finite element approximations satisfies a relation of the same form as Eq. (3.105) in two dimensions; hence trilinear elements produce a second-order method. Interface and interelement conditions can also be found to obey the same relations derived in Section 3.7.4 for two-dimensional elements.

3.9 CLOSURE

In this chapter we have established most of the basic concepts involved in the finite element discretization of time-independent boundary value problems. In fact, we are already in a position to attack some fairly complex problems, such as the one defined by Eqs. (3.1) – (3.5) and shown in Fig. 3.2. Only the most necessary mathematical concepts have been introduced in order to establish the method, and it is hoped that the intricate link between the mathematics and the physical aspects of heat conduction problems has become transparent. At this point we are still restricted in many ways, e.g., only polygonal regions where all sides are parallel to one coordinate axis have been considered, and we have dealt strictly with linear problems. In the subsequent chapters most of these restrictions will be lifted, and we will end up with very general and powerful finite element tools.

REFERENCES

Allaire, P. E. (1985) *Basics of the Finite Element Method.* Dubuque, Iowa: Wm. C. Brown Publishers.

Anderssen, R. S. and Mitchell, A. R. (1979) "The Petrov Galerkin Method," *Math. Methods Appl. Sci.*, Vol. 1, pp. 3–15.

Carey, G. F. (1982) "Derivative Calculation from Finite Element Solutions," *Comput. Methods Appl. Methods Eng.*, Vol. 35, pp. 1–14.

Carey, G. F. and Oden, J. T. (1983) *Finite Elements: A Second Course.* Englewood Cliffs, N.J.: Prentice-Hall.

Cheung, Y. K. (1976) *Finite Strip Method in Structural Analysis.* New York: Pergamon Press.

Douglas, Jr., J., Dupont, T. and Wahblin, L. (1975) "Optimal L_∞ Error Estimates for Galerkin Approximations to Solutions of Two-Point Boundary Value Problems," *Math. Comput.*, Vol. 29, pp. 475–483.

Finlayson, B. A. (1972) *The Method of Weighted Residuals and Varational Principles.* New York: Academic Press.

Fletcher, C. A. J. (1982) *The Galerkin Weighted Residual Method.* New York: Springer-Verlag.

Fletcher, C. A. J. (1984) *Computational Galerkin Methods.* New York: Springer-Verlag.

Gottlieb, D. and Orzsag, S. A. (1977) *Numerical Analysis of Spectral Methods: Theory and Applications.* Philadelphia, PA: SIAM.

Greenberg M. D. (1978) *Foundations of Applied Mathematics.* Englewood Cliffs, N.J.: Prentice-Hall.

Hall, C. A. and Porsching, T. A. (1990) *Numerical Analysis of Partial Differential Equations.* Englewood Cliffs, N.J.: Prentice-Hall.

Hughes, T. J. R. (1987) *The Finite Element Method.* Englewood Cliffs, N.J.: Prentice-Hall.

Kreith, F. (1958) *Principles of Heat Transfer.* Scranton, PA: International Textbooks.

Marshall, R. S., Heinrich, J. C. and Zienkiewicz, O. C. (1978) "Natural Convection in Square Enclosure by a Finite Element, Penalty Function Formulation Using Primitive Fluid Variables," *Numer. Heat Transfer*, Vol. 1, pp. 315–330.

Mitchell, A. R. and Wait, R. (1977) *The Finite Element Method in Partial Differential Equations*. New York: John Wiley and Sons.

Nitsche, J. A. (1979) "L_∞-Error Analysis for Finite Elements," pp. 173–186 in *The Mathematics of Finite Elements and Applications III* (J. P. Whiteman, editor). London: Academic Press.

Oden, J. T. and Carey, G. F. (1983) *Finite Elements: Mathematical Aspects*, Vol. 4. Englewood Cliffs, N.J.: Prentice-Hall.

Pepper, D. W. and Heinrich, J. C. (1992) *The Finite Element Method: Basic Concepts and Applications*. Washington, D.C.: Hemisphere (Taylor and Francis).

Strang, G. and Fix, G. J. (1973) *An Analysis of the Finite Element Method*. Englewood Cliffs, N.J.: Prentice-Hall.

Zienkiewicz, O. C. (1977) *The Finite Element Method*, 3rd ed. London: McGraw-Hill.

EXERCISES

3.1 Given a function u(**x**) defined over the disk $|\mathbf{x}| \leq 1$, find the normal derivative $\partial u / \partial n$ along the boundary using (a) Cartesian coordinates and (b) polar coordinates.

3.2 Repeat Exercise 3.1 for a sphere of radius 1, using both Cartesian and spherical coordinate systems.

3.3 Show that the function $f(x) = \lim_{n \to \infty} f_n(x)$, where

$$f_n(x) = \begin{cases} \sqrt{n} & -\dfrac{1}{n} \leq x \leq \dfrac{1}{n} \\ 0 & \text{otherwise} \end{cases}$$

is in $L^2(-1,1)$. However, the function $g(x) = \lim_{n \to \infty} g_n(x)$, where

$$g_n(x) = \begin{cases} n & -\dfrac{1}{n} \le x \le \dfrac{1}{n} \\ 0 & \text{otherwise} \end{cases}$$

is not in $L^2(-1,1)$.

3.4 If $p_n(\mathbf{x})$ is a polynomial of degree n, show that for any interval [a,b], $p_n(x)$ is in the space $H^1[a,b]$. Use this result to argue that if Ω is any finite, connected two-dimensional region, the polynomial function $p_n(\mathbf{x})$ is in $H^1(\Omega)$.

3.5 Show that function $f(x) = x^\alpha$ is in $L^2([0,1])$ if $\alpha > -1/2$ but it is not in $H^2([0,1])$ unless $\alpha > 1/2$.

3.6 What is the space $H_0^1(\Omega)$ for the problem defined by Eqs. (3.6) – (3.8)?

3.7 Find the solution to the differential equation

$$-\frac{d^2 u}{dx^2} = \delta(x) \qquad -1 < x < 1$$

$$u(-1) = u(1) = 0$$

The above is a particular case of Eq. (3.23) with $\zeta = 0$. *hint*: Write two separate solutions for $-1 \le x \le 0$ and $0 \le x \le 1$, and determine the extra constants imposing interface conditions at x = 0. The above equation models the small static deflection of an elastic string under unit tension subject to a point load at x = 0. The solution is called a *fundamental* solution. This kind of solution constitutes the basis for the method of Green's function.

3.8 Show that the solution obtained for Exercise 3.7 is an element of the Sobolev space $H_0^1([-1,1])$.

3.9   Repeat the development leading to Eq. (3.35) (and subsequently to an expression of the form Eq. (3.28)) for the case when L is the Laplace operator in axisymmetric coordinates.

3.10   Use Gauss's divergence theorem to obtain the weak formulation for the three-dimensional Laplace's equation in Cartesian coordinates.

3.11   Using integration by parts, re-derive the weak weighted residual formulation for Eqs. (3.6) - (3.8).

3.12   Derive the weak weighted residual form for the axisymmetric heat conduction problem.

$$-\frac{1}{r}\frac{d}{dr}\left(rk\frac{dT}{dr}\right) = Q \qquad\qquad r_1 < r < r_2$$

$$-k\frac{dT}{dr} + h_c(T - T_\infty) = 0 \qquad\qquad r = r_1$$

$$T = T_L \qquad\qquad r = r_2$$

3.13   Show that the functions $\phi_i(x) = \sin i\pi x$ are in the space $H^1([0,1])$.

3.14   Find the solution to Eq. (3.44) using the functions ϕ_i given by Eq. (3.45) with n = 3.

3.15   Use Eq. (3.46) with n = 1 to find Eq. (3.47).

3.16   Find the exact solution to the two-point boundary value problem

$$\frac{d^2u}{dx^2} = 1 \qquad\qquad 0 < x < 1$$

$$u(0) = u(1) = 0$$

Compare it to the Galerkin approximation using the functions in Eq. (3.45) for n = 1, 2, and 3.

3.17   Find the Galerkin approximation to the solution of Eqs. (3.40) – (3.42) using $\phi_i = x^i$ for n = 3 and n = 4. Can you show that $a_i = 0$ for all n > 4?

3.18   Show that the boundary value problem

$$\frac{d}{dx}\left(k\frac{du}{dy}\right) = f(x) \qquad a < x < b$$

$$u(a) = p, \ u(b) = q$$

can be transformed into the problem

$$\frac{d}{dx}\left(k\frac{dv}{dx}\right) = g(x) \qquad a < x < b$$

$$v(a) = 0, \ v(b) = 0$$

Substitute u(x) = v(x) + u$_p$(x), where u$_p$(x) is any known function that satisfies the boundary conditions on u. Hence the approximation function can be written in the form $v_n(x) = (x - a)(x - b)(a_1 + a_2 + \dots a_n x^n)$. What are the weighting functions in the Galerkin formulation?

3.19   Consider the two-dimensional boundary value problem

$$-\left(\frac{\partial^2 u}{\partial x^2} + \frac{\partial^2 u}{\partial y^2}\right) = 1 \qquad 0 < x, y < 1$$

$$u = 0 \qquad \text{on } x = 0, \ x = 1, \ y = 0, \ y = 1$$

(a)   Show that the functions $\phi_{i,j}(x,y) = \sin(2i - 1)\pi x \ \sin(2j - 1)\pi y$, i,j = 1,2,3,... are admissible functions.

(b)   Find the Galerkin approximations to the solution u for i = j = 1 and i,j = 1,2.

(c)   Can you find a recurring relation valid for all values of i, j?

3.20   Show that for any value of the parameters b and c, the function u given by Eq. (3.49) is in $H_0^1([0,1])$. Be careful when showing that the derivative of u is in $L^2([0,1])$.

3.21   Show that if f(x) is a piecewise linear function between $0 \le x \le 1/2$ and $1/2 \le x \le 1$, then f(x) can be written as

$$f(x) = N_1(x)f_1 + N_2(x)f_2 + N_3(x)f_3$$

where $x_1 = 0, x_2 = 1/2$ and $x_3 = 1$.

3.22   Integrate and solve Eqs. (3.53) and (3.54); then use Eq. (3.52) to find Eq. (3.55).

3.23   If the error at a point in the domain is defined as $e(x) = |u^*(x) - u^h(x)|$ where u^* is the exact solution and u^h is the finite element approximation, for a uniform mesh of size h_1, show that the convergence rate is given by the slope of the line defined by the points (ln h, ln e). Hint: assume that $e = Ch^p$, where C is a constant and p is the rate of convergence. Given (h_1,e_1) and (h_2,e_2), find an expression for p, using $h_2 = h_1/2$.

3.24   Use the error estimate given in Eq. (3.9) to find the expression $e^1 = 0.3h$ in Example 3.5 (find the constant using results for h = 1/2).

3.25   Show that if the interval [0,1] is subdivided into three equal elements, the contribution to the right–hand side vector derived from

$$\int_0^1 w f \, dx$$

using linear interpolation for the function f defined in Fig. 3.15(b) is identical to the one obtained using the exact piecewise constant function f and is given by

$$\frac{1}{6}\begin{bmatrix} f_1 \\ 2f_1 \\ f_1 + f_2 \\ f_2 \end{bmatrix}$$

3.26   Find the solution to Eqs. (3.87) – (3.88) and discuss it in relation to what you would expect the solution to look like from physical considerations.

3.27   Derive Eq. (3.95) from Eqs. (3.93) and (3.94). Then find Eq. (3.96) using the data of Fig. 3.20.

3.28   Find the shape function for node number 19 in Fig. 3.18 if the coordinates of nodes 8, 9, 18, 19, and 26 are as shown in the table below.

Node	x	y
8	0.2	1.0
9	0.7	0.0
18	1.0	0.25
19	1.0	0.725
26	1.5	0.25

3.29   Fill in the details that lead to Eqs. (3.97) and (3.98). Then construct the remaining element equations and assemble them to find Eq. (3.99).

3.30   Repeat the solution of the problem in Example 3.8 using the same parameters except for $q = -1$ along $y = b$. Compare the results with Eq. (3.104) and discuss the solution.

3.31   In Example 3.9, set $\alpha = 10$, and add the interface term along $x = a/2$ to Eq. (3.103). Solve the new system and compare the solution to that of Example 3.8. Discuss the differences from a physical point of view.

3.32   Show that if $\mathbf{n}$ in Fig. 3.24 is defined as $\mathbf{n} = \mathbf{i}$ instead of $\mathbf{n} = \mathbf{i}$, Eq. (3.120) remains unchanged.

3.33   Solve the problem

$$-\left(\frac{\partial^2 u}{\partial x^2}+\frac{\partial^2 u}{\partial y^2}\right)=0 \qquad\qquad 0<x,y<1$$

$$u(0,y)=\ln\left[1.21+(y-1.5)^2\right]$$

$$u(1,y)=\ln\left[0.01+(y-1.5)^2\right]$$

$$u(x,0)=\ln\left[(x-1.1)^2+2.25\right]$$

$$u(x,1)=\ln\left[(x-1.1)^2+0.25\right]$$

using regular meshes of 4, 16, and 64 elements. Calculate the error at the midpoint $(x,y)=(0.5,0.5)$ using the exact solution

$$u(x,y)=\ln\left[(x-1.1)^2+(y-1.5)^2\right]$$

and plot the error for $u(x,y)$ and its first partial derivatives at the midpoint in $\ln h$ versus $\ln e$ to find the convergence rate in the same way as was done in Fig. 3.13.

3.34   Discretize the region in Fig. 3.2 using an appropriate finite element mesh according to what we have discussed in this chapter. Solve using the following parameters.

AlO_2 ceramic matrix:	$K = 40$ W/m K
	$h = .015$ W/m^2 K
Copper:	$k = 400$ W/m^2 K
Silicon chip:	$k = 150$ W/m K
	$h = .015$ W/m^2 K
Geometric dimensions:	$a = 2.5$ cm
	$b = 1$ cm
	$c = 0.5$ cm
	$d = 0.25$ cm
	$e = 0.01$ cm

Other quantities: $\quad\quad\quad\quad\quad T_a = 25° \, C \quad T_b = 5° \, C \quad Q = 1550 \, W/m^2$

Once you have found a solution, refine the mesh and solve again. Compare the answers to see if your solution is close to converged.

3.35   In Example 3.4, generate the plot for the error in the flux at $x = 1.0$ shown in the figure below. Use $h = 0.2, 0.1,$ and 0.05, and calculate both the derivative of the interpolant and the boundary residuals.

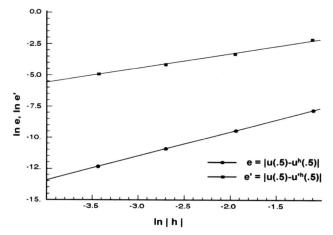

3.36   Solve the problem in Example 3.11 using uniform finite element meshes of 10 x 10 and 20 x 20 bilinear elements. Examine the error and rate of convergence at the point (0,1).

3.37   Derive Eq. (3.96).

3.38   Derive Eq. (3.98).

3.39   Derive Eq. (3.127).

3.40   Using bilinear elements, find a finite element solution to the problem

$$\nabla^2 u = 2 \quad\quad\quad \text{in } \Omega$$

$$u = 0 \quad\quad\quad \text{in } \Gamma_1$$

$$\frac{\partial u}{\partial u} = 0 \quad\quad\quad \text{in } \Gamma_2$$

where $\Omega, \Gamma_1,$ and Γ_2 are shown in the following figure.

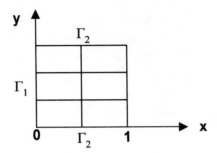

3.41   Derive the equivalent to Eq. (3.121) in three dimensions.

FOUR

HIGHER ORDER ELEMENTS

4.1 OVERVIEW

As stated in Section 3.6, the finite element method (FEM) is born when we choose the shape functions in a Galerkin weighted residuals formulation to be defined locally over subdomains of Ω that we identify as elements. In this chapter we will develop families of elements in one and two dimensions and introduce the most basic three-dimensional elements. We will restrict ourselves to the ones applicable to fluid flow and heat transfer. Hence we will not venture into the development of some classes of elements of importance in structural analysis such as C^1 and hierarchical elements or hybrid elements. However, occasionally the need arises to construct a special element, and therefore we will include the theory of blending function interpolation, which provides us with an easy method to construct elements with practically any behavior for second-order boundary value problems.

Two-dimensional quadrilateral elements will first be defined over a rectangular domain. Then we will introduce the concept of isoparametric transformations and apply it to all elements. We will follow with the section on blending function interpolation, which can be used to construct elements with special, predetermined behavior, and close with a section on three-dimensional elements.

Because the concept of shape functions is so fundamental to finite element methodology, every book on the method must devote a significant portion to it and references are difficult to choose. The paper by Gallagher (1980) is recommended for those seeking complementary reading

4.2 ONE-DIMENSIONAL ELEMENTS

In Chapter 3 we introduced the linear element, which is the lowest order C^o element. This is, of course, one of infinitely many choices, since we could choose polynomials of any degree to interpolate the unknown function over an element. To do this, however, we must introduce more nodes (or degrees of freedom) in the elements. For example, if we desire to use quadratic polynomials over an element $e=\{x\ /0 \leq x \leq h\}$, a function $u(x)$ will be approximated as

$$u(x) \cong a + bx + cx^2 \qquad 0 \leq x \leq h \tag{4.1}$$

which contains three unknown parameters. To determine the shape functions we place three nodes within the element, one at each end of the interval and one at the midpoint. Setting the nodes at $x_1 = 0$, $x_2 = h/2$, and $x_3 = h$, the shape functions become

$$N_1(x) = 1 - \frac{3x}{h} + \frac{2x^2}{h^2}$$

$$N_2(x) = \frac{4x}{h}\left(1 - \frac{x}{h}\right) \tag{4.2}$$

$$N_3(x) = \frac{x}{h}\left(\frac{2x}{h} - 1\right)$$

when $0 \leq x \leq h$ and zero otherwise.

Figure 4.1 shows the local and global quadratic shape functions. These functions interpolate a quadratic polynomial exactly over the element, and hence we can expect that a finite element approximation based on quadratic elements will be more accurate than one based on linear elements. In fact, why stop here? We can go further and construct cubic elements with two interior nodes located at a distance of $h/3$ to each end, or a quartic element with three interior nodes, etc.

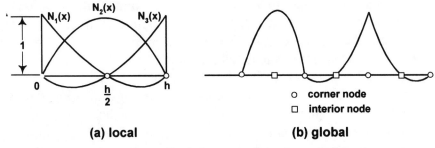

Figure 4.1 One-dimensional quadratic shape function, (a) local, and (b) global.

If we choose to approximate the function u with a polynomial of degree n, so that our element e = {x /0 ≤ x ≤ h}has n + 1 nodes located at $x_i = (i-1)h/n$, i = 1, 2, ..., n + 1, the shape functions can be readily obtained using Lagrange's formula and are given by

$$N_i(x) = \begin{cases} \dfrac{(x-x_1)(x-x_2)\cdots(x-x_{i-1})(x-x_{i+1})\cdots(x-x_{n+1})}{(x_i-x_1)(x_i-x_2)\cdots(x_i-x_{i-1})(x_i-x_{i+1})\cdots(x_i-x_{n+1})} & 0 \le x \le h \\[2mm] 0 \text{ otherwise} & i = 1, \cdots, n+1 \end{cases}$$

(4.3)

In particular, it is easy to recover the expressions in Eq. (3.50) for linear shape functions when n = 1 and Eqs. (4.2) for quadratic elements when n = 2 (Exercise 4.1).

Detailed examples utilizing one-dimensional heat conduction and quadratic elements have been given by Pepper and Heinrich (1992). A detailed treatment of one-dimensional problems with a stronger orientation towards structural analysis is given by Reddy (1984). Here we will only discuss the most important aspects of the use of higher order one-dimensional elements.

REMARKS

1. There is no particular reason for the nodes in higher order elements to be always equidistant; in fact, it is well known that Chebyshev interpolation points optimize the error distribution over an

element in the maximum norm (Isaacson and Keller, 1966). However, Chebyshev points never include the end points of the interval. Hence we cannot define conformal C^0 piecewise polynomial functions using the Chebyshev points as the nodal locations.

The location of interior nodes can also be used to achieve some particular behavior. For example, by moving the midnodes of a two-dimensional quadratic element a quarter of the way toward a corner node, a special singular element is produced with a singularity in the derivative at the corner node (Barsoum, 1976) that can be used to model crack propagation in fracture mechanics.

2. In problems of fluid flow or heat transfer, special elements are very rarely used, and only Lagrangian elements with equidistant nodes will be considered here. For such elements, the following error estimates are available (Becker et al., 1981).

Let u* be the exact solution to the boundary value problem

$$-\frac{d}{dx}\left(p(x)\frac{du}{dx}\right)+q(x)u=f(x) \qquad \Omega=\{x/a \leq x \leq b\} \quad (4.4)$$

with $p(x) \geq p_0 > 0$ and $q(x) \geq 0$, and such that u* has square integrable derivatives of order s in Ω. Let u^h be the finite element approximation to u* using a mesh of elements of size h and Lagrangian shape functions of degree n. Then

$$\left\| u^*-u^h \right\|_0 \leq C_1 h^{\alpha+1} \qquad (4.5)$$

where $\alpha = \min(n,s)$ and C_1 is a constant independent of h.

Notice that the use of higher order elements will increase the rate of convergence only if u* has square integrable derivatives of high enough order. For example, if u* is in $H^1(\Omega)$ but does not possess square integrable second-order derivatives, the use of quadratic or cubic elements will still yield an $O(h^2)$ approximation, the same as linear elements.

3. There are two ways in which a finite element approximation to any problem can be improved. The first consists in increasing the number of elements used in the mesh, therefore decreasing the size of h and, consequently, the error, in accordance with Eq. (4.5). This is called

the h-method and relies on decreasing the size of the mesh to achieve better accuracy, utilizing always the same element.

The second possibility is to keep the number of elements fixed and to increase the degree of the interpolation polynomials in the elements. In this way the number of nodes is increased and so is the order of the element. This is called the p-method. Of course, a combination of both can also be used, and this is referred to as the h-p method.

In fluid flow and heat transfer the p-method is simply not practical, and we will use the h-method exclusively. Moreover, we will use only the simplest linear or, at most, quadratic elements.

4. Let us consider heat conduction in an axisymmetric tube (Reddy, 1984). The governing equation is

$$\frac{d}{dr}\left(kr\frac{dT}{dr}\right) = 0 \qquad\qquad r_1 < r < r_2 \qquad\qquad (4.6)$$

with boundary conditions

$$T(r_1) = T_1 \quad T(r_2) = T_2 \qquad\qquad (4.7)$$

for which the exact solution is given by

$$T(x) = T_1 - (T_1 - T_2)\frac{\ln(r/r_1)}{\ln(r_2/r_1)} \qquad\qquad (4.8)$$

We choose $r_1 = 1.75$ inches, $r_2 = 3.25$ inches and the inner and outer temperatures are $T_1 = 400°F$ and $T_2 = 80°F$. The solution is approximated using six linear elements, three quadratic elements, and two cubic elements to keep the total number of nodes constant. The results using these finite element meshes are presented in Table 4.1. The maximum error using linear elements is 4.38%, quadratic elements 0.003%, and cubic elements 0.0027%.

Table 4.1 FEM approximations to axisymmetric heat conduction

r	Linear (6 elements)	Quadratic (3 elements)	Cubic (2elements)	Exact
1.75	400.0000	400.0000	400.0000	400.0000
2.00	331.0090	330.9848	330.9645	330.9736
2.25	270.1346	270.0892	270.0863	270.0879
2.50	215.6680	215.6293	215.6240	215.6242
2.75	166.3887	166.3560	166.3525	166.3553
3.00	121.3946	121.3789	121.3734	121.3767
3.25	80.0000	80.0000	80.0000	80.0000

The values in Table 4.1 clearly illustrate that there is a very significant improvement going from linear to quadratic elements. However, the gains going from quadratic to cubic elements are marginal. On the other hand, the calculation cost increases considerably as we increase the order of the elements. To obtain the element stiffness matrices requires significantly more operations for the higher order elements. However, more important is the fact that the bandwidth of the coefficient matrix becomes larger with higher order elements. For linear elements the stiffness matrix is tridiagonal, for quadratic elements pentadiagonal, and for cubic elements heptadiagonal. Therefore the solution of the resulting linear system of equations becomes significantly more difficult and more costly as the order of the elements increases.

The behavior of the error in our example is typical of what we can expect in the finite element solution of second-order self-adjoint equations. In many cases a few quadratic elements will yield solutions of much better accuracy than a much larger number of linear elements, and their use is therefore desirable. However, the use of cubic or higher order elements is rarely justified. Furthermore, in two dimensions the bandwidth produced by quadratic elements severely limits its application in many flow problems. The use of cubic elements is out of the question in two dimensions. Therefore we will only utilize linear and quadratic elements in this book.

5. The Lagrangian elements discussed in this section are $C^0(\Omega)$ elements. Hence, even though they may be of any degree of smoothness in the interior of the elements, their derivatives are discontinuous at interelement boundaries. However, the derivatives are indeed approximated more accurately the higher the element order, provided

that the solution has enough smoothness. If the conditions for Eq. (4.5) hold, we also have

$$\left\| \frac{d}{dx}(u*-u^h) \right\|_0 \leq C_2 h^\alpha \qquad (4.9)$$

where C_2 is a constant independent of h.

     6. There are other important families of elements based on piecewise polynomial interpolation that are not of the Lagrangian type. An example is Hermite polynomials, which are based on interpolating derivatives as well as the function at the nodes. The cubic Hermite two-noded element (Exercise 4.3) is a $C^1(\Omega)$ element for which both the function and its first derivative are continuous over the entire domain Ω. This element is widely used in finite element simulations of beam bending in solid mechanics (Reddy, 1984). It has also been successfully used to simulate transport in aquifers (Pinder and Gray, 1977). Smooth finite element approximations can also be obtained using splines as shape functions (de Boor, 1978). However, these are rarely used in problems of interest to us.

     Elements need not be based on polynomial approximations. Hemker (1977) proposed the use of exponential two-noded elements with shape functions of the form

$$N_1(x) = \frac{e^\gamma - e^{\gamma x/h}}{e^\gamma - 1} \qquad (4.10a)$$

$$N_2(x) = \frac{e^{\gamma\left(\frac{x}{h}-1\right)} - e^\gamma}{1 - e^{-\gamma}} \qquad (4.10b)$$

where γ is a parameter that depends on the problem data, over an element $e = \{x \ /0 \leq x \leq h\}$, and successfully used them to treat singular perturbation equations. Wachpress (1975) developed elements based on rational functions. Indeed, we can construct elements with any kind of functional variation we wish within the elements. We need only make sure that they will produce converging approximation sequences of solutions.

7. The Lagrangian elements based on Eq. (4.3) satisfy the following:

$$\text{(i)} \quad \sum_{i=1}^{n+1} N_i(x) \equiv 1 \tag{4.11}$$

$$\text{(ii)} \quad \sum_{i=1}^{n+1} \frac{dN_i}{dx} \equiv 0 \tag{4.12}$$

$$\text{(iii)} \quad x = \sum_{i=1}^{n+1} N_i(x) x_i \tag{4.13}$$

(iv) A Lagrangian element of degree n interpolates *exactly* any polynomial of degree k, with k = 0, 1, 2, ..., n.

To prove the above properties is not difficult and is left to Exercise 4.5; in particular property iii is important in the context of isoparametric transformations defined later.

4.3 TWO-DIMENSIONAL ELEMENTS

A very important difference between the solution of one-dimensional and two- or three-dimensional problems is the necessity to deal with geometry in the latter cases. The type, size, and density (number) of elements in two-dimensional domains become noticeably more important and significant in the overall solution; three-dimensional problems are considerably more troublesome. In Chapter 3 we introduced the bilinear quadrilateral element as a natural extension of the one-dimensional linear element to two dimensions. However, from a topological point of view, triangular elements are simpler than quadrilaterals, and we will discuss those first. We will begin with the 3-node linear triangular element, followed by higher order triangles utilizing Pascal's triangle.

4.3.1 Triangular Elements

The simplest two-dimensional figure that defines an area is the triangle. The simplest triangular element is obtained defining a linear interpolation field of the form

$$u(x,y) \cong a + bx + cy \qquad (4.14)$$

and placing the nodes at the corners of the triangle. The shape functions can be written in terms of the nodal coordinates as

$$N_1(x,y) = \frac{1}{2A}\left[x_2 y_3 - x_3 y_2 + (y_2 - y_3)x + (x_3 - x_2)y\right]$$

$$N_2(x,y) = \frac{1}{2A}\left[x_3 y_1 - x_1 y_3 + (y_3 - y_1)x + (x_1 - x_3)y\right] \qquad (4.15)$$

$$N_3(x,y) = \frac{1}{2A}\left[x_1 y_2 - x_2 y_1 + (y_1 - y_2)x + (x_2 - x_1)y\right]$$

where the area A is given by

$$2A = (x_2 y_3 - x_3 y_2) + (x_3 y_1 - x_1 y_3) + (x_1 y_2 - x_2 y_1) \qquad (4.16)$$

and the nodes are numbered counterclockwise as in Fig. 4.2, where the shape function $N_1(x,y)$ is also shown. These elements have been discussed extensively by Pepper and Heinrich (1992).

Higher order triangular elements can be easily developed by association with Pascal's triangle, as shown in Fig. 4.3. In the triangle, line n contains the terms that must be added to a complete polynomial of degree n − 1 in order to obtain a complete polynomial of degree n.

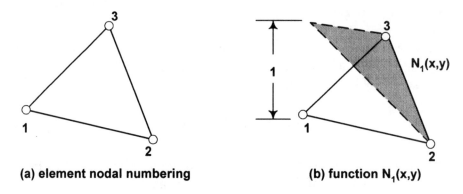

(a) element nodal numbering   **(b) function $N_1(x,y)$**

Figure 4.2 Linear triangular element and shape function $N_1(x,y)$.

Notice that in Fig. 4.3, we start from n = 0 for a constant polynomial. Furthermore, we can visualize the location of the nodes in the triangular element from the location of the terms in Pascal's triangle as shown in Fig. 4.3. Notice that the total number of nodes for an interpolation polynomial of order n is given by $n_e = 1/2(n + 1)(n + 2)$.

The shape functions can be more easily obtained if we use area coordinates. If we join any point P in the triangle to the vertices of the triangle, we define three areas, A_1, A_2, and A_3, as shown in Fig. 4.4.

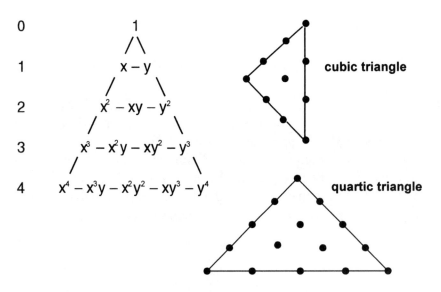

Figure 4.3 Pascal's triangle and corresponding triangular elements.

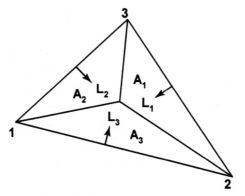

Figure 4.4 Area natural coordinates for a triangle.

A coordinate system that uniquely represents every point in the triangle is given by

$$L_i = \frac{A_i}{A} \qquad i = 1,2,3 \qquad (4.17)$$

If the nodes are uniformly distributed along the element sides, the shape functions are easily constructed in this coordinate system, also called the *natural* coordinate system for the triangle.

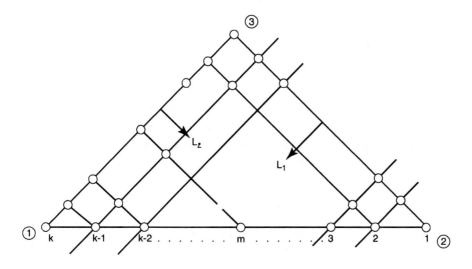

Figure 4.5 General triangle with k nodes along each side.

Suppose we want to construct an element with k nodes on a side, i.e., the one corresponding to n = k−1 in Pascal's triangle, as shown in Fig. 4.5. It is easy to show (Exercise 4.6) that the coordinates L_i, i = 1, 2, 3 are the normalized distances from the point P to side i, i.e., in Fig. 4.4, L_1 is the orthogonal distance from point P to side 2-3 divided by the orthogonal distance from node 1 to side 2-3. Hence, if for example, a node lies on line m parallel to side 1, its shape function must vanish on lines parallel to sides 2 and 3 except for those that contain the node. Therefore, if we observe that the coordinates L_1 that define each of the lines parallel to side 1 are given by

$$L_1 = 0, \frac{1}{k-1}, \frac{2}{k-1}, \cdots, 1 \text{ or } \frac{j-1}{k-1}, \quad j = 1, 2, \cdots k$$

the shape function for, say, node m in Fig. 4.5 must contain factors of the form

$$L_1, L_2, \left(L_1 - \frac{1}{k-1}\right)\left(L_2 - \frac{1}{k-1}\right), \text{ etc.}$$

Let us illustrate this further using the following example.

Example 4.1

Find the shape functions for the cubic triangular element. The node numbering is shown in Fig. 4.6(a); the coordinate locations for the nodes are given by 0, 1/3, 2/3, and 1.

Consider first the corner nodes. It is easily seen that these shape functions will depend only on one coordinate, the one corresponding to the node, because they must vanish along lines parallel to their own side. Therefore the shape functions are given by Lagrange's formula

$$N_i = \frac{L_i\left(L_i - \frac{1}{3}\right)\left(L_i - \frac{2}{3}\right)}{(1-0)\left(1-\frac{1}{3}\right)\left(1-\frac{2}{3}\right)} = \frac{1}{2}L_i(3L_i - 1)(3L_i - 2)$$

$$(4.18)$$

$$i = 1, 2, 3$$

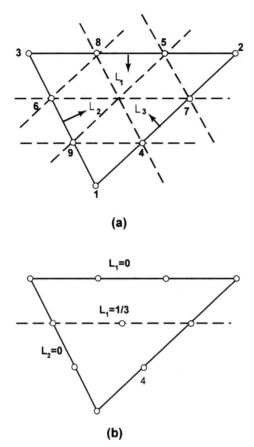

Figure 4.6 Cubic triangle, (a) nodal numbering, (b) lines along which $N_4(L_1, L_2, L_3)$ must vanish.

For the nodes lying on the sides of the triangle, such as node 4, we have

$$N_4 = \frac{L_1\left(L_1 - \dfrac{1}{3}\right)}{\dfrac{2}{3}\left(\dfrac{2}{3} - \dfrac{1}{3}\right)}\frac{L_2}{\left(\dfrac{1}{3}\right)} = \frac{9}{2}L_1L_2(3L_1 - 1) \qquad (4.19a)$$

where it is suggested that the shape function is obtained as a combination of those partial Lagrangian polynomials that include the lines parallel to sides 1 and 2 that lie between the node for which we wish to determine the shape function and the sides that do not contain the node under consideration, sides 1 and 2 in this case. Notice that those factors representing lines beyond the point, such as $(L_2 - 2/3)$ in Fig. 4.6 need not appear. The shape function already vanishes at nodes 5 and 7 because of the factor L_1, and $(L_1 - 2/3)$. The same is true for $(L_2 - 1)$, and $(L_1 - 1)$ need not appear because of the presence of the factor L_2 (as shown in Fig. 4.6b). Similarly we can obtain the rest of the shape functions for nodes 5 through 9. These turn out to be

$$N_5 = \frac{9}{2} L_2 L_3 (3 L_2 - 1) \tag{4.19b}$$

$$N_6 = \frac{9}{2} L_1 L_3 (3 L_3 - 1) \tag{4.19c}$$

$$N_7 = \frac{9}{2} L_1 L_2 (3 L_2 - 1) \tag{4.19d}$$

$$N_8 = \frac{9}{2} L_2 L_3 (3 L_3 - 1) \tag{4.19e}$$

$$N_9 = \frac{9}{2} L_1 L_3 (3 L_1 - 1) \tag{4.19f}$$

Shape functions corresponding to nodes on a side on the triangle depend only on the two coordinates corresponding to the other two sides.

Finally, for interior nodes such as node 10 in our case, the same conditions as in the previous case apply to all three sides. That is, the shape function must contain the factors that make it vanish at every line between the node and the side corresponding to the coordinate. Hence

$$N_{10} = \frac{L_1}{\left(\frac{1}{3}\right)} \frac{L_2}{\left(\frac{1}{3}\right)} \frac{L_3}{\left(\frac{1}{3}\right)} = 27 \, L_1 L_2 L_3 \tag{4.20}$$

Shape functions for interior nodes depend on all three natural coordinates. The construction of the shape functions for other triangular elements is left as an exercise. ∎

REMARKS

1. It is easily found (Exercise 4.7) that the shape functions for the linear triangle are $N_i = L_i$, $i = 1,2,3$, and this element must be capable of interpolating linear functions exactly over the element. In particular, it interpolates the functions $f(x,y) = x$ and $g(x,y) = y$; therefore we must have

$$x = L_1x_1 + L_2x_2 + L_3x_3$$

$$(4.21)$$

$$y = L_1y_1 + L_2y_2 + L_3y_3$$

plus we have $L_1 + L_2 + L_3 = 1$. Replacing this last expression into Eqs. (4.21) and re-writing in matrix form, we have

$$\begin{bmatrix} x_1 - x_3 & x_2 - x_3 \\ y_1 - y_3 & y_2 - y_3 \end{bmatrix} \begin{bmatrix} L_1 \\ L_2 \end{bmatrix} = \begin{bmatrix} x - x_3 \\ y - y_3 \end{bmatrix} \qquad (4.22)$$

the 2 x 2 matrix is nonsingular as long as the three points are not colinear. Hence it can be inverted to obtain L_1, L_2, and L_3 as a function of x and y.

2. The shape functions obtained in Eqs. (4.18) – (4.20) help us justify the nodal numbering used in the cubic triangle. Besides the fact that we want to be able to easily associate the corner nodes with the coordinate directions, by assigning a cyclic numbering to the nodes, the shape functions taken in groups of 3 always have the exact same form.

Notice also that the condition $N_i(x_j, y_j) = \delta_{ij}$ is satisfied, since the shape functions are of the Lagrangian type in the natural triangular coordinates.

3. The shape functions in natural coordinates are independent of the shape of the triangle. This is particularly appealing when we are dealing with highly irregular geometries that may require a large variety of very differently shaped triangles to discretize it properly. In fact, the ability of the triangle to discretize any kind of geometric figure with relative ease is the main reason for the wide use of triangular elements. Moreover, from this point of view, triangular elements are always better than quadrilateral elements. Very powerful mesh generators have been

developed based on the triangular geometry that can automatically discretize extremely complex regions. Although in the last years much progress has been made in this area using quadrilateral elements, these mesh generators still lack the versatility and degree of automation of those based on triangles.

On the other hand, the use of triangular elements can be difficult, especially when applied to incompressible flows. The advantages and drawbacks of triangular elements will be discussed in more detail as we progress further.

4.3.2 Rectangular Elements

A rectangular (and more generally quadrilateral) element is defined by four corner points and therefore is no longer linear, which makes it more complex than a triangular element. However, there is a greater variety of quadrilateral elements, and in general, they can offer many advantages over the use of triangles. We will start by presenting the most common families of rectangular elements. These will be extended to general quadrilaterals in the next section on isoparametric transformations.

The simplest way to obtain rectangular elements consists in taking the product (also referred to as the tensor product) of one-dimensional elements. In this fashion we generate the family of Lagrangian C^o elements that are bilinear, biquadratic, etc., and contain 2^2, 3^2, 4^2, ... nodes as shown in Fig. 4.7. To obtain the shape functions we only need to know the form of the shape functions, in one dimension, and the two-dimensional function at a node is obtained as the product of the one-dimensional functions that would correspond to that node in the x and y directions, respectively.

Example 4.2

As an example, let us find the shape functions for the nine-noded biquadratic element. In Fig. 4.8 we have established the axial system,

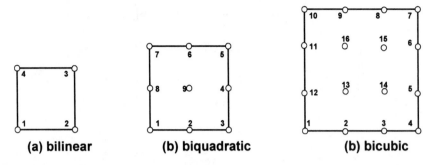

Figure 4.7 2-D quadrilateral elements, (a) bilinear, (b) biquadratic, and (c) bicubic.

and we show the quadratic one-dimensional functions in each direction. The shape functions are then given by

$$N_1(x,y) = \overline{N}_1(x)\overline{N}_1(y) = \frac{1}{4a^2b^2}xy(x-a)(y-b)$$

$$N_2(x,y) = \overline{N}_2(x)\overline{N}_1(y) = \frac{1}{2a^2b^2}(a^2-x^2)y(y-b)$$

$$N_3(x,y) = \overline{N}_3(x)\overline{N}_1(y) = \frac{1}{4a^2b^2}xy(x+a)(y-b)$$

$$N_4(x,y) = \overline{N}_3(x)\overline{N}_2(y) = \frac{1}{2a^2b^2}x(x+a)(b^2-y^2)$$

$$N_5(x,y) = \overline{N}_3(x)\overline{N}_3(y) = \frac{1}{4a^2b^2}xy(x+a)(y+b)$$

$$N_6(x,y) = \overline{N}_2(x)\overline{N}_3(y) = \frac{1}{2a^2b^2}(a^2-x^2)(y+b) \qquad (4.23)$$

$$N_7(x,y) = \overline{N}_1(x)\overline{N}_3(y) = \frac{1}{4a^2b^2}(x-a)(y+b)$$

$$N_8(x,y) = \overline{N}_1(x)\overline{N}_2(y) = \frac{1}{2a^2b^2}x(x-a)(b^2-y^2)$$

$$N_9(x,y) = \overline{N}_2(x)\overline{N}_2(y) = \frac{1}{a^2b^2}(a^2-x^2)(b^2-y^2)$$

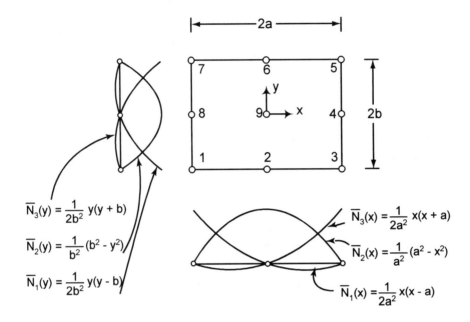

$$\overline{N}_3(y) = \frac{1}{2b^2}\, y(y + b)$$

$$\overline{N}_2(y) = \frac{1}{b^2}\, (b^2 - y^2)$$

$$\overline{N}_1(y) = \frac{1}{2b^2}\, y(y - b)$$

$$\overline{N}_3(x) = \frac{1}{2a^2}\, x(x + a)$$

$$\overline{N}_2(x) = \frac{1}{a^2}\, (a^2 - x^2)$$

$$\overline{N}_1(x) = \frac{1}{2a^2}\, x(x - a)$$

Figure 4.8 Biquadratic element and one-dimensional quadratic shape functions.

Another important family of rectangular elements is known as the *serendipity* elements. These elements differ from the Lagrangian family just discussed in that they do not contain any interior nodes. Examples of such elements are the eight-noded quadratic and the 12-noded cubic elements shown in Fig. 4.9.

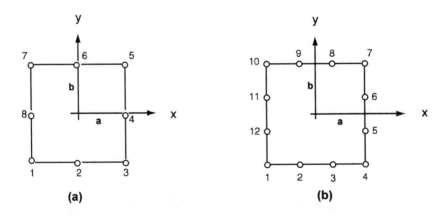

Figure 4.9 Serendipity elements for (a) quadratic and (b) cubic.

The shape functions for these elements are not so easily obtained, and initially, they were found by a trial and error process, thus the name serendipity. However, with the development of blending function interpolation techniques, a systematic way to construct the shape functions is now available that will be explained later. For the eight-noded quadratic element of Fig. 4.9(a) the shape functions are given by

$$N_1(x,y) = \frac{-1}{4a^2b^2}(a-x)(b-y)(bx+ay+ab)$$

$$N_2(x,y) = \frac{1}{2ab^2}\left(a^2-x^2\right)(b-y)$$

$$N_3(x,y) = \frac{1}{4a^2b^2}(a+x)(b-y)(bx-ay-ab)$$

$$N_4(x,y) = \frac{1}{2a^2b}(a+x)\left(b^2-y^2\right)$$

$$N_5(x,y) = \frac{1}{4a^2b^2}(a+x)(b+y)(bx+ay-ab) \qquad (4.24)$$

$$N_6(x,y) = \frac{1}{2ab^2}\left(a^2-x^2\right)(b+y)$$

$$N_7(x,y) = \frac{-1}{4a^2b^2}(a-x)(b+y)(bx-ay+ab)$$

$$N_8(x,y) = \frac{1}{2a^2b}(a-x)\left(b^2-y^2\right)$$

The shape functions for the 12-noded cubic element can be found in many finite element textbooks, e.g., Reddy (1984). We will return to them in Section 4.5, where we will see that we can construct rectangular elements with virtually any number of nodes along each side.

4.3.3 Error bounds for Two-Dimensional Elements

The following result is valid for all the two-dimensional elements discussed above (Strang and Fix, 1973; Oden and Carey, 1983). If u* is the exact solution to Eqs. (3.9) – (3.12), where L is given by Eq. (3.15) and u^h is the finite element approximation to u^* using elements that contain a complete polynomial of order k, then u^h satisfies

$$\left\| u^* - u^h \right\|_0 \le C_1 h^{k+1} \left\| f \right\|_0 \qquad (4.25)$$

$$\left\| \frac{\partial}{\partial x_i} \left(u^* - u^h \right) \right\|_0 \le C_2 h^k \left\| u^* \right\|_k \qquad (4.26)$$

the k-th norm in the inequality Eq. (4.26) is defined in Appendix A, and C_1 and C_2 are constants independent of h.

REMARKS

1. Notice that inequality Eq. (4.25) is identical to Eq. (3.107) for the case of linear elements, when k = 1. These error estimates are, in fact, valid for one, two, or three dimensions, and they are "optimal" in the sense that we cannot expect to have faster convergence in any norm. The technique used to obtain these error estimates is due to Aubin (1967) and Nitsche (1968) and is called the "Aubin-Nitsche trick." An important additional result of this analysis is that "the average of the error in a finite element approximation is much smaller than the typical error at a point." This fact is expressed as

$$\left| \int_\Omega \left(u^* - u^h \right) d\Omega \right| \sim h^{k+2} \qquad (4.27)$$

and what it means in practice is that the error must alternate very rapidly in sign. Therefore there must be points within each element where the solution is exceptionally accurate (error = 0), and we would like to find those points. We will exploit this result further later on in the context of the evaluation of the derivatives.

2. The requirement that the element shape functions must contain the *complete* polynomial of order k means that the shape functions must contain every term in the k-th line of Pascal's triangle as shown in Fig. 4.3. Hence the linear triangle and bilinear quadrilateral exhibit exactly the same rate of convergence because the bilinear quadrilateral contains

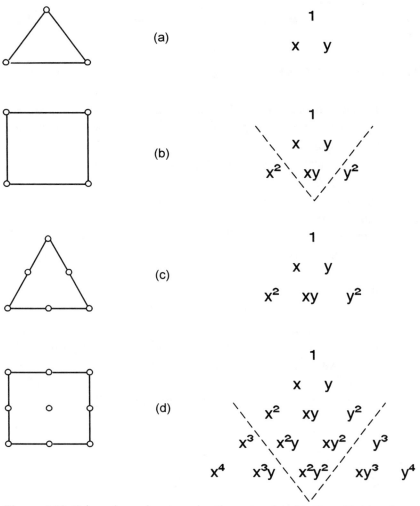

Figure 4.10 Triangular and rectangular elements related to Pascal's triangle.

only a complete polynomial of degree k = 1. The additional term xy does not change the rate of convergence in the bilinear element because the other two quadratic terms, and x^2 and y^2 are missing, as shown in Fig. 4.10. A similar situation applies to quadratic elements.

3. In practice, we usually obtain better *accuracy* using quadrilateral elements than triangular elements, even though, as is the case for the linear triangle and the bilinear rectangle, both elements exhibit the same rate of convergence. The difference between accuracy

and rate of convergence will be explored in more detail in Chapter 5. Superconvergence will also be investigated further in subsequent chapters.

4.4 ISOPARAMETRIC ELEMENTS

In the previous section we constructed rectangular elements and argued that they usually offer advantages over the use of triangular elements. However, the rectangular geometry is very restrictive and we would like to be able to build general quadrilateral elements such as depicted in Fig. 4.11, in order to be able to deal with more general geometry. In fact, we would also like to be able to construct elements with curved sides to better match curved boundaries.

There are two important difficulties that must be overcome in order to achieve these goals. These are as follows.

4.4.1 Difficulty 1.

It becomes impossible to find the shape functions for an element such as depicted in Fig. 4.11 because the shape functions will no longer be bilinear. We can always find bilinear functions of the form $u(x,y) = a + bc + cy + dxy$ that interpolate u at the four corner nodes for elements e_1 and e_2 in Fig. 4.11. However, those shape functions will not coincide

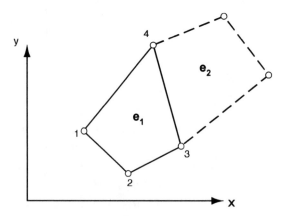

Figure 4.11 General quadrilateral elements in a global frame of reference.

along side 3-4, and conformity will be lost (Exercise 4.9). Hence the shape functions cannot be bilinear.

Example 4.3

The shape functions for the nonrectangular element shown in Fig. 4.12 are given by

$$N_1(x,y) = \frac{1}{4ab}(a - x + cy)(b - y)$$

$$N_2(x,y) = \frac{1}{4ab}(a + x - cy)(b - y)$$

$$N_3(x,y) = \frac{1}{4ab}(a + x - cy)(b + y)$$

$$N_4(x,y) = \frac{1}{4ab}(a - x + cy)(b + y)$$

(4.28)

where $c = \cot\theta = b/L$. To show that the shape functions in Eq. (4.28) also satisfy the conditions in Remark 6 of Section 4.2 is left to Exercise 4.10.

Notice that the shape functions, even for this simple generalization of a rectangular element, are not bilinear; they contain quadratic terms in y. ∎

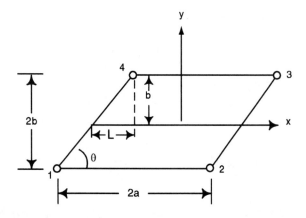

Figure 4.12 Nonrectangular rhomboidal element.

For elements with a more general quadrilateral shape, the shape functions become too complicated to find.

4.4.2 Difficulty 2.

Even if we resolve the difficulty of finding the shape functions, we will be faced next with the problem of integrating over arbitrary quadrilateral regions such as shown in Fig. 4.11. This would be extremely difficult to program if it is indeed possible to devise general formulas. For those readers who may not appreciate off hand the magnitude of this problem, a simple illustrative example is provided by Exercise 4.11.

Both of these difficulties can be resolved using transformations of coordinates. The novelty in finite element methodology is that the transformations will be defined locally, element wise, and hence can be applied to arbitrary geometries. At the same time, they will introduce the need for numerical integration, since as we will see, the integrals in the weak weighted residuals formulation will no longer be simple polynomial functions.

To resolve the above-mentioned difficulties in a convenient and very powerful way, the concept of *isoparametric* transformations was introduced by Taig (1961) for the bilinear rectangular element and later generalized by Irons (1966) to general rectangular elements. The idea is based on performing a local (element by element) transformation between a general quadrilateral element in the global coordinate system and a "parent" rectangular element defined in a $\xi - \eta$ coordinate system in the square $-1 \leq \xi, \eta \leq 1$ as depicted in Fig. 4.13 for a four-noded element.

We can now transform our integrals so that

$$\iint_e F\left(u(x,y), \frac{\partial u}{\partial x}, \frac{\partial u}{\partial y}\right) dx\, dy$$

$$= \int_{-1}^{1} \int_{-1}^{1} F\left(u(x(\xi,\eta), y(\xi,\eta)), \frac{\partial u}{\partial x}, \frac{\partial u}{\partial y}\right) |\det \mathbf{J}| d\xi\, d\eta \qquad (4.29)$$

where $\mathbf{J}$ is the Jacobian of the transformation, which will be discussed shortly (Amazigo and Rubenfeld, 1980).

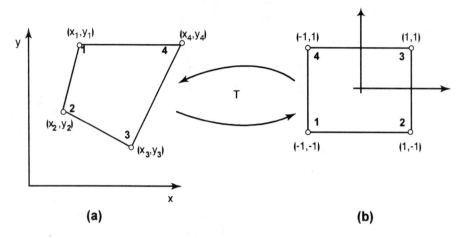

Figure 4.13 Bilinear isoparametric elements, (a) actual element and (b) parent element.

The isoparametric transformation itself is easily obtained, recalling Eq. (4.13),

$$T\begin{bmatrix} \xi \\ \eta \end{bmatrix} = \begin{bmatrix} x \\ y \end{bmatrix} = \sum_{i=1}^{N_e} N_i(\xi, \eta) \begin{bmatrix} x_i \\ y_i \end{bmatrix} \qquad (4.30)$$

where N_e is the number of nodes in the element and $N_i(\xi,\eta)$ are the shape functions for the corresponding parent element, in the square $-1 \leq \xi, \eta \leq 1$. Actually, this transformation is the inverse of what we really need, since it maps the parent element, not the actual element. However, it has some very important advantages.

1. It is clearly defined once we know the coordinates of the nodes (x_i, y_i) in the global system and the shape functions $N_i(\xi,\eta)$ in the square parent system of coordinates. These data are essential to all finite element models and are immediately available.

2. Given any function, $\phi(x,y)$ we can find the derivatives of ϕ with respect to x and y in the parent coordinate system immediately; however, we need the derivatives in the x–y plane. To obtain these, we apply the chain rule,

$$\frac{\partial \phi}{\partial \xi} = \frac{\partial \phi}{\partial x}\frac{\partial x}{\partial \xi} + \frac{\partial \phi}{\partial y}\frac{\partial y}{\partial \xi}$$

$$\frac{\partial \phi}{\partial \eta} = \frac{\partial \phi}{\partial x}\frac{\partial x}{\partial \eta} + \frac{\partial \phi}{\partial y}\frac{\partial y}{\partial \eta}$$

written in matrix form

$$\begin{bmatrix} \dfrac{\partial \phi}{\partial \xi} \\[2ex] \dfrac{\partial \phi}{\partial \eta} \end{bmatrix} = \begin{bmatrix} \dfrac{\partial x}{\partial \xi} & \dfrac{\partial y}{\partial \xi} \\[2ex] \dfrac{\partial x}{\partial \eta} & \dfrac{\partial y}{\partial \eta} \end{bmatrix} \begin{bmatrix} \dfrac{\partial \phi}{\partial x} \\[2ex] \dfrac{\partial \phi}{\partial y} \end{bmatrix} \tag{4.31}$$

where

$$\mathbf{J} = \begin{bmatrix} \dfrac{\partial x}{\partial \xi} & \dfrac{\partial y}{\partial \xi} \\[2ex] \dfrac{\partial x}{\partial \eta} & \dfrac{\partial y}{\partial \eta} \end{bmatrix}$$

is the Jacobian matrix of the transformation. Assuming for the time being that $\mathbf{J}$ is nonsingular, we have

$$\begin{bmatrix} \dfrac{\partial \phi}{\partial x} \\[2ex] \dfrac{\partial \phi}{\partial y} \end{bmatrix} = \mathbf{J}^{-1} \begin{bmatrix} \dfrac{\partial \phi}{\partial \xi} \\[2ex] \dfrac{\partial \phi}{\partial \eta} \end{bmatrix} = \frac{1}{\det \mathbf{J}} \begin{bmatrix} \dfrac{\partial y}{\partial \eta} & -\dfrac{\partial y}{\partial \xi} \\[2ex] -\dfrac{\partial x}{\partial \eta} & \dfrac{\partial x}{\partial \xi} \end{bmatrix} \begin{bmatrix} \dfrac{\partial \phi}{\partial \xi} \\[2ex] \dfrac{\partial \phi}{\partial \eta} \end{bmatrix} \tag{4.32}$$

and the derivatives are, in expanded form,

$$\frac{\partial \phi}{\partial x} = \frac{1}{\det \mathbf{J}} \left(\frac{\partial y}{\partial \eta} \frac{\partial \phi}{\partial \xi} - \frac{\partial y}{\partial \xi} \frac{\partial \phi}{\partial \eta} \right)$$

$$\frac{\partial \phi}{\partial y} = \frac{1}{\det \mathbf{J}} \left(-\frac{\partial x}{\partial \eta} \frac{\partial \phi}{\partial \xi} + \frac{\partial x}{\partial \xi} \frac{\partial \phi}{\partial \eta} \right)$$

(4.33)

Example 4.4

Let us consider the Laplace equation over an arbitrary element e:

$$\frac{\partial^2 u}{\partial x^2} + \frac{\partial^2 u}{\partial y^2} = 0$$

The weak form of the equation is

$$I = \iint_e \left(\frac{\partial w}{\partial x} \frac{\partial u}{\partial x} + \frac{\partial w}{\partial y} \frac{\partial u}{\partial y} \right) dx \, dy = 0$$

We want to transform the integral into an integral in the parent coordinate system as in Eq. (4.21). Using Eqs. (4.32) and (4.33) we obtain

$$I = \int_{-1}^{1} \int_{-1}^{1} \frac{1}{\det \mathbf{J}} \left[\left(\frac{\partial y}{\partial \eta} \frac{\partial w}{\partial \xi} - \frac{\partial y}{\partial \xi} \frac{\partial w}{\partial \eta} \right) \left(\frac{\partial y}{\partial \eta} \frac{\partial u}{\partial \xi} - \frac{\partial y}{\partial \xi} \frac{\partial u}{\partial \eta} \right) \right.$$

$$\left. + \left(-\frac{\partial x}{\partial \eta} \frac{\partial w}{\partial \xi} + \frac{\partial x}{\partial \xi} \frac{\partial w}{\partial \eta} \right) \left(-\frac{\partial x}{\partial \eta} \frac{\partial u}{\partial \xi} + \frac{\partial x}{\partial \xi} \frac{\partial u}{\partial \eta} \right) \right] d\xi \, d\eta$$

where u an w are expressed in terms of shape functions in the ξ–η plane and the derivatives $\partial x/\partial \xi$, $\partial y/\partial \xi$, $\partial x/\partial \eta$, and $\partial y/\partial \eta$ can be calculated from Eq. (4.30). ∎

We now concentrate on the bilinear quadrilateral as shown in Fig. 4.13, where the set of local element nodes and the nodes in the parent element must both be numbered in the same direction (counterclockwise). The isoparametric transformation is given by

$$\begin{bmatrix} x \\ y \end{bmatrix} = \begin{bmatrix} \sum_{i=1}^{4} N_i(\xi,\eta)x_i \\ \sum_{i=1}^{4} N_i(\xi,\eta)y_i \end{bmatrix} \tag{4.34}$$

The shape functions in the parent element are

$$N_1 = \frac{1}{4}(1-\xi)(1-\eta) \tag{4.35a}$$

$$N_2 = \frac{1}{4}(1+\xi)(1-\eta) \tag{4.35b}$$

$$N_3 = \frac{1}{4}(1+\xi)(1+\eta) \tag{4.35c}$$

$$N_4 = \frac{1}{4}(1-\xi)(1+\eta) \tag{4.35d}$$

Clearly,

$$\begin{bmatrix} x(-1,-1) \\ y(-1,-1) \end{bmatrix} = \begin{bmatrix} x_1 \\ y_1 \end{bmatrix}$$

Hence node 1 in the parent element is mapped to node 1 in the actual element, and similarly for nodes 2, 3, and 4. Next we examine the effect of the transformation on the element sides. Let us set $\eta = -1$, which gives us the line between nodes 1 and 2 in the parent element, and the transformation reduces to

$$x = \frac{1}{2}(1-\xi)x_1 + \frac{1}{2}(1+\xi)x_2$$

$$y = \frac{1}{2}(1-\xi)y_1 + \frac{1}{2}(1+\xi)y_2$$

eliminating ξ from the above equations, we obtain

$$y = \frac{y_2 - y_1}{x_2 - x_1} x + \frac{x_2 y_1 - x_1 y_2}{x_2 - x_1}$$

which is the equation of the straight line through the points (x_1, y_1) and (x_2, y_2). Hence each of the sides of the parent elements is mapped to the corresponding side in the actual element. Finally, the point $(\xi, \eta) = (0,0)$ is mapped to the centroid of the actual element $((x_1 + x_2 + x_3 + x_4)/4, (y_1 + y_2 + y_3 + y_4)/4)$; hence the entire interior of the parent element must be mapped into the interior of the actual element.

The Jacobian of the transformation and derivatives of the shape functions are also easily obtained. We have

$$\mathbf{J} = \begin{bmatrix} \sum_{i=1}^{4} \frac{\partial N_i}{\partial \xi} x_i & \sum_{i=1}^{4} \frac{\partial N_i}{\partial \xi} y_i \\ \sum_{i=1}^{4} \frac{\partial N_i}{\partial \eta} x_i & \sum_{i=1}^{4} \frac{\partial N_i}{\partial \eta} y_i \end{bmatrix} \tag{4.36}$$

so that

$$\det \mathbf{J} = \left(\sum_{i=1}^{4} \frac{\partial N_i}{\partial \xi} x_i \right) \left(\sum_{i=1}^{4} \frac{\partial N_i}{\partial \eta} y_i \right) - \left(\sum_{i=1}^{4} \frac{\partial N_i}{\partial \xi} y_i \right) \left(\sum_{i=1}^{4} \frac{\partial N_i}{\partial \eta} x_i \right) \tag{4.37}$$

and hence

$$\frac{\partial}{\partial x} = \frac{1}{\det \mathbf{J}} \left(\sum_{i=1}^{4} \frac{\partial N_i}{\partial \eta} y_i \right) \frac{\partial}{\partial \xi} - \frac{1}{\det \mathbf{J}} \left(\sum_{i=1}^{4} \frac{\partial N_i}{\partial \xi} y_i \right) \frac{\partial}{\partial \eta}$$

$$\tag{4.38}$$

$$\frac{\partial}{\partial y} = \frac{-1}{\det \mathbf{J}} \left(\sum_{i=1}^{4} \frac{\partial N_i}{\partial \eta} x_i \right) \frac{\partial}{\partial \xi} + \frac{1}{\det \mathbf{J}} \left(\sum_{i=1}^{4} \frac{\partial N_i}{\partial \xi} x_i \right) \frac{\partial}{\partial \eta}$$

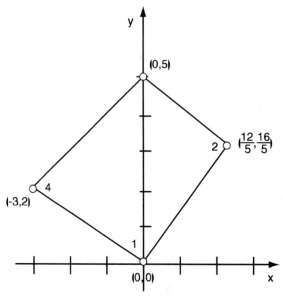

Figure 4.14 Isoparametric element for Example 4.5.

Example 4.5

Let us formulate the integral I of Example 4.4 for the Laplacian operator over the element shown in Fig. 4.14. From Eq. (4.34), we obtain

$$x = \frac{1}{20}\left(-3 + 27\xi - 27\eta + 3\xi\eta\right)$$

$$y = \frac{1}{20}\left(51 + 31\xi + 19\eta - \xi\eta\right)$$

the Jacobian of the transformation is

$$\mathbf{J} = \begin{bmatrix} \dfrac{1}{20}(27 + 3\eta) & \dfrac{1}{20}(31 - \eta) \\ \dfrac{1}{20}(-27 + 3\xi) & \dfrac{1}{20}(19 - \xi) \end{bmatrix}$$

with determinant

$$\det \mathbf{J} = \frac{1}{200}(675 - 60\xi + 15\eta) = d(\xi, \eta)$$

Notice that det $\mathbf{J}$ is always positive in $1 \leq \xi, \eta \leq 1$; hence the transformation is invertible.

The derivatives are given by

$$\frac{\partial}{\partial x} = \frac{19 - \xi}{20\, d(\xi, \eta)} \frac{\partial}{\partial \xi} - \frac{31 - \eta}{20\, d(\xi, \eta)} \frac{\partial}{\partial \eta}$$

$$\frac{\partial}{\partial y} = -\frac{3\xi - 27}{20\, d(\xi, \eta)} \frac{\partial}{\partial \xi} + \frac{27 + 3\eta}{20\, d(\xi, \eta)} \frac{\partial}{\partial \eta}$$

Setting $w_i = N_i(\xi, \eta)$ and $u = \sum_{j=1}^{4} N_j(\xi, \eta) u_j$, we obtain

$$I = \sum_{j=1}^{4} \frac{1}{4000} \int_{-1}^{1} \int_{-1}^{1} \frac{1}{675 - 60\xi + 15\eta} \left\{ \left[(19 - \xi)\frac{\partial N_i}{\partial \xi} - (31 - \eta)\frac{\partial N_i}{\partial \eta} \right] \right.$$

$$\left. \bullet \left[(19 - \xi)\frac{\partial N_j}{\partial \xi} - (31 - \eta)\frac{\partial N_j}{\partial \eta} \right] u_j + \left[(27 - 3\xi)\frac{\partial N_i}{\partial \xi} + (27 + 3\eta)\frac{\partial N_j}{\partial \eta} \right] \right.$$

$$\left. \bullet \left[(27 - 3\xi)\frac{\partial N_j}{\partial \xi} + (27 + 3\eta)\frac{\partial N_j}{\partial \eta} \right] u_j \right\} d\xi\, d\eta$$

Notice that the integral is no longer a polynomial function of ξ and η but a rational function of the form $P(\xi,\eta)/Q(\xi,\eta)$, and hence it is no longer possible to integrate it analytically in a convenient way. For this reason, numerical integration will be used to evaluate it. Finally, notice that the shape functions in the x–y coordinate system are not needed. ∎

Just as we have defined an isoparametric transformation for the bilinear element using Eq.(4.22), we can do it for triangular and higher order elements. With higher order transformations we can produce elements with parabolic sides as illustrated in Fig. 4.15. The only differences are that now $N_e > 4$ in Eq. (4.22) and the shape functions must be replaced by those for the corresponding element. Table 4.2 gives the shape functions for the eight-noded quadratic and the biquadratic isoparametric elements shown below.

Table 4.2 Shape functions for bilinear and eight-noded quadratic isoparametric elements

$N_1(\xi,\eta) = 1/4\,\xi\,\eta(\xi-1)(\eta-1)$	$N_1(\xi,\eta) = -1/4(1-\xi)(1-\eta)(1+\xi+\eta)$
$N_2(\xi,\eta) = 1/2\,\eta(1-\xi^2)(\eta-1)$	$N_2(\xi,\eta) = 1/2(1-\xi^2)(1-\eta)$
$N_3(\xi,\eta) = 1/4\,\xi\,\eta(\xi+1)(\eta-1)$	$N_3(\xi,\eta) = 1/4(1+\xi)(1-\eta)(\xi-\eta-1)$
$N_4(\xi,\eta) = 1/2\,\xi\,(\xi+1)(1-\eta^2)$	$N_4(\xi,\eta) = 1/2(1+\xi)(1-\eta^2)$
$N_5(\xi,\eta) = 1/4\,\xi\,\eta(\xi+1)(\eta+1)$	$N_5(\xi,\eta) = 1/4(1+\xi)(1+\eta)(\xi+\eta-1)$
$N_6(\xi,\eta) = 1/2\,\eta(1-\xi^2)(\eta+1)$	$N_6(\xi,\eta) = 1/2(1-\xi^2)(1+\eta)$
$N_7(\xi,\eta) = 1/4\,\xi\,\eta(\xi-1)(\eta+1)$	$N_7(\xi,\eta) = 1/4(1-\xi)(1+\eta)(1+\xi-\eta)$
$N_8(\xi,\eta) = 1/2\,\xi(\xi-1)(1-\eta^2)$	$N_8(\xi,\eta) = 1/2(1-\xi)(1-\eta^2)$
$N_9(\xi,\eta) = (1-\xi^2)(1-\eta^2)$	

Similarly, cubic isoparametric elements can be defined; however, in applications of interest to us, quadratic isoparametric elements are used much less than bilinear elements and cubics are hardly ever considered. The reader interested in further details and applications of higher order isoparametric elements can consult the texts by Gallagher (1975), Zienkiewicz (1977), Becker et al. (1981), or Hughes (1987) and the papers by Ergatoudis et al (1968), McLeod and Mitchell (1972) and Gallagher (1980).

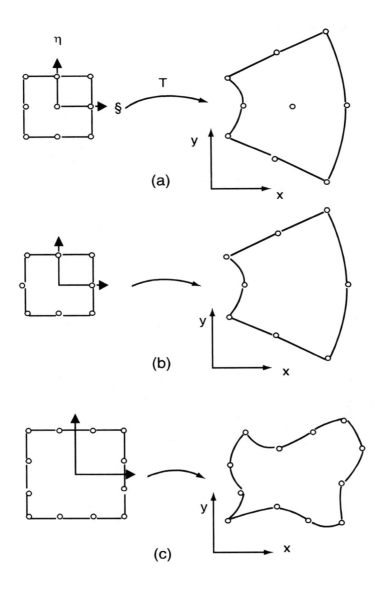

Figure 4.15 Higher order isoparametric quadrilateral elements for (a) biquadratic, (b) eight-noded quadratic, and (c) 12-noded cubic.

For triangular elements the concepts described above are applied using the right triangle shown in Fig. 4.16 as the parent element; in this case we use $0 \leq \xi \leq 1$ and $0 \leq \eta \leq 1$. The shape functions for the linear and quadratic parent elements are not difficult to find (Exercise 4.14). For the linear element these are

$$\begin{aligned}
N_1(\xi, \eta) &= 1 - \xi - \eta \\
N_2(\xi, \eta) &= \xi \\
N_3(\xi, \eta) &= \eta
\end{aligned} \tag{4.39}$$

and for the quadratic triangle,

$$\begin{aligned}
N_1(\xi, \eta) &= \left[2(1 - \xi - \eta) - 1\right](1 - \xi - \eta) \\
N_2(\xi, \eta) &= \xi(2\xi - 1) \\
N_3(\xi, \eta) &= \eta(2\eta - 1) \\
N_4(\xi, \eta) &= 4\xi\eta \\
N_5(\xi, \eta) &= 4\eta(1 - \xi - \eta) \\
N_6(\xi, \eta) &= 4\xi(1 - \xi - \eta)
\end{aligned} \tag{4.40}$$

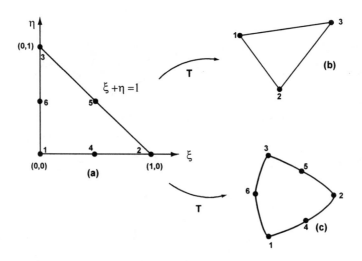

Figure 4.16 Triangular isoparametric elements for (a) parent element, (b) linear element, and (c) curved quadratic triangle.

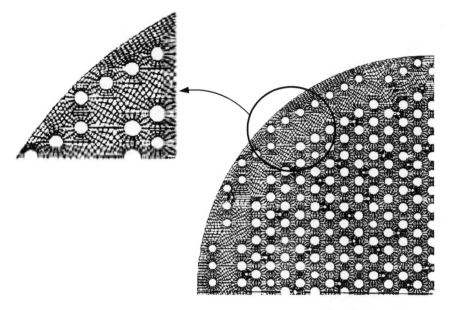

Figure 4.17 Computational mesh for an array consisting of 600 individually heated tubes (from Pepper and Dyne, 1993).

REMARKS

1. Isoparametric elements are essential in making the finite element method a powerful numerical tool. The fact that the transformations are performed locally is crucial because it frees us from having to deal with the entire geometry. When the global geometry is very complicated, it is impossible to find global transformations using conformal mapping that will map the entire domain into a simpler rectangular domain. However, isoparametric elements can always be used to define an appropriate mesh, as shown in Fig. 4.17, which depicts the partial computational domain for convective flow within a 600–tube array (Pepper and Dyne, 1993).

At this point we begin asking what is the best way to define a mesh, how to produce algorithms for automatic mesh generation, and how to be able to adapt meshes, etc. We deal with these important subjects in Chapter 11.

2. The invertibility of a transformation is usually difficult to determine, and intuitively, we can imagine that the accuracy will decrease if we distort the elements too much. Mathematically, the

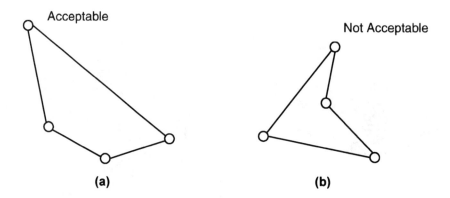

Figure 4.18 Examples of (a) permissible quadrilateral elements and (b) illegal quadrilateral elements.

condition for invertibility is that det $| \mathbf{J} | > 0$ at all points in the parent element. This condition leads to different restrictions for every set of shape functions and must be examined for each individual element. For the bilinear transformation (Strang and Fix, 1973) it can be shown that det $\mathbf{J} \neq 0$ if and only if the quadrilateral is convex, that is, all its interior angles must be less than π. Figure 4.18 shows sample permissible and nonpermissible elements.

For higher order elements these conditions are more difficult to determine and are very sensitive to the location of midside and interior nodes. Serendipity elements do not contain interior nodes, and for this reason the quadratic serendipity element, in particular, has been a very popular element to deal with curved boundaries. In most practical applications, only one side (or at most two) needs to be curved in order to match a boundary. In this case it can be proved that isoparametric transformations are invertible under most practical circumstances (Strang and Fix, 1973).

3. In second-order differential equations the continuity of the interpolation functions after the isoparametric transformation is guaranteed when transformations are defined by Eq. (4.22). Hence no overlaps or gaps are created along interelement boundaries and convergence of finite element approximations is assured.

Isoparametric transformations owe their name to the fact that the same shape functions used to interpolate the dependent variables are

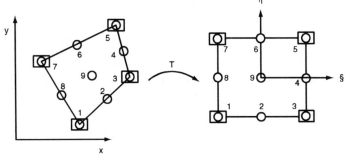

$$T:\begin{Bmatrix}x\\y\end{Bmatrix} = \frac{1}{4}(1-\xi)(1-\eta)\begin{Bmatrix}x_1\\y_1\end{Bmatrix} + \frac{1}{4}(1+\xi)(1-\eta)\begin{Bmatrix}x_3\\y_3\end{Bmatrix} + \frac{1}{4}(1+\xi)(1+\eta)\begin{Bmatrix}x_5\\y_5\end{Bmatrix} + \frac{1}{4}(1-\xi)(1+\eta)\begin{Bmatrix}x_7\\y_7\end{Bmatrix}$$

Figure 4.19 Example of a subparametric transformation on a biquadratic element. Circles denote interpolation nodes and squares geometric transformation nodes.

used in the geometric transformation. This is, of course, not always necessary, and we could define a *subparametric* transformation using lower order interpolation functions for the geometric transformations than we use for interpolation. For example, as shown in Fig. 4.19, we can define a biquadratic element associated with a bilinear transformation if the sides of the elements are always straight lines. Indeed, this is a very commonly used element. In the same note, we can define *superparametric* elements in which the geometric transformation is of higher order than the interpolation. These are hardly justifiable though and can lead to nonconformity, therefore we do not discuss them any further. A simple example is shown in Fig. 4.20.

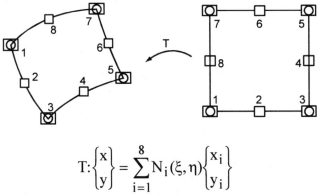

$$T:\begin{Bmatrix}x\\y\end{Bmatrix} = \sum_{i=1}^{8} N_i(\xi,\eta)\begin{Bmatrix}x_i\\y_i\end{Bmatrix}$$

Figure 4.20 Example of a superparametric transformation on a bilinear element. Circles denote interpolation nodes, squares geometric transformation nodes. The shape functions $N_i(\xi,\eta)$, $i = 1,...8$, are given in Table 4.2.

4. We now turn our attention to the question of accuracy of the approximation using isoparametric transformations. Here we need to distinguish between order of convergence, which was shown by Ciarlet and Raviart (1972) to be preserved by the isoparametric transformation, and the accuracy of the calculation on a particular mesh. This may appear confusing at first sight, but it can be simply explained by the following result.

Suppose that a bilinear element is used in the $\xi-\eta$ plane and that the solution u^* is in $H^2(\Omega)$. Then the finite element approximation u^h to u^* satisfies the estimate

$$\left\| u^* - u^h \right\|_0 \leq C' h^2 \left\| u \right\|_2 \tag{4.41}$$

This is just another form of Eq. (3.77). The important difference is that when isoparametric elements are used, the constant C' has the form

$$C' = C'' \left(\frac{\max\left(\det \mathbf{J} \right)}{\min\left(\det \mathbf{J} \right)} \right)^{\frac{1}{2}} \left\| \mathbf{T} \right\|_2 \left\| \mathbf{T}^{-1} \right\|_2 \tag{4.42}$$

where C'' is independent of the transformation. Clearly, when the elements are too distorted and $\det \mathbf{J}$ becomes very small, C' can be a very large number. Hence, even though the method still converges as $O(h^2)$, we may require a very large number of elements (or a very small h) for the error to be small. Therefore it is important to keep in mind that elements should be distorted as little as possible to achieve the best allowed accuracy.

When curved boundaries are fitted, curved isoparametric elements can produce enormous improvements in some classes of problems (Zlamal, 1974), and in problems governed by self-adjoint operators it is possible to take full advantage of curved elements. For non-self-adjoint problems such as convective flows and heat and mass transfer, such techniques must be applied carefully. The reader interested in a more rigorous mathematical treatment of the errors associated with isoparametric elements should consult the works by Ciarlet and Raviart (1972), McLeod and Mitchell (1972), Strang and Fix (1973), Zlamal (1974), Wachpress and McLeod (1979), Carey and Oden (1983) and references therein.

5. As seen in Example 4.5, the use of isoparametric elements requires *numerical integration* because the determinant of the Jacobian transforms the integral into a rational function, even though the shape functions are polynomials. It will also become important to assess the effect of numerical integration in the convergence and accuracy of the results. This is done in the next chapter.

We will now turn our attention to a very powerful technique to construct any kind of C^o element.

4.5 BLENDING FUNCTION INTERPOLATION

We will now develop a general methodology to construct rectangular elements. The techniques can also be applied to triangular elements (Marshall, 1975). However, theoretical and practical difficulties arise in the case of triangular elements that significantly reduce the advantages of the use of blending function interpolation in this case. The ideas stem from the work of Coons (1967) in the context of representing complex car body shapes. The following development is based on the work of Gordon and Hall (1973 a, b).

Let us consider a rectangular element of size $0 \le x \le h$ and $0 \le y \le k$ with the origin of coordinates as shown in Fig. 4.21. To introduce the basic concepts, we will assume that a continuous $C^o(\Omega)$ function is defined over the element, and we will choose a set of one-dimensional shape functions, which can be linear, quadratic, cubic, etc. We will denote this set of shape functions by $\phi_i(x)$, $i = 1,\ldots,N_e$, where N_e is the number of nodes in the element. It is important to recall that we *choose* these functions.

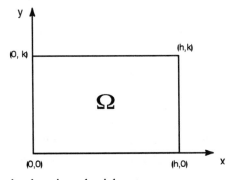

Figure 4.21 Rectangular domain and axial system.

Assuming for the time being that the functions $\phi_i(x)$ are the linear shape functions,

$$\phi_1(x) = \left(1 - \frac{x}{h}\right)$$

$$\phi_2(x) = \frac{x}{h} \tag{4.43}$$

we define an operator P_x of the form $P_x[u] = \sum_{i=1}^{2} \phi_i(x)u(x_i,y)$, i.e.,

$$P_x[u] = \left(1 - \frac{x}{h}\right)u(0,y) + \left(\frac{x}{h}\right)u(h,y) \tag{4.44}$$

P_x exhibits the properties of a geometric projector, which are

(i) It is linear, i.e., if α and β are any two real numbers then $P_x[\alpha u + \beta v] = \alpha P_x[u] + \beta P_x[v]$.

(ii) If we apply it twice to a function u, we obtain the same result as if we apply it only once, i.e.,

$$P_x\big[P_x(u)\big] = P_x(u) \tag{4.45}$$

this property is called *idempotence*. Geometrically it means that if we are looking for the projection of a vector in a plane, and the vector is contained in the plane, then its projection is the vector itself.

The projector P_x maps the functions u(x,y) in $C^0(\Omega)$ to functions that are linear in the x-direction. Moreover, at x = 0

$$P_x\big[u(0,y)\big] \equiv u(0,y)$$

and at x = h,

$$P_x\big[u(h,y)\big] \equiv u(h,y)$$

Hence $P_x[u]$ interpolates u(x,y) at every point along the lines x = 0 and x = h. On the other hand, $P_x[u]$ is a linear function of x for each value of y.

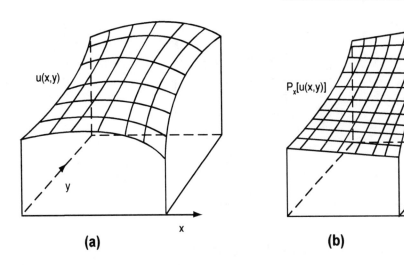

Figure 4.22 Effect of the projection P_x : (a) original function $u(x,y)$ and (b) projected function $P_x[u(x,y)]$.

The effect of P_x on u is illustrated in Fig. 4.22. Because $P_x[u]$ interpolates u at an infinite number of points, it is called a *transfinite interpolant* of u.

The above definition of $P_x[u]$ is immediately extended to any Lagrangian one-dimensional element, that is

$$P_x[u] = \sum_{i=1}^{\ell} u(x_i, y) \phi_i(x) \qquad (4.46)$$

where $0 = x_1 < x_2 \ldots < x_l = h$ and the shape functions $\phi_i(x)$, $i = 1,2,\ldots,l$ are given by Lagrange's formula, Eq. (4.3).

Example 4.6

Let us choose $l = 3$, that is, quadratic shape functions with nodes $x_1 = 0$, $x_2 = h/2$, and $x_3 = h$. The projection of u is

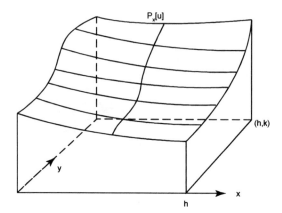

Figure 4.23 Schematic of a quadratic blending projection.

$$P_x[u] = \phi_1(x)u(0,y) + \phi_2(x)u\left(\frac{h}{2},y\right) + \phi_3(x)u(h,y)$$

$$= \left(\frac{2x^2}{h^2} - \frac{3x}{h} + 1\right)u(0,y) + \left(\frac{4x}{h} - \frac{4x^2}{h^2}\right)u\left(\frac{h}{2},y\right)$$

$$+ \left(\frac{2x^2}{h^2} - \frac{x}{h}\right)u(h,y) \tag{4.47}$$

where the functions $\phi_i(x)$ are the quadratic shape functions given in Eq. (4.2). This projection, depicted in Fig. 4.23, interpolates $u(x,y)$ exactly along $x = 0$, $x = h/2$, and $x = h$ and is quadratic in the x-direction. ∎

The functions $\phi_i(x)$ are called *blending functions*, the projector $P_x[u]$ interpolates $u(x,y)$ exactly along the lines $x = x_i$, $i = 1,...,l$, and varies as a polynomial of degree $l-1$ in the x-direction.

Further generalizations such as the use of Hermite polynomials (Watkins, 1976) or transcendental functions can be immediately visualized. However, let us now return to using Lagrangian shape functions as blending functions and similarly define a projector P_y as

$$P_y[u] = \sum_{i=1}^{m} u(x,y_i)\psi_i(y) \tag{4.48}$$

which is the y-direction counterpart of $P_x[u]$ defined in Eq. (4.44), where now $\psi_i(y)$ are the Lagrangian blending functions and $0 = y_1 < y_2 < ... < y_m = k$. Hence, for $m = 2$ we obtain the linear projector

$$P_y[u] = \left(1 - \frac{y}{k}\right)u(x,0) + \left(\frac{y}{k}\right)u(x,k)$$

(4.49)

which enjoys the same properties as P_x in Eq. (4.36).

It is easily established that for the projectors P_x and P_y given by Eqs. (4.44) and (4.40) the product operator defined as

$$P_x P_y[u] \equiv P_x\Big[P_y[u]\Big]$$

(4.50)

commutes, i.e., $P_x P_y = P_y P_x$. This property is necessary if we are to construct elements in order for the elements to be uniquely defined, and in this case it is easily verified (Exercise 4.22). However, in general, it has to be shown to hold for every combination of operator P_x and P_y individually.

In the linear case of Eqs. (4.44) and (4.49), we have

$$P_x P_y[u] \equiv P_x\left[\left(1 - \frac{y}{k}\right)u(x,0) + \left(\frac{y}{k}\right)u(x,k)\right]$$

$$= \left(1 - \frac{x}{h}\right)\left[\left(1 - \frac{y}{k}\right)u(0,0) + \left(\frac{y}{k}\right)u(0,k)\right]$$

$$+ \left(\frac{x}{h}\right)\left[\left(1 - \frac{y}{k}\right)u(h,0) + \left(\frac{y}{k}\right)u(h.k)\right]$$

or

$$P_x P_y[u] = \left(1 - \frac{x}{h}\right)\left(1 - \frac{y}{k}\right)u(0,0) + \left(1 - \frac{x}{h}\right)\left(\frac{y}{k}\right)u(0,k)$$

$$+ \left(\frac{x}{h}\right)\left(1 - \frac{y}{k}\right)u(h,0) + \left(\frac{x}{h}\right)\left(\frac{y}{k}\right)u(h,k)$$

(4.51)

which is the bilinear interpolant of u, i.e.,

$$u(x,y) \cong P_x P_y[u] = \sum_{i=1}^{2}\sum_{j=1}^{2}\phi_i(x)\psi_j(y)u(x_iy_j) = \sum_{i=1}^{4}N_i(x,y)u_i$$

(4.52)

and hence the projectors can be utilized to find the shape functions of various elements. For $l = m = 3$ we obtain the biquadratic element (Exercise 4.22). For $l = m = 4$ we get the bicubic element, etc.

An immediate generalization can be obtained for $l \neq m$, in this case we can construct *bipolynomial* Lagrangian elements of the form

$$P_x P_y[u] = \sum_{i=1}^{\ell}\sum_{j=1}^{m}\phi_i(x)\psi_j(y)u(x_iy_j)$$

(4.53)

Example 4.7

We can find the shape functions for the six-noded element of Fig. 4.24 using $l = 3$ and $m = 2$ in Eq. (4.53). To show that the operators still commute is left to Exercise 4.23. The resulting shape functions follow from

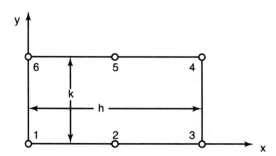

Figure 4.24 Six-noded bipolynomial, quadratic-linear element.

$$P_x P_y[u] = \left(\frac{2x^2}{h^2} - \frac{3x}{h} + 1\right)\left(1 - \frac{y}{k}\right)u(0,0) + \left(\frac{2x^2}{h^2} - \frac{3x}{h} + 1\right)\left(\frac{y}{k}\right)u(0,k)$$

$$+ \left(\frac{4x}{h} - \frac{4x^2}{h^2}\right)\left(1 - \frac{y}{k}\right)u\left(\frac{h}{2},0\right) + \left(\frac{4x}{h} - \frac{4x^2}{h^2}\right)\left(\frac{y}{k}\right)u\left(\frac{h}{2},k\right)$$

$$+ \left(\frac{2x^2}{h^2} - \frac{x}{h}\right)\left(1 - \frac{y}{k}\right)u(h,0) + \left(\frac{2x^2}{h^2} - \frac{x}{h}\right)\left(\frac{y}{k}\right)u\left(\frac{h}{k}\right)$$

$$(4.54)$$

and can be numbered according to Fig. 4.24. Notice that this element has a quadratic behavior in the x-direction and linear in the y-direction. ■

Clearly, many more elements can be constructed this way. A great variety is shown by Gordon and Hall (1973a).

There is another important way to combine the projectors P_x and P_y called the *Boolean* sum, defined as

$$P_x \oplus P_y \equiv P_x + P_y - P_x P_y \qquad (4.55)$$

which is also a projector if P_x and P_y commute. Let us examine the Boolean sum for the case when the blending functions in Eqs. (4.46) and (4.48) are linear, i.e., $l = m = 2$:

$$P_x \oplus P_y[u] = P_x[u] + P_y[u] - P_x P_y[u]$$

$$= \left(1 - \frac{x}{h}\right)u(0,y) + \left(\frac{x}{h}\right)u(h,y) + \left(1 - \frac{y}{k}\right)u(x,0)$$

$$+ \left(\frac{y}{k}\right)u(x,k) - \left(1 - \frac{x}{h}\right)\left(1 - \frac{y}{k}\right)u(0,0) - \left(\frac{x}{h}\right)\left(1 - \frac{y}{k}\right)u(h,0)$$

$$- \left(\frac{x}{h}\right)\left(\frac{y}{k}\right)u(h,k) - \left(1 - \frac{x}{h}\right)\left(\frac{y}{k}\right)u(0,k)$$

$$(4.56)$$

Notice that

$$P_x \oplus P_y[u](x,0) \equiv u(x,0)$$

$$P_x \oplus P_y[u](x,k) \equiv u(x,k)$$

$$P_x \oplus P_y[u](0,y) \equiv u(0,y)$$

$$P_x \oplus P_y[u](h,y) \equiv u(h,y)$$

Hence the Boolean projector matches the function *exactly* along the element boundaries. This property is depicted in Fig. 4.25, where the Boolean projection of the function $u(x,y) = \sin[\pi(x^2 + y^2)/(2h^2)]$ over a square element of size h is shown. We refer to the Boolean projectors as the *blended function interpolants,* or simply blended interpolants, since we can visualize them as interpolants in which the boundary data are blended into the interior of the element.

The Boolean sum operation can be applied to the general projector of Eqs. (4.46) and (4.48) for any value of l and m, e.g., for $l = m = 3$, we obtain a function that matches u(x,y) exactly along the lines x = 0, h/2, and h, and the lines y = 0, k/2, and k. This situation is sketched in Fig. 4.26, where we also show the case $l = 3$, m = 2, which yields a function that matches along x = 0, h/2, h and y = 0,k.

The procedure just described allows us to construct interpolants capable of matching the original function exactly along any number of

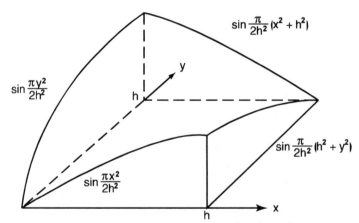

Figure 4.25 Blended function interpolant of sin $[\pi(x^2 + y^2)/2h^2]$ over the square for element for $0 \le x, y \le h$.

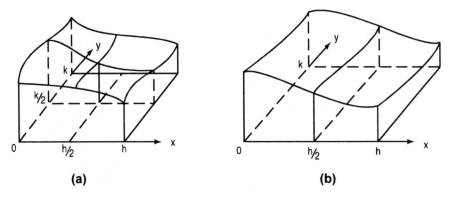

Figure 4.26 Higher order blending function interpolation (a) $l{=}m{=}3$ and (b) $l{=}3,m{=}2$.

lines parallel to the coordinate axis. However, this information is generally not available, and in order to produce elements that will be useful in finite element analysis, we must perform a second level of interpolation. We will illustrate some applications to the construction of blended finite elements through a series of examples.

Example 4.8

Consider the linear blended interpolant in Eq. (4.56). In order to construct $P_x \oplus P_y[u]$ we would need to know the functions $u(0,y)$, $u(h,y)$, $u(x,0)$, and $u(x,k)$, which are not known if this is an element. However, we can approximate those functions using other interpolants if we assume that the four functions are linear, e.g.,

$$u(x,0) = \left(1 - \frac{x}{h}\right)u(0,0) + \left(\frac{x}{h}\right)u(h,0)$$

and similarly for $u(x,k)$, $u(0,y)$, and $u(h,y)$. Substituting them back into Eq. (4.56), we will recover the bilinear element (Exercise 4.25). Let us now use quadratic interpolation, i.e.,

$$u(x,0) \cong \left(\frac{2x^2}{h^2} - \frac{3x}{h} + 1 \right) u(0,0) + \left(\frac{4x}{h} - \frac{4x^2}{h^2} \right) u\left(\frac{h}{2},0 \right)$$

$$+ \left(\frac{2x^2}{h^2} - \frac{x}{h} \right) u(h,0)$$

$$u(x,k) \cong \left(\frac{2x^2}{h^2} - \frac{3x}{h} + 1 \right) u(0,k) + \left(\frac{4x}{h} - \frac{4x^2}{h^2} \right) u\left(\frac{h}{2},k \right)$$

$$+ \left(\frac{2x^2}{h^2} - \frac{x}{h} \right) u(h,k)$$

$$u(0,y) \cong \left(\frac{2y^2}{k^2} - \frac{3y}{k} + 1 \right) u(0,0) + \left(\frac{4y}{k} - \frac{4y^2}{k^2} \right) u\left(0,\frac{k}{2} \right)$$

$$+ \left(\frac{2y^2}{k^2} - \frac{y}{k} \right) u(0,k)$$

$$u(h,y) \cong \left(\frac{2y^2}{k^2} - \frac{3y}{k} + 1 \right) u(h,0) + \left(\frac{4y}{k} - \frac{4y^2}{k^2} \right) u\left(h,\frac{k}{2} \right)$$

$$+ \left(\frac{2y^2}{k^2} - \frac{y}{k} \right) u(h,k)$$

Substituting into Eq. (4.56) and simplifying, we obtain

$$u(x,y) \cong \left[\left(1-\frac{y}{k}\right)\left(1-\frac{3x}{h}+\frac{2x^2}{h^2}-\frac{2y}{k}+\frac{2xy}{hk}\right)\right]u_1$$

$$+\left[\left(\frac{4x}{h}\right)\left(1-\frac{y}{k}\right)\left(1-\frac{x}{h}\right)\right]u_2 +\left[\left(\frac{x}{h}\right)\left(-1-\frac{y}{k}+\frac{2y^2}{k^2}+\frac{2x}{h}-\frac{2xy}{hk}\right)\right]u_3$$

$$+\left[\left(\frac{4xy}{hk}\right)\left(1-\frac{y}{k}\right)\right]u_4 +\left[\left(\frac{xy}{hk}\right)\left(\frac{2x}{h}+\frac{2y}{k}-3\right)\right]u_5 +\left[\left(\frac{4xy}{hk}\right)\left(1-\frac{x}{h}\right)\right]u_6$$

$$+\left[\left(\frac{y}{k}\right)\left(-1+\frac{x}{h}+\frac{2y}{k}-\frac{2xy}{hk}+\frac{2x^2}{h^2}\right)\right]u_7 +\left[\left(1-\frac{x}{h}\right)\left(1-\frac{3y}{k}+\frac{2y^2}{k^2}\right)\right]u_8$$

which is an eight-noded quadratic serendipity element. If cubic interpolation is used, we obtain the 12-noded cubic serendipity element (Exercise 4.26). Higher order blending functions combined with any order of interpolation in each coordinate direction can be used to construct *macro elements,* such as the 35-noded element of Fig. 4.27

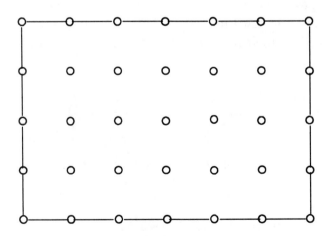

Figure 4.27 Thirty-five-noded macroelement.

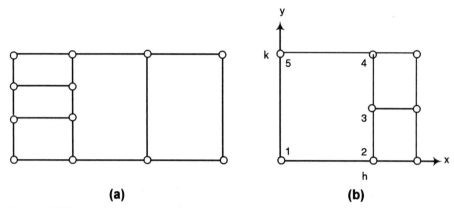

(a)   **(b)**

Figure 4.28 Local mesh refinement using blending interpolation (a) transition from four to two nodes and (b) mesh refinement by one half.

which was obtained using $l = 4$ and $m = 3$ in Eqs. (4.46) and (4.48), respectively, combined with sixth-order interpolation in the x-direction and fourth degree interpolation in the y-direction (Gordon and Hall, 1973a). ∎

Example 4.9

Blending techniques can be used to construct transition elements for local mesh refinement, as shown in Fig. 4.28 using bilinear elements. To construct the transition element in Fig. 4.28b, we first substitute $u(x,0)$, $u(x,k)$, and $u(0,y)$ by their linear interpolants into Eq. (4.56). This yields

$$u(x,y) \cong \left(1 - \frac{x}{h}\right)\left(1 - \frac{y}{k}\right)u_1 + \left(1 - \frac{x}{h}\right)\frac{y}{k}u_5 + \frac{x}{h}u(h,y) \qquad (4.57)$$

We now approximate the function $u(h,y)$ using a piecewise linear interpolant,

$$u(h,y) \cong \phi_2(y)u_2 + \phi_3(y)u_3 + \phi_4(y)u_4$$

where the interpolation functions $\phi_i(y)$, $i = 2,3,4$ are given by

$$\phi_2(y) = \begin{cases} \left(1 - \dfrac{2y}{k}\right) & 0 \le y \le \dfrac{k}{2} \\ 0 & \text{otherwise} \end{cases}$$

$$\phi_3(y) = \begin{cases} \dfrac{2y}{k} & 0 \le y \le \dfrac{k}{2} \\ \left(2 - \dfrac{2y}{k}\right) & \dfrac{k}{2} \le y \le k \\ 0 & \text{otherwise} \end{cases}$$

$$\phi_4(y) = \begin{cases} \left(\dfrac{2y}{k} - 1\right) & \dfrac{k}{2} \le y \le k \\ 0 & \text{otherwise} \end{cases}$$

Replacing into Eq. (4.49), we can write

$$u(x,y) \cong \sum_{i=1}^{5} N_i(x,y) u_i$$

where the shape functions $N_i(x,y)$ are

$$N_1(x,y) = \left(1 - \frac{x}{h}\right)\left(1 - \frac{y}{k}\right)$$

$$N_2(x,y) = \begin{cases} \left(\dfrac{x}{h}\right)\left(1 - \dfrac{2y}{k}\right) & 0 \le y \le \dfrac{k}{2} \\ 0 & y > \dfrac{k}{2} \end{cases}$$

$$N_3(x,y) = \begin{cases} \dfrac{2xy}{hk} & 0 \le y \le \dfrac{k}{2} \\ \left(\dfrac{x}{h}\right)\left(2 - \dfrac{2y}{k}\right) & \dfrac{k}{2} \le y \le k \end{cases}$$

$$N_4(x,y) = \begin{cases} 0 & 0 \le y \le \dfrac{k}{2} \\ \left(\dfrac{x}{h}\right)\left(\dfrac{2y}{k} - 1\right) & \dfrac{k}{2} \le y \le k \end{cases}$$

$$N_5(x,y) = \left(1 - \dfrac{x}{h}\right)\left(\dfrac{y}{k}\right)$$

Cavendish (1975) introduced the idea of using blending function interpolation to construct elements that can be piecewise linear along any side, so that they can have more than two nodes on one side and still be compatible with bilinear elements. ∎

Example 4.10

Another direct application of blending interpolation to the construction of elements is derived if we would like to incorporate the boundary data exactly into our finite element scheme. We can build elements that match the Dirichlet conditions along the boundary exactly, while remaining linear along their interior sides, as shown in Fig. 4.29, which shows an element whose side x = 0 lies along the boundary of the domain and can incorporate the function u(0,y), and a corner element whose sides x = 0 and y = k lie in the boundary and can match u(0,y) and u(x,k) exactly. These ideas were introduced by Marshall and Mitchell (1973), and extended to the approximation of Neumann conditions by Hall and Heinrich (1978). To illustrate their application, let us look at the stochastic equation in probabilistic design known as the Pontriagin-Vitt equation for a randomly accelerated particle,

$$\pi S_o \frac{\partial T}{\partial y^2} + y \frac{\partial T}{\partial x} = 1 \qquad (4.58)$$

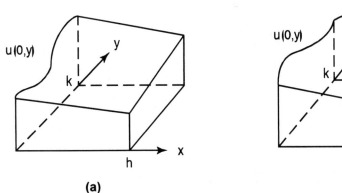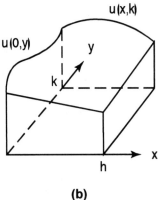

Figure 4.29 Blended elements that incorporate boundary data exactly (a) side element and (b) corner element.

where S_o is the magnitude of the spectral density of the excitation and T is the zeroth moment of time to first passage (Bergman and Heinrich, 1981). The equation is defined over the infinite strip depicted in Fig. 4.30 with boundary conditions

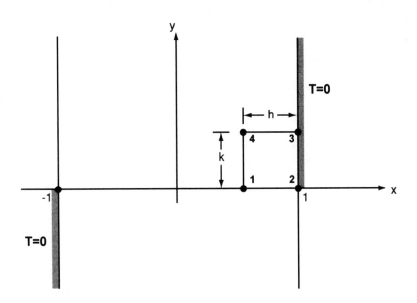

Figure 4.30 Computational domain and one of the four elements with singular behavior along $y = 0$, for the solution of the Pontriagin-Vitt equation.

$$T(-1,y) = 0 \text{ for } y < 0$$

$$T(1,y) = 0 \text{ for } y > 0$$

$$T(x,y) \to 0 \text{ as } |y| \to \infty.$$

The analytical solution can be obtained along the axis $y = 0$ and is given by

$$T(x,0) = \left(\frac{3^{11/16}}{2\Gamma\left(\frac{1}{3}\right)} \right) (1-x^2)^{\frac{1}{6}} \left[F\left(-\frac{1}{3},1,\frac{7}{6};\frac{1}{2}(1-x)\right) \right.$$

$$\left. - F\left(-\frac{1}{3},1,\frac{7}{6};\frac{1}{2}(1+x)\right) \right] \tag{4.59}$$

where Γ is the gamma function and F is the hypergeometric function. The term $(1 - x^2)^{1/6}$ carries the singular part of the solution, and it is desirable to construct an element that will reproduce that behavior $u(x,0) = [1 - (x/h)^2]^{1/6}$. Using Eq. (4.56), we have

$$P_x \oplus P_y[T] = \left(1 - \frac{x}{h}\right)\left[\left(1 - \frac{y}{k}\right)T_1 + \frac{y}{k}T_4\right]$$

$$+ \left(\frac{x}{h}\right)\left[\left(1 - \frac{y}{k}\right)T_2 + \left(\frac{y}{k}\right)T_3\right] + \left(1 - \frac{y}{k}\right)\left(1 - \left(\frac{x}{h}\right)^2\right)^{\frac{1}{6}}T_1$$

$$+ \left(\frac{y}{k}\right)\left[\left(1 - \frac{x}{h}\right)T_4 - \left(\frac{x}{h}\right)T_3\right] - \left(1 - \frac{x}{h}\right)\left(1 - \frac{y}{k}\right)T_1$$

$$- \left(\frac{x}{h}\right)\left(1 - \frac{y}{k}\right)T_2 - \left(\frac{x}{h}\right)\left(\frac{y}{k}\right)T_3 - \left(1 - \frac{x}{h}\right)\left(\frac{y}{k}\right)T_4$$

Simplifying and using the fact that $T_2 = T_3 = 0$, we finally obtain

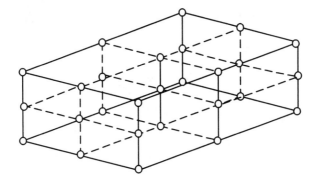

Figure 4.31 Triquadratic element of 27 nodes.

$$P_x \oplus P_y[T] = \left(1-\frac{y}{k}\right)\left(1-\left(\frac{x}{h}\right)^2\right)^{\frac{1}{6}}T_1 + \left(1-\frac{x}{h}\right)\left(\frac{y}{k}\right)T_4 \qquad (4.60)$$

which is an interpolant to $T(x,y)$ over the element with the desired behavior along $y = 0$ and is compatible with bilinear elements along $x = 0$ and $y = k$. ∎

It is not difficult to imagine further applications of blending interpolation. In fact, it has also been used as a method of mesh generation (Gordon and Hall, 1973).

4.6 THREE-DIMENSIONAL ELEMENTS

In section 3.8 we introduced the trilinear element as the natural extension to three dimensions of the one- and two-dimensional Lagrangian linear and bilinear elements. Evidently, higher order hexagonal Lagrangrian elements such as triquadratic, tricubic, etc., can be constructed by direct multiplication of the corresponding one-dimensional functions in each coordinate direction. A triquadratic element, for example, will contain 27 nodes, and these can be corner, side, or interior nodes, as shown in Fig. 4.31.

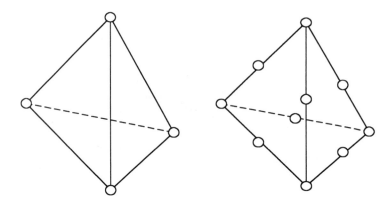

Figure 4.32 Linear and quadratic tetrahedral elements.

Three-dimensional tetrahedral elements are the natural extension of two-dimensional triangles. The simplest of these elements linear and quadratic tetrahedrons depicted in Fig. 4.32. These elements can also be related to the three-dimensional Pascal tetrahedron that gives the complete polynomial of corresponding degree (Carey and Oden, 1983). A natural coordinate system is defined for tetrahedral elements by means of the four internal volumes determined when any interior point is connected with the four vertices of the tetrahedron – in a manner similar to that used in two dimensions to define the area coordinates. In this case we refer to them as volume coordinates. These have been discussed in detail by Pepper and Heinrich (1992).

Isoparametric transformations are readily defined for the three-dimensional elements as

$$
\mathbf{T}\begin{bmatrix} \xi \\ \eta \\ \zeta \end{bmatrix} = \begin{bmatrix} x \\ y \\ z \end{bmatrix} = \sum_{i=1}^{N_e} N_i(\xi,\eta,\zeta) \begin{bmatrix} x_i \\ y_i \\ z_i \end{bmatrix}
\tag{4.61}
$$

The Jacobian matrix is

$$
\mathbf{J} = \begin{bmatrix} \dfrac{\partial x}{\partial \xi} & \dfrac{\partial y}{\partial \xi} & \dfrac{\partial z}{\partial \xi} \\[2ex] \dfrac{\partial x}{\partial \eta} & \dfrac{\partial y}{\partial \eta} & \dfrac{\partial z}{\partial \eta} \\[2ex] \dfrac{\partial x}{\partial \zeta} & \dfrac{\partial y}{\partial \zeta} & \dfrac{\partial z}{\partial \zeta} \end{bmatrix}
\tag{4.62}
$$

and the derivatives of the shape functions are obtained from

$$
\begin{bmatrix} \dfrac{\partial N_i}{\partial x} \\[2ex] \dfrac{\partial N_i}{\partial y} \\[2ex] \dfrac{\partial N_i}{\partial z} \end{bmatrix} = \mathbf{J}^{-1} \begin{bmatrix} \dfrac{\partial N_i}{\partial \xi} \\[2ex] \dfrac{\partial N_i}{\partial \eta} \\[2ex] \dfrac{\partial N_i}{\partial \zeta} \end{bmatrix}
\tag{4.63}
$$

in the same manner as was done in the two-dimensional case.

As far as the elements themselves are concerned, no new concepts arise when we move from two to three dimensions. However, important differences appear due to the difficulty to visualize three-dimensional domains and to define the finite element meshes. These are discussed in Chapter 11. Another important aspect of three-dimensional calculations will be the solution of the resulting linear system of equations, which now become extremely large and usually prohibitive even for modern supercomputers.

4.7 CLOSURE

In this chapter we have constructed several families of elements for second-order differential equations in one and two dimensions. Extensions to three dimensions are straightforward in all cases except for the Boolean sum operator, which takes a slightly more complicated form. We have established, hopefully, that the possibilities for constructing elements are limitless. Therefore any additional knowledge about the physics of the problem or the expected behavior of the solution we are seeking can always be built into the elements.

In Section 4.4 we introduced the concept of isoparametric elements. This concept is crucial for the ability of finite elements to model irregular geometries. Isoparametric Lagrangian or serendipity elements are the elements you will find in every commercial finite element analysis software package, and in most of our application to heat and mass transfer, we will use the bilinear isoparametric Lagrangian element.

REFERENCES

Amazigo, J. C. and Rubenfeld, L. A. (1980) *Advanced Calculus.* New York: John Wiley and Sons.

Aubin, J. P. (1967) "Behavior of the Error of the Approximate Solution of Boundary-Value Problems for Linear Elliptic Operators by Galerkin's and Finite Difference Methods," *Ann. Scuola Normale Pisa, Ser. 3*, Vol. 21, pp. 599–637.

Barsoum, R. (1976) "On the Use of Isoparametric Finite Elements in Linear Fracture Mechanics," *Int. J. Numer. Methods. Eng.,* Vol. 10, pp. 25–37.

Becker, E. B., Carey, G. F. and Oden, J. T. (1981) *Finite Elements: An Introduction,* Vol. 1. Englewood Cliffs, N.J.: Prentice-Hall.

Bergman, L. A. and Heinrich, J. C. (1981) "Petrov-Galerkin Finite Element Solution for the First Passage Probability and Moments of First Passage Time of the Randomly Accelerated Free Particle," *Comput. Methods. Appl. Mech. Eng.,* Vol. 27, pp. 345–362.

Carey, G. F. and Oden, J. T. (1983) *Finite Elements,* Vol. II, *A Second Course.* Englewood Cliffs, N.J.: Prentice-Hall.

Cavendish, J. C. (1975) "Local Mesh Refinement Using Rectangular Blended Elements," *J. Comput. Phys.,* Vol. 19, pp. 211–288.

Ciarlet, P. and Raviart, P. (1972) "Interpolation Theory over Curved Element with Applications to Finite Element Methods," *Comput. Methods Appl. Mech. Eng.,* Vol. 1, pp. 217–249.

Coons, S. A. (1967) "Surfaces for Computer Aided Design of Space Forms," Project MAC, Design Div., Dept. of Mech. Eng. MIT Report MAC-TR-41.

de Boor, C. (1978) "A Practical Guide to Splines," in *Applied Mathematical Sciences Series*, No. 27. New York: Springer-Verlag.

Ergatoudis, J. G., Irons, B. M. and Zienkiewicz, O. C. (1968) "Curved Isoparametric 'Quadrilateral' Elements for Finite Element Analysis," *Int. J. Solids Struct.*, Vol. 4, pp. 31–42.

Gallagher, R. H. (1975) *Finite Element Analysis: Fundamentals.* Englewood Cliffs, N. J.: Prentice-Hall.

Gallagher, R. H. (1980) "Shape Functions," Chapter 3 in *Finite Elements in Electrical and Magnetic Field Problems* (M. B. K. Chari and P. Silvester, editors). New York: John Wiley and Sons.

Gordon, W. J. and Hall, C. A. (1973a) "Transfinite Element Methods: Blending Function Interpolation over Arbitrary Curved Element Domains," *Numer. Math.* Vol. 21, pp. 109–129.

Gordon, W. J. and Hall, C. A. (1973b) "Construction of Curvilinear Co-ordinate Systems and Applications to Mesh Generation," *Int. J. Numer. Methods Eng.*, Vol .7, pp. 461–477.

Hall, C. A. and Heinrich, J. C. (1978) "A Finite Element That Satisfies Natural Boundary Conditions Exactly," *J. Inst. Methods Appl.* Vol. 21, pp. 237–250.

Hemker, P. W. (1977) *A Numerical Study of Stiff Two-Point Boundary Problems.* Amsterdam: Mathematisch Centrum.

Hughes, T. J. R. (1987) *The Finite Element Method, Linear Static and Dynamic Analysis*, Vol. 11, pp. 347–362, Englewood Cliffs, N. J.: Prentice-Hall.

Irons, B. M. (1966) "Engineering Applications of Numerical Integration in Stiffness Methods," *AIAA J.,* Vol. 4, pp. 2035–2037.

Isaacson, E. and Keller, H. B. (1966) *Analysis of Numerical Methods.* New York: John Wiley and Sons.

Kawahara, M. and Hasegawa, K. (1978) "Periodic Galerkin Finite Element Method of Tidal Flow," *Int. J. Numer. Methods Eng.*, Vol. 12, pp. 115–120.

Marshall, J. A. and Mitchell, A. R. (1973) "An Exact Boundary Technique for Improved Accuracy in the Finite Element Method," *J. Inst. Mats. Appl.,* Vol. 12, pp. 355–362.

Marshall, J. A. (1975) "Some Applications of Blending Function Techniques to Finite Element Methods," Ph.D. thesis, Univ. of Dundee.

McLeod, R. and Mitchell, A. R. (1972) "The Construction of Basis Functions for Curved Elements in the Finite Element Method," *J. Inst. Mats. Appl.*, Vol. 10, pp. 382–393.

Nitsche, J. (1968) "Ein Kriterium für die Quasi-Optimalität des Ritzschen Verfahrens," *Numer. Math.*, Vol. 11, pp. 346–348.

Oden, J. T. and Carey, G. F. (1983) *Finite Elements*, Vol. IV, *Mathematical Aspects*. Englewood Cliffs, N.J.: Prentice-Hall.

Pepper, D. W. and Dyne, B. R. (1995) "Numerical Simulation of Forced Convective Cooling Within a Massive Array of Heated Tubes." *AIAA J. Thermophysics and Heat Transfer*, Vol. 12, pp. 123–456.

Pepper, D. W. and Heinrich, J. C. (1992) *The Finite Element Method: Basic Concepts and Applications.* Washington, D.C.: Hemisphere (Taylor and Francis).

Pinder, G. F. and Gray, G. (1977) *Finite Element Simulations in Surface and Subsurface Hydrology.* New York: Academic Press.

Reddy, J. N. (1984) *An Introduction to the Finite Element Method.* New York: McGraw-Hill.

Strang, G. and Fix, G. J. (1973) *An Analysis of the Finite Element Method.* Englewood Cliffs, N. J.: Prentice-Hall.

Taig, I. C. (1961) *Structural Analysis by the Matrix Displacement Method.* English Electric Aviation Report S017.

Wachspress, E. L. (1975) *A Rational Finite Element Basis.* New York: Academic Press.

Wachpress, E. and McLeod, R. editors (1979) "Curved Finite Elements," Special issue of *Comput. Methods Appl. Mech. Eng.*, Vol. 5, No. 4.

Watkins, D. S. (1976) "On the Construction of Conforming Rectangular Plate Elements," *Int. J. Numer. Methods Eng.*, Vol. 10, pp. 925–933.

Zienkiewicz, O. C. (1977) *The Finite Element Method*, 3rd ed. London: McGraw-Hill.

Zlamal, M. (1974) "Curved Element in the Finite Element Method," *SIAM J. Numer. Anal.*, Vol. 123, pp. 123–456.

EXERCISES

4.1   Use Eq. (4.3) with n = 1 to derive Eq. (3.59) for linear shape functions and n = 2 to derive Eq. (4.2) for quadratic shape functions.

4.2   For a one-dimensional element of size h, find the shape functions for interpolation using quartic polynomials, i.e., n = 4, and sketch them.

4.3   Assuming that the shape functions are cubic polynomials of the form $N_i(x) = a + bx + cx^2 + dx^3$ and imposing the conditions that one of $N_i(0)$, $N_i(h)$, $N_i'(0)$, or $N_i'(h)$ is equal to 1, and the other three are equal to zero, find the cubic Hermite shape functions for an element e = {x /0 $\leq x \leq h$}.

4.4   Using the shape functions in Exercise 4.3, find the cubic Hermite interpolant of f(x) = sin x between 0 and π using two elements of the same size, and show that both the function and its derivative are continuous at x = $\pi/2$.

4.5   Find the element stiffness matrices for Eq. (4.4) using quadratic and cubic shape functions and assuming that p and q are constant.

4.6   Show that if in Fig. 4.4 we denote by h_i the orthogonal distance from node i to its opposite side, and if the area coordinates of a point P in the triangle are (L_1, L_2, L_3), then

$$L_i = \frac{p_i}{h_i} \qquad\qquad i = 1,2,3$$

where p_i, i = 1,2,3, are the orthogonal distances from the point P to the corresponding sides of the triangle.

4.7   Using the method described in Section 4.3.1, show that the shape functions for the linear and quadratic triangles are given by

$$N_i = L_i \qquad\qquad i = 1,2,3$$

and

$$N_i = L_i(2L_i - 1)$$
$$N_{i+3} = 4L_iL_{i+1}$$
$$i = 1,2,3$$

respectively, where $L_4 \equiv L_1$ and the nodes in the quadratic element are numbered in a manner consistent with Fig. 4.6a.

4.8   Find the shape functions for the triangular element of fourth degree in natural area coordinates.

4.9   Consider the two elements in the figure below, and find the bilinear interpolants over e_1 and e_2 given by

$$u_{e_1}(x,y) = a_1 + b_1 x + c_1 y + d_1 xy$$
$$u_{e_2}(x,y) = a_2 + b_2 x + c_2 y + d_2 xy$$

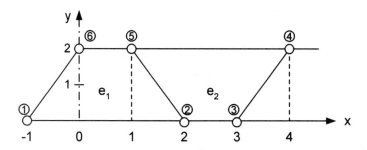

Write them in the form $u_{e_j}(x,y) = \sum_{i=1}^{4} N_{ji}(x,y) u_i$, $j = 1,2$. Show that if a function $u(x,y)$ defined over the two elements is approximated over each element using the interpolants above, the interpolating function will not be continuous along the side common to both elements. Show that, in particular, the shape functions obtained satisfy the relation $N_i(x_j,y_j) = \delta_{ij}$, but N_1, N_3, N_4, and N_6 do not vanish along the line 2-5.

4.10   Show that the shape functions given by Eq. (4.20) for the element of Fig. 4.12 satisfy the condition $N_i(x_j,y_j) = \delta_{ij}$ and Eqs. (4.11), (4.12), and (4.13).

4.11 Find the integral of the function $f(x,y) = x^2 + y^2$ over element e_1 in Exercise 4.9. Would you like to try to repeat this exercise over the general element of Fig. 4.13?

4.12 In the element of Fig. 4.14, move node number 4 to the position $(x_4,y_4) = (-12/5, 9/5)$ and find the element stiffness matrix for the Laplace operator using the isoparametric transformation.

4.13 Find the shape functions for the cubic serendipity element of Figure 4.15c in the parent $\xi-\eta$ plane.

4.14 Derive the shape functions given in Eqs. (4.31) and (4.32) for the linear and quadratic isoparametric triangles.

4.15 Show that the isoparametric transformation for biquadratic and eight-noded quadratic elements reduce to a bilinear transformation when the element sides remain straight lines.

4.16 Repeat Exercise 4.15 for a 12-noded cubic element with (a) parabolic sides, and (b) straight-line sides.

4.17 Show that the condition for the bilinear isoparametric transformation to be invertible is that all interior angles be less than π. (See figure below). Hint: write

$$x(\xi,\eta) = x_1 + (x_2 - x_1)\xi + (x_3 - x_1)\eta + (x_4 - x_3 - x_2 + x_1)\xi\eta$$
$$y(\xi,\eta) = y_1 + (y_2 - y_1)\xi + (y_3 - y_1)\eta + (y_4 - y_3 - y_2 - y_1)\xi\eta$$

calculate det $\mathbf{J}$, evaluate it at $(0,0)$ and use the cross product formula for $(\mathbf{x_4} - \mathbf{x_1}) \times (\mathbf{x_2} - \mathbf{x_1})$.

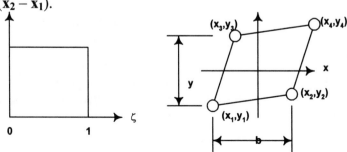

4.18   Given the weak weighted residuals form

$$\int_0^a \left(\int_0^b \left[\frac{\partial w}{\partial x} \frac{\partial u}{\partial x} + cwu - wf \right] dy \right) dx = 0$$

use an isoparametric transformation based on quadratic elements to rewrite the integral in terms of the variables ξ, η in the parent element.

4.19   Construct the shape functions for the six-noded element shown in the figure below using products of Lagrangian element shape functions in each direction. Determine if the shape functions satisfy conditions Eqs. (4.11) through (4.13).

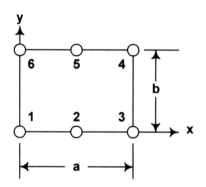

4.20   What is the order of convergence for the element in Exercise 4.19?

4.21   Show that the operator P_x defined by Eq. (4.36) is linear and idempotent.

4.22   Show that if $l = m = 3$, the product operator $P_x P_y[u]$ is the biquadratic interpolant of u.

4.23   Show that if $l = 3$ and $m = 2$ in Eqs. (4.38) and (4.40) the operators P_x and P_y commute. If you feel confident enough, try the next problem.

4.24   Show that if in Eqs. (4.38) and (4.40) $l \neq m$ the operators P_x and P_y always commute.

4.25   Show that, if the functions $u(x,0)$, $u(x,k)$, $u(0,y)$, and $u(h,y)$ in Eq. (4.56) are replaced by their linear interpolants, we obtain the bilinear element.

4.26   Repeat Exercise 4.25 using cubic interpolation to obtain the 12–noded cubic serendipity element.

4.27   We can construct a five-noded element as shown in the figure below in the $-1 \leq \xi,\eta \leq 1$ domain by adding a term of the form $\xi^2\eta^2$ to the bilinear interpolant. Find the shape functions for such element.

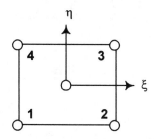

4.28   Consider a three-noded, one-dimensional element with nodes at $x_1 = -1$, $x_2 = 0$, $x_3 = 1$. Assuming interpolation functions of the form

$$N_i(x) = \sin\left[\pi\left(a_i + b_i x + c_i x^2\right)\right]$$

determine the lowest order polynomials such that the functions $N_i(x)$ satisfy the conditions $N_i(x_j) = \delta_{ij}$.

4.29   (a) In the interval $-1 \leq \xi \leq 1$, construct a one-dimensional element capable of interpolating the functional values at the nodes $x = \pm 1$ plus the first derivative at the node $x = -1$. Sketch the shape functions for this three-degrees-of-freedom element.
        (b) Extend to the two-dimensional $-1 \leq \xi,\eta \leq 1$ square. Is this extension unique?

(c) Construct an element in $-1 \le \xi, \eta \le 1$ that reduces to the element in Exercise 4.29a when $y = -1$, i.e., it interpolates the function at the points $(\pm 1, -1)$ and the derivative only at $(-1, -1)$.

(d) Are the elements in Exercises 4.29a and 4.29b conformal?

4.30 Use blending function interpolation to find the shape functions of the 10-noded element in the figure below.

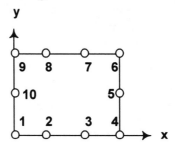

4.31 For the six-noded element in the following figure, construct the shape functions and discuss compatibility.

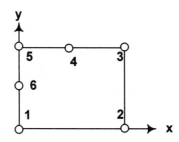

NUMERICAL INTEGRATION

5.1 OVERVIEW

In the previous chapter we introduced the concept of isoparametric elements, which allows us to use elements of very general shapes. The integrands in our finite element formulations are changed from simple polynomials to rational functions that can no longer be integrated analytically (in the general case). To be able to automate the evaluation of the integrals over each element, numerical integration is required. At the same time, new issues arise that make numerical integration in finite element methodology a subject in itself. In this chapter we will discuss the most basic issues that arise with the use of numerical integration; other more specialized aspects will be introduced later when needed. First, we present some of the most commonly used families of quadrature rules, then we discuss the degree of precision necessary to guarantee convergence and optimal accuracy of the finite element approximations. The concept of reduced integration will be introduced, and we will finish the chapter with a discussion on how to evaluate the gradients of the dependent variables once the solution has been obtained.

5.2 QUADRATURE FORMULAE

We summarize the basic principles of numerical integration starting with the one-dimensional case. We seek an appropriate way to numerically approximate the value of an integral

$$I(g) = \int_a^b g(x)\,dx \tag{5.1}$$

In finite element approximations using isoparametric elements, the integrals are always performed over an element at a time in the parent coordinate system $-1 \le \xi \le 1$; hence we are also especially interested in integrals of the form

$$I(f) = \int_{-1}^1 f(\xi)\,d\xi \tag{5.2}$$

We will consider approximations of the form

$$I_n(g) \cong \sum_{i=1}^n w_i g(x_i) \tag{5.3}$$

which are called *numerical quadrature* or *numerical integration formulae.* The points x_i are called *quadrature points,* and the coefficients w_i are called *quadrature weights.*

We then need to choose the quadrature points and weights in such a way that $I_n(g)$ is a good approximation to $I(g)$, i.e., so that the error $E_n(g)$ defined by

$$E_n(g) = I(g) - I_n(g) \tag{5.4}$$

is small.

In order to investigate the errors associated with quadrature formulae we define the *degree of precision* of an integration formula as the maximum integer m such that

$$E\left(x^n\right) = 0 \qquad n = 0,1,2,\ldots,m$$

but

$$E\left(x^{m+1}\right) \ne 0$$

Hence a quadrature formula with degree of precision m integrates polynomials of degree $\leq$ m exactly. To estimate the errors is not very difficult. The proofs are based on Taylor series expressions and the theory of orthogonal polynomials.[*] Here we will state the types of quadratures most frequently used in finite element analysis and their order of accuracy.

The formulae most commonly used in finite element approximations fall into the category of *interpolatory quadratures*. They are found by subdividing the interval $a \leq x \leq b$ into $n + 1$ points, i.e.,

$$a \leq x_1 < x_2 \ldots < x_{n+1} \leq b$$

and constructing the interpolating polynomial $P_n(x)$ of degree n such that

$$g(x_i) \equiv P_n(x_i) \qquad i = 1,2,\ldots,n+1$$

The integral $I(g)$ is then approximated by

$$I_{n+1}(g) \cong \int_a^b P_n(x)dx \qquad (5.5)$$

which is readily evaluated using the Lagrangian form of the shape functions in Eq. (4.3), i.e.,

$$P_n(x) = \sum_{j=1}^{n+1} N_j^n(x)g(x_j) \qquad (5.6)$$

and

---

[*]Here we present the main results as they apply to finite element methodology. For a thorough mathematical analysis, see Isaacson and Keller (1966, Chapter 7, and references therein).

$$N_j^n(x) = \frac{\displaystyle\prod_{\substack{k\neq j \\ k=1}}^{n+1}(x - x_k)}{\displaystyle\prod_{\substack{k\neq j \\ k=1}}^{n+1}(x_j - x_k)} \tag{5.7}$$

Substituting Eq. (5.6) into Eq. (5.5), we obtain

$$I_{n+1}(g) = \sum_{i=1}^{n+1} w_i^{n+1} g(x_i) \tag{5.8}$$

with

$$w_i^{n+1} = \int_a^b N_i^n(x)\,dx \tag{5.9}$$

Hence the quadrature weights w_i^{n+1} are determined completely by the points $x_1, x_2, \ldots, x_{n+1}$.

The error $E_{n+1}(g)$ can be calculated using the results of interpolation theory. We can prove that an interpolatory quadrature using $n+1$ points has degree of precision *at least n*.

5.2.1 Newton-Cotes Formulae

If the quadrature points are equally spaced and $x_1 = a$, $x_{n+1} = b$, the interpolatory quadratures are known as *closed Newton-Cotes* formulae, and the error $E_{n+1}(g)$ in Eq. (5.4) can be expressed in the form

$$E_{n+1}(g) = \frac{M_n}{(n+2)!} h^{n+3} \left(\frac{d^{n+1}g}{dx^{n+1}}\right)_{x=\zeta} \qquad \text{if n is even} \quad (5.10a)$$

and

$$E_{n+1}(g) = \frac{M_n}{(n+1)!} h^{n+2} \left(\frac{d^{n+1}g}{dx^{n+1}}\right)_{x=\zeta} \qquad \text{if n is odd} \quad (5.10b)$$

where

$$M_n \equiv \int_0^n y^2 (y-1)(y-2)...(y-n)dy \text{ for n even}$$

$$M_n \equiv \int_0^n y (y-1)(y-2)...(y-n)dy \text{ for n odd}$$

Also, h = b − a and ζ is a point such that a < ζ < b. These estimates, of course, are only valid if the function g has continuous derivatives of order n + 1.

It follows (Exercise 5.1) that if n = 2k (n is even) and we use an odd number of interpolation points, the error is of the same order as that of the next higher quadrature (n = 2k+1), which contains an even number of interpolation points. Therefore formulae using n even are more efficient, and the number of integration points should be increased in pairs in order to increase the accuracy of the approximation.

The resulting closed Newton-Cotes formulae for the first four values of n are

$$n = 1 \qquad \int_a^b g(x)dx = \frac{h}{2}(g_1 + g_2) + O(h^3) \qquad (5.11)$$

which is the *trapezoidal* rule

$$n = 2 \qquad \int_a^b g(x)dx = \frac{h}{6}(g_1 + 4g_2 + g_3) + O(h^5) \qquad (5.12)$$

which is *Simpson's* rule, and

$$n = 3 \qquad \int_a^b g(x)dx = \frac{h}{8}(g_1 + 3g_2 + 3g_3 + g_4) + O(h^5) \qquad (5.13)$$

$$n = 4 \qquad \int_a^b g(x)dx = \frac{h}{90}(7g_1 + 32g_2 + 12g_3 + 32g_4 + 7g_5) + O(h^7) \qquad (5.14)$$

In the above formulae we have used the notation $g_i \equiv g(x_i)$.

It can be seen that the trapezoidal rule is not efficient when compared to Simpson's rule and that the order of the approximation does not improve going from n = 2 to n = 3, but it improves by two orders going from n = 2 to n = 4.

If the end points of the interval are not included as interpolation points, a new family of quadrature formulae is obtained known as the *open Newton-Cotes* formulae. The only one of interest to us here is the one using one integration point, which is the midpoint of the interval, also know as the *midpoint rule*. This is

$$\int_a^b g(x)dx = hg_1 + O\left(h^3\right) \qquad (5.15)$$

For the open Newton-Cotes formulae, the error estimates given in Eqs. (5.10a) and (5.10b) are valid with slightly modified coefficients M_n (Isaacson and Keller, 1966). However, if the end points of the interval are not used as integration points, we will prefer to use Gauss quadratures presented in the next section.

The composite Newton-Cotes formulae are obtained by subdividing the interval [a,b] into subintervals and applying the formulae over each of them. Here we adopt the point of view that the interval [a,b] constitutes an element; thus we are not interested in subdividing it further to find the finite element integrals. Therefore, only the basic form of the formulae is considered.

5.2.2 Gaussian Quadrature

We now ask ourselves the following question: What is the maximum degree of precision that we can achieve with a quadrature rule that uses n integration points? The answer is that the maximum possible degree of precision that we can obtain using n integration points is m = 2n − 1, and the quadrature formulae that achieve this accuracy are known as Gaussian quadratures.

When Gauss quadratures are used, it is convenient to map the interval [a,b] into the parent interval [-1,1] by means of the transformation

$$\xi = \frac{2}{b-a}x - \frac{a+b}{b-a} \qquad (5.16)$$

The integrals are then performed over the interval $-1 \leq \xi \leq 1$ as in Eq. (5.2), and the integration formulae are referred to as Gauss-Legendre quadratures.

The integration points are given by the zeroes of the (orthogonal) Legendre polynomials of degree n over the interval [-1,1]* and the weights are obtained from the formula

$$w_j = \frac{1}{P_n'(\xi_j)} \int_{-1}^{1} \frac{P_n(\xi)}{\xi - \xi_j} d\xi \qquad\qquad j = 1,2,\ldots,n \qquad (5.17)$$

where $P_n(\xi)$ is the Legendre polynomial of degree n in the interval $-1 \leq \xi \leq 1$ and ξ_j, j = 1, ..., n, are the integration points. The error incurred when using an n–point Gauss quadrature to approximate $\int_{-1}^{1} f(\xi) d\xi$ is given by

$$E_n(f) = \frac{2}{(2n+1)!} \left[\frac{2^n (n!)^2}{2n!} \right]^2 \left. \frac{d^{2n} f}{d\xi^{2n}} \right|_{\xi = \bar{\xi}} \qquad (5.18)$$

where $\bar{\xi}$ is a point in the interval $-1 < \xi < 1$. Notice that this error estimate holds only if the function f has continuous derivatives of order 2n in the interval.

To put the error in Eq. (5.18) in terms of powers of h = b − a is not possible, hence a qualitative comparison of the order of convergence of Gauss quadratures with the Newton-Cotes formulae eludes us. However, numerical comparisons show that if f possesses enough continuous derivatives, the use of Gauss integration produces an incredible improvement in accuracy when compared with other integration methods.

---

*The fundamentals on the theory of orthogonal polynomials can be found in most applied mathematics books; in particular, the interested reader may consult Greenberg (1978) or O'Neil (1983).

Example 5.1

Let us approximate the integral (Yakowitz and Szidarovszky, 1989)

$$\int_0^{10} 10\ \sin(1 - 0.1x)\,dx$$

using the midpoint, trapezoidal and Simpson's rules and compare with Gaussian integration using 2, 3, and 5 integration points. The errors for the different quadrature rules are shown in Table 5.1.

Table 5.1 Accuracy of different quadrature rules

Number	Quadrature			
of points	Midpoint	Trapezoidal	Simpson	Gauss
2	-4.8×10^{-1}	3.9	N/A	1.1×10^{-2}
3	-2.1×10^{-1}	9.6×10^{-1}	1.6×10^{-2}	2.4×10^{-5}
5	-7.7×10^{-2}	2.4×10^{-1}	-1.0×10^{-3}	-1.9×10^{-11}

■

The above example serves to illustrate the impressive power of Gauss quadrature as well as the inefficiency of the trapezoidal rule. Examples like this can be found in most numerical methods books such as the one mentioned above. However, accuracy and efficiency are not the only reasons to use Gauss quadrature in finite element approximations. As we will see in the following section, there are other properties of Gauss integration formulae that can make them important to use in the calculations.

We close this section with a list of the integration points and weights for Gauss quadratures up to order n = 5 (Table 5.2). Higher order formulae are given in texts with an emphasis in structural analysis, where higher order elements are commonly used, such as that of Zienkiewicz (1977).

5.3 MULTIPLE INTEGRALS

In two and three dimensions, we need to evaluate double and triple integrals over areas and volumes, respectively. Let us consider the two-dimensional integral

Table 5.2. Sampling points and weight coefficients for Gaussian
quadrature formulae

$$\int_{-1}^{1} f(\xi)\,d\xi \cong \sum_{i=1}^{n} w_i f(\xi_i)$$

n	ξ_i	w_i
1	0.0	2.0
2	-0.57735026918963	1.0
	0.57735026918963	1.0
3	-0.77459666924148	0.55555555555556
	0.0	0.88888888888889
	0.77459666924148	0.55555555555556
4	-0.86113631159405	0.34785484513745
	-0.33998104358486	0.65214515486255
	0.33998104358486	0.65214515486255
	0.86113631159405	0.34785484513745
5	-0.90617984593866	0.23692688505619
	-0.53846931010568	0.47862867049937
	0.0	0.56888888888889
	0.53846931010568	0.47862867049937
	0.90617984593866	0.23692688505619

$$I = \int_{-1}^{1} \int_{-1}^{1} f(\xi,\eta)\,d\xi\,d\eta \tag{5.19}$$

The easiest way to extend the concepts already established in the one-dimensional case to two (and three) dimensions is to evaluate the variables one by one in succession, keeping the others constant, or "separation of variables." In Eq. (5.19) we can fix the independent variable h and define

$$F(\eta) \equiv \sum_{i=1}^{n} w_i\, f(\xi_i,\eta) \cong \int_{-1}^{1} f(\xi,\eta)\,d\xi \tag{5.20}$$

Next we can approximate

$$\int_{-1}^{1} F(\eta) d\eta \cong \sum_{j=1}^{m} w_j \, F(\eta_j) = \sum_{j=1}^{m} w_j \left(\sum_{i=1}^{n} w_i \, f(\xi_i, \eta_j) \right) \quad (5.21)$$

Hence we have

$$I_{nm} = \sum_{i=1}^{n} \sum_{j=1}^{m} w_i w_j \, f(\xi_i, \eta_j) \quad (5.22)$$

where m and n are not necessarily equal.

These are also interpolatory quadratures, and the results in one dimension can be extended to obtain error estimates. If the domain is rectangular, the degree of precision of the quadrature given by Eq. (5.22) is at least the minimum of those for Eqs. (5.20) and (5.21). If the domain is not rectangular, this is no longer true. We do not consider such cases, however, since we assume that we always use isoparametric transformations if the elements are not rectangular.

Example 5.2

A quadrature formula with degree of precision 3 is obtained if we use a Gauss quadrature with $n = 2$ in each direction, that is,

$$\int_{-1}^{1} \int_{-1}^{1} f(\xi, \eta) d\xi \, d\eta \cong I_{22} = \sum_{i=1}^{2} \sum_{j=1}^{2} w_i w_j \, f(\xi_i, \eta_j)$$

where the sampling points ξ_i, η_j in each direction and the weights w_i are given in Table 5.2 for $n = 2$. Figure 5.1 shows the approximate location of the Gauss points in two and three dimensions for $n = m = 2$ and $n = m = 3$. The extension of Gauss quadratures to three dimensions is left as an exercise. ∎

5.4 MINIMUM AND OPTIMAL ORDER OF INTEGRATION

It is evident from the previous sections that increasing the number of quadrature points in an integration formula also increases the accuracy of the numerical approximation to the exact integral. The problem is that by

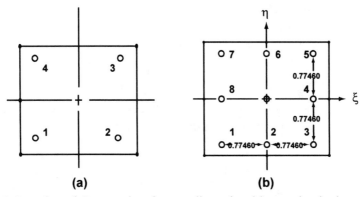

Figure 5.1 Location of Gauss points for two-dimensional integration in the square $1 \leq \xi, \eta \leq 1$ for (a) 2 x 2 quadrature and (b) 3 x 3 quadrature.

doing so, we also increase the computational cost, which can be a significant consideration in large calculations. From this point of view, we would like to use as few integration points as possible to evaluate the stiffness matrices and right-hand side vectors. The use of quadratures with a degree of precision that is too low also causes difficulties. In fact, in some cases, we may lose convergence altogether; in other cases, we may obtain stiffness matrices that are singular, even though the equations are well posed and the approximation spaces are admissible. It is therefore natural to think that there will be a *minimum* order of integration that guarantees convergence of the finite element approximations, and perhaps an *optimal* one that guarantees the highest possible rate of convergence and accuracy while minimizing the computational cost. To illustrate these ideas, let us consider the following example, which shows how the number of integration points affect the final stiffness matrices.

Example 5.3

Find the solution to the heat conduction problem given in Eqs. (3.40)–(3.62), that is,

$$\frac{d^2u}{dx^2} = 1 \quad 0 < x < 1$$

with boundary conditions

$$u(0) = 0 \qquad \frac{du}{dx}\bigg|_{x=1} = 1$$

We will apply the finite element method using one quadratic element. Therefore the weak Galerkin form can be written as

$$\int_0^1 \left(\frac{dN_i}{dx} \frac{du}{dx} + N_i \right) dx - N_i(1) = 0 \qquad i = 1,2,3$$

with $u(x) = \sum_{j=1}^{3} N_j(x) u_j$. From Eq. (4.2) the shape functions and their derivatives are

$$N_1(x) = 1 - 3x + 2x^2 \qquad \frac{dN_1}{dx} = -3 + 4x$$

$$N_2(x) = 4x(1-x) \qquad \frac{dN_2}{dx} = 4 - 8x$$

$$N_3(x) = x(2x-1) \qquad \frac{dN_3}{dx} = 4x - 1$$

Hence the system of three equations is

$$\left[\int_0^1 (-3+4x)^2 dx \right] u_1 + \left[\int_0^1 (-3+4x)(4-8x) dx \right] u_2$$
$$+ \left[\int_0^1 (-3+4x)(4x-1) dx \right] u_3 = -\int_0^1 (1-3x+2x^2) dx \qquad (5.23a)$$

$$\left[\int_0^1 (4-8x)(-3+4x) dx \right] u_1 + \left[\int_0^1 (4-8x)^2 dx \right] u_2$$
$$+ \left[\int_0^1 (4-8x)(4x-1) dx \right] u_3 = -\int_0^1 4x(1-x) dx \qquad (5.23b)$$

$$\left[\int_0^1 (4x-1)(-3+4x)dx\right]u_1 + \left[\int_0^1 (4x-1)(4-8x)dx\right]u_2$$

$$+\left[\int_0^1 (4x-1)^2 dx\right]u_3 = -\int_0^1 x(2x-1)dx + 1 \qquad (5.23c)$$

We will now integrate Eqs. (5.23) numerically using Newton-Cotes formulae for simplicity.

a) Using the midpoint rule given by Eq. (5.15), we obtain the linear system of equations

$$\begin{bmatrix} 1 & 0 & -1 \\ 0 & 0 & 0 \\ -1 & 0 & 1 \end{bmatrix} \begin{bmatrix} u_1 \\ u_2 \\ u_3 \end{bmatrix} = \begin{bmatrix} 0 \\ -1 \\ 1 \end{bmatrix}$$

It is easily verified that the stiffness matrix has rank 1 (or that the degree of singularity of the matrix is 2). Therefore, only one linearly independent equation is obtained, and the reduced 2 x 2 matrix that results after imposing the Dirichlet boundary condition u(0) = 0 remains singular. Furthermore, the matrix augmented by the right-hand side has rank 2, and hence the system is incompatible – a solution does not exist. Obviously, our numerical integration formula was too inaccurate, and as a result, we obtained a system of equations without solutions.

(b) Let us now use the two-point trapezoidal rule (given in Eq. (5.11)) for integration. This yields

$$\begin{bmatrix} 5 & -8 & 3 \\ -8 & 16 & -8 \\ 3 & -8 & 5 \end{bmatrix} \begin{bmatrix} u_1 \\ u_2 \\ u_3 \end{bmatrix} = \begin{bmatrix} -1/2 \\ 0 \\ 1/2 \end{bmatrix}$$

This time the stiffness matrix has rank 2 (a degree of singularity 1). The effect of adding one more quadrature point is to increase the number of linearly independent equations. Now a nonsingular reduced system results after incorporating the Dirichlet boundary condition. Thus

$$\begin{bmatrix} 16 & -8 \\ -8 & 5 \end{bmatrix} \begin{bmatrix} u_1 \\ u_2 \end{bmatrix} = \begin{bmatrix} 0 \\ 1/2 \end{bmatrix}$$

with solution $u_2 = 1/2$, $u_3 = 1/2$. This solution is very inaccurate (recall that a correct finite element approximation should yield the exact solution at the nodes) and shows a 300% error at node 2. This is a direct consequence of the poor accuracy of the trapezoidal rule that has degree of precision m = 1 and therefore cannot integrate quadratic functions with enough accuracy over an interval of this size.

(c) If we now use the three-point Simpson rule, which has degree of precision m = 2, the integrations are exact and yield

$$\begin{bmatrix} 14 & -16 & 2 \\ -16 & 32 & -16 \\ 2 & -16 & 14 \end{bmatrix} \begin{bmatrix} u_1 \\ u_2 \\ u_3 \end{bmatrix} = \begin{bmatrix} -1 \\ -4 \\ 5 \end{bmatrix}$$

which gives the exact solution at the nodal points, $u_2 = 1/8$, $u_3 = 1/2$. Therefore, if a quadrature with four (or more) points is used, we cannot improve on this result, and we have achieved both the minimum and optimal order of integration for this case. ∎

Had we used Gauss quadratures instead of Newton-Cotes in the above example, the same result would have been obtained for a one-point quadrature, which is identical to the midpoint rule. However, a two-point Gauss quadrature yields the same result as Simpson's rule since it has degree of precision 3 and integrates quadratic polynomials exactly.

The general result (Strang and Fix, 1973) can be stated as follows:

"The minimum order of numerical integration required for the finite element approximations to second order differential equations to be convergent is the one needed to integrate exactly the first derivative of all shape functions, and produce a positive definite stiffness matrix."

In the previous example, the derivatives are linear, and hence a quadrature formula with degree of precision m = 1 should suffice to guarantee convergence according to the above result. However, these low orders of integration can be impractical and lead to anomalous behavior, as shown in our example, where the one-point quadrature produced a singular system and the two-point quadrature was too inaccurate.[*] The reader interested in further details should consult the work of Fix (1972) and Fried (1974). We will abandon the use of this condition in favor of the following criteria, which are sufficient to ensure that we obtain the optimal rate of convergence for second-order elliptic equations:

"The squares of the first derivatives of all shape functions must be integrated exactly by the quadrature rule."

Hence in one dimension or if triangular elements are used in two dimensions, with shape functions that are polynomials of degree k, we need a quadrature formula with degree of precision m = 2(k − 1). If rectangular elements are used we normally require more precision owing to the extra terms of order higher than the highest complete polynomial contained in the shape functions.

Example 5.4

Consider the linear triangle and bilinear quadrilateral applied to the equation

$$\frac{\partial^2 u}{\partial x^2} + \frac{\partial^2 u}{\partial y^2} + a(x,y) u = f(x,y)$$

The above criterion states that the integrals

$$\int_\Omega \left(\frac{\partial N_i}{\partial x} \frac{\partial N_j}{\partial x} + \frac{\partial N_i}{\partial y} \frac{\partial N_j}{\partial y} \right) d\Omega$$

---

[*] The additional condition that the contribution to the stiffness matrix of the second-order diffusion term to be a positive definite matrix was not satisfied when one integration point was used, resulting in a singular matrix.

must be computed exactly to achieve the optimal rate of convergence. In the case of the linear triangle, the shape functions are of the form

$$N_i(x,y) = a + bx + cy$$

Hence all partial derivatives are constant, and a one-point quadrature suffices.

For bilinear elements, on the other hand, the shape functions are of the form

$$N_i(x,y) = a + bx + cy + dxy \qquad (5.24)$$

and their derivatives are linear functions of x or y. Therefore we must integrate quadratic polynomials exactly, and a two-point Gauss quadrature is required. A similar situation arises for quadratic triangles and biquadratic rectangular elements (Exercise 5.15). ∎

For triangular elements, convenient integration formulae were first derived by Hammer et al. (1956). Later, Cowper (1973) added integration formulae for triangles with degree of precision up to 7. Table 5.3 gives the sampling points and weights for integration formulae with degree of precision up to 5 are given. The positions of the quadrature points are shown in Fig. 5.2.

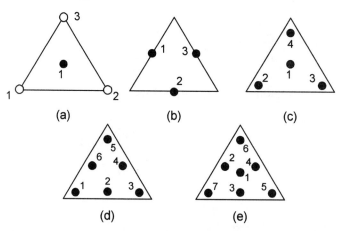

Figure 5.2 Schematic of the location of the integration points in triangles for the quadrature given in Table 5.3 (a) one-point rule, (b) three-point rule, (c) four-point rule, (d) six-point rule, and (e) seven-point rule.

Table 5.3 Quadrature formulae for triangles*

Number of points	Degree of precision	Point number i	Area coordinates of quadratures			Weight w_i
			$(L_1)_i$	$(L_2)_i$	$(L_3)_i$	
1	1	1	1/2	1/2	1/2	1
3	2	1	1/2	1/2	0	1/3
		2	0	1/2	1/2	1/3
		3	1/2	0	1/2	1/3
4	3	1	1/3	1/3	1/3	-27/48
		2	11/15	2/15	2/15	25/48
		3	2/15	11/15	2/15	25/48
		4	2/15	2/15	11/15	25/48
6	4	1	a_1	a_2	a_2	b_1
		2	a_3	a_3	a_4	b_2
		3	a_2	a_1	a_2	b_1
		4	a_4	a_3	a_3	b_2
		5	a_2	a_2	a_1	b_1
		6	a_3	a_4	a_3	b_2
7	5	1	c_1	c_1	c_1	d_1
		2	c_2	c_3	c_3	d_2
		3	c_4	c_4	c_5	d_3
		4	c_3	c_2	c_3	d_2
		5	c_5	c_4	c_4	d_3
		6	c_3	c_3	c_2	d_2
		7	c_4	c_5	c_4	d_3

$a_1 = 0.81684757298046$ $c_2 = 0.79742698535309$

$a_2 = 0.09157621350977$ $c_3 = 0.10128650732346$

$a_3 = 0.44594849091597$ $c_4 = 0.47014206410512$

$a_4 = 0.10810301816807$ $c_5 = 0.05971587178977$

$b_1 = 0.10995174365532$ $d_1 = 0.22500000000000$

$b_2 = 0.22338158967801$ $d_2 = 0.12593918054483$

$c_1 = 0.33333333333333$ $d_3 = 0.13239415278851$

*Integration point numbers are indicated in Fig. 5.2.

The question now is how to determine the necessary order of integration when isoparametric elements are used and the functions to be integrated are no longer polynomials. It was found by Irons (1966) (also see Strang and Fix, 1973) that it is sufficient to be able to evaluate the area $\int_\Omega d\Omega$ correctly. Therefore, if an isoparametric transformation is used, we must be able to integrate exactly the expression

$$\int_{-1}^{1} \int_{-1}^{1} \det J\, d\xi\, d\eta \qquad (5.25)$$

Because the determinant of the transformation contains products of the derivatives of the shape functions as given in Eq. (4.29), the criteria stated above will still be applicable in the case of isoparametric elements.

REMARK

In Example 5.4, we must also integrate the term $\int_\Omega a(x)N_i N_j d\Omega$ which contributes to the stiffness matrix. Consider one-dimensional linear elements (or triangles) and assume that a(x) is a constant function, without loss of generality. The products of the shape functions in the above integral are quadratic polynomials and are not integrated exactly by a one-point quadrature. Nevertheless, our result states that this is not necessary, and the method will still converge at an optimal rate. It is therefore important to clarify the difference between *rate of convegence* and *accuracy*, since the first is often confused with the second. To do this let us examine the following simple one-dimensional example using linear elements.

Example 5.5

Consider the equation

$$-\frac{d^2u}{dx^2} + u = 0 \qquad 0 \le x \le 1 \qquad (5.26)$$

with the boundary conditions chosen so that the exact solution is u*(x) = 100 sinh (x). Because the derivatives of the shape functions are constant, a one-point quadrature is sufficient to guarantee optimal rate of convergence. The element stiffness equations are

$$\left(\frac{1}{h}\begin{bmatrix} 1 & -1 \\ -1 & 1 \end{bmatrix} + \frac{h}{4}\begin{bmatrix} 1 & 1 \\ 1 & 1 \end{bmatrix} \right) \begin{bmatrix} u_1 \\ u_2 \end{bmatrix} = \begin{bmatrix} 0 \\ 0 \end{bmatrix}$$

On the other hand, using a two-point Gauss quadrature, we integrate the second term in the equation exactly, and this changes the element stiffness equations to

$$\left(\frac{1}{h}\begin{bmatrix} 1 & -1 \\ -1 & 1 \end{bmatrix} + \frac{h}{6}\begin{bmatrix} 2 & 1 \\ 1 & 2 \end{bmatrix} \right) \begin{bmatrix} u_1 \\ u_2 \end{bmatrix} = \begin{bmatrix} 0 \\ 0 \end{bmatrix}$$

Clearly, the two approximations will differ – in fact, the second one is much more accurate than the first. This is shown in Table 5.4, which gives the maximum relative error in finite element approximations obtained using 5, 10, and 20 elements.

Table 5.4 Maximum error in the solution of Eq. (5.26)

Number of elements	Maximum error with one Gauss point	Maximum error with two Gauss points
5	-0.236 %	0.050 %
10	-0.059 %	0.013 %
20	-0.013 %	0.003 %

It is evident from Table 5.4 that the use of two Gauss integration points produces a significantly more *accurate* solution. However, the *rate of convergence* of the finite element method is the same for both integration schemes. This is shown in Fig. 5.3 for ln(h) versus ln(|e|), where h is the mesh size and |e| is the magnitude of the relative error at x = 0.2. It can be observed that the two lines have the same slope, and hence the schemes have the same rate of convergence. But the results using a two-point Gauss quadrature are always more accurate. A similar situation arises with higher order elements (Exercise 5.16). ■

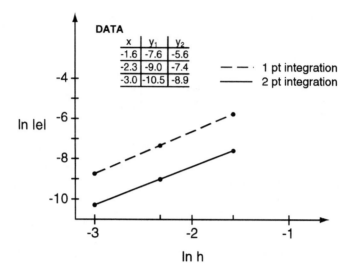

Figure 5.3 Convergence rates at x = 0.2 using linear elements with one- and two-point Gauss quadrature.

If quadrilateral elements are used instead, this situation does not arise because both terms in the weak form involve polynomials of the same order in each coordinate direction. That is, the degree of precision required by our integration criterion gives m = 2 for bilinear elements, and m = 4 for biquadratic elements. Therefore a 2 x 2 Gauss quadrature is needed for bilinear elements, and a 3 x 3 Gauss rule for biquadratics.

5.5 REDUCED INTEGRATION, EVALUATING GRADIENTS

In the previous section we established that the order of integration required to guarantee an optimal rate of convergence must have degree of precision of at least $m = 2l$ in each coordinate direction. In this instance, l is the polynomial degree of the derivatives of the shape functions. This condition occurs regardless of whether the derivatives contain the complete polynomial of degree l. We will now use this result to introduce other concepts.

5.5.1 Reduced and Selective Integration

We will refer to the order of integration applied to rectangular elements as *full integration* when a Gauss quadrature with at least the number of points needed to guarantee the required degree of precision $2l$, is used. If a Gauss quadrature with degree of precision less than $2l$ is used, we will call it *reduced integration*. If full integration requires n Gauss points in each coordinate direction, normally, reduced integration will use one point less, i.e., n − 1 Gauss points in each coordinate direction.

Example 5.6

(a) For bilinear rectangular elements, l = 1. Hence the 2 x 2 Gauss quadrature is necessary for full integration. The reduced integration quadrature will use 2 − 1 = 1 Gauss points in each direction.

(b) A biquadratic element, for which l = 2, requires the 3 x 3 Gauss quadrature for full integration. Fourth-degree polynomials must be integrated exactly, and a degree of precision of at least 4 is required. The reduced integration formula is the 2 x 2 Gauss quadrature, which has degree of precision m = 3.

These are illustrated in Fig. 5.4, and further examples are left to Exercise 5.12. ∎

We can, of course, integrate different terms in the weighted residuals expressions using different quadrature rules. When this is done, we refer to it as *selective reduced integration*. Selective integration is often used in finite element analysis for specific purposes, especially in solid mechanics. We will deal mainly with two applications of reduced and selective integration. First, we will use reduced integration to evaluate the gradients of finite element solutions. Later, in Chapter 10,

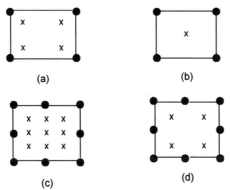

(a) (b) (c) (d)

Figure 5.4 Gauss quadrature points denoted by for (a) bilinear element, full integration, (b) bilinear element, reduced integration, (c) biquadratic element, full integration, and (d) biquadratic element, reduced integration.

we will use selective reduced integration in penalty function algorithms for the solution of the Navier-Stokes equations.

5.5.2 Evaluating Gradients

In Chapter 3, Sections 3.6.4 and 3.7.2, we discussed the evaluation of boundary fluxes using the boundary residual equations. We showed that calculating the gradients in the direction normal to the boundary using this method yields approximations that are more accurate than those obtained by differentiation of the finite element solution.

It is also of great interest to evaluate the gradients of the finite element solution in the interior of the computational domain accurately. This is particularly important in stress analysis, where the stresses are a function of the derivatives of the displacements, and in heat transfer analysis, when the heat fluxes are the quantities of interest. As was the case for boundary fluxes, increased accuracy can be obtained in the evaluation of gradients in the interior of the computational domain, with a little extra work. In this case, the extra accuracy is achieved through the use of reduced numerical integration. To illustrate the basic concepts, let us look first at a simple one-dimensional example.

Example 5.7

Consider the function $f(x) = e^x$ over the interval $-1 \leq x \leq 1$, where we define one quadratic element. The finite element interpolant to $f(x)$ is given by

$$g(x) = \sum_{i=1}^{3} N_i(x) f_i = \frac{1}{2}x(x-1)e^{-1} + \left(1 - x^2\right) + \frac{1}{2}x(x+1)e \quad (5.27)$$

The derivative $f'(x)$ is approximated by the derivative of $g(x)$, which is

$$g'(x) = \left(x - \frac{1}{2}\right)e^{-1} - 2x + \left(x + \frac{1}{2}\right)e \quad (5.28)$$

where $f'(x)$ and $g'(x)$ are shown in Fig. 5.5.

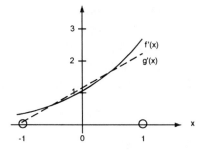

Figure 5.5 Derivative of the function $f(x) = e^x$ and its finite element interpolant $g'(x)$ in $-1 \leq x \leq 1$.

We will now calculate the error $e(x) = |(f'(x)-g'(x))/f'(x)|$ at different points in the interval.

(a) If we evaluate the derivatives at the nodes, we obtain the results shown in Table 5.5. The approximation is very poor, especially if we consider that the maximum error in the approximation of the function is about 3%. However, Fig. 5.5 suggests that at two points the derivative is exact, and we should try to get closer to those values.

Table 5.5 Derivatives of the quadratic interpolant to $f(x) = e^x$ evaluated at the nodes

x	f'(x)	g' (x)	e (x) (%)
-1.0	0.367879	0.089030	98
0.0	1.000000	1.175200	18
1.0	2.71828	2.26136	17

(b) Let us now evaluate the derivatives at the three points corresponding to full Gauss integration. The results, shown in Table 5.6, show a significant improvement, but the error remains fairly high.

Table 5.6 Derivatives of the quadratic interpolant to $f(x) = e^x$ evaluated at the three full integration Gauss points

x	f'(x)	g' (x)	e (x) (%)
-0.774597	0.460880	0.333864	28
0.00000	1.00000	1.17520	18
0.774597	2.6972	2.01654	7

(c) If we now evaluate the derivatives at the two reduced integration Gauss points, the results given in Table 5.7 show a dramatic improvement. In fact, at these points the derivatives are superconvergent. The general result will be stated in a moment. Meanwhile, the reader can verify that the same is true for the linear interpolant in this example (Exercise 5.18). ■

Table 5.7 Derivatives of the quadratic interpolant to $f(x) = e^x$ at two reduced integration Gauss points

x	$f'(x)$	$g'(x)$	e (x) (%)
-0.577350	0.561384	0.548106	2
0.577350	1.78131	1.80230	1

To explain why we obtained such improved accuracy at the reduced integration Gauss points, we need the following results:

If $p(x)$ is a polynomial of order k, the least squares polynomial fit of order k−1 to $p(x)$ intersects $p(x)$ at the Gauss points, of the k-point Gauss quadrature. Let us illustrate this for the quadratic case.

Example 5.8

Consider a quadratic polynomial, $k = 2$ in the interval $-1 \le x \le 1$, i.e., $p(x) = a + bx + cx^2$. The linear polynomial $g(x)$, which is the least squares fit to $p(x)$ over this interval, is easily found to be $q(x) = (a + c/3) + bx$. Hence $p(x) - q(x) = c(x^2 - 1/3) = 0$ gives $x = \pm \sqrt{1/3}$, which are the Gauss points for the two-point quadrature as given in Table 5.2. To show that if $p(x)$ is linear the constant least squares fit intersect at $x = 0$ is left to Exercise 5.19. ■

The argument is this: it can be proved (Herrmann, 1972; Moan, 1973) that if the differential operator is self-adjoint, then the finite element solution to the second-order equation is equivalent to finding the least squares approximation to the derivatives of the dependent variable in the weighted residuals sense. Therefore, at the reduced integration Gauss points, the values of the derivatives, as evaluated from the finite element solution, are the same as the values that we would obtain if we interpolate the derivatives using the shape functions. Hence the approximation to the derivatives (at these points) is of the same order as for the function itself. That is, at the reduced integration points the error in the derivatives is $O(h^{k+1})$ and the derivatives are superconvergent.

In general, this result is exact in one dimension and approximate in two and three dimensions, depending on the boundary conditions (Barlow, 1976).

Let us now illustrate this behavior in the context of finite element solutions.

Example 5.9

First we will consider the one-dimensional heat transfer problem of Example 3.4:

$$-\frac{d^2u}{dx^2} + u = 1 \qquad 0 < x < 1$$

$$u(0) = 0, \quad u(1) = 1 \tag{5.29}$$

For which the exact solution is

$$u^* = \frac{1}{e^2 - 1}\left(e^x - e^{2-x}\right) + 1 \tag{5.30}$$

with derivatives

$$\frac{du^*}{dx} = \frac{1}{e^2 - 1}\left(e^x + e^{2-x}\right) \tag{5.31}$$

A finite element solution of Eq. (5.29) using five linear finite elements yields a maximum relative error at the nodes of 0.1%. The derivatives evaluated at the center of each element, which correspond to the reduced integration point, yield a 0.2% maximum error. The calculated derivatives and relative errors are shown in Table 5.8. ∎

The gradients, in the interior of the computational region, should be calculated at the reduced integration Gauss points for maximum accuracy. This translates into using one Gauss point for linear, bilinear, or trilinear elements, and two Gauss points for quadratic, 2 x 2 points for biquadratic, and 2 x 2 x 2 points for triquadratic elements.

Table 5.8 Error in the approximation of the derivatives in Example 5.9 at the reduced integration Gauss point

x	Exact	Finite elements	Relative error
0.1	1.21944	1.22215	-0.0022
0.3	1.06805	1.07003	-0.0019
0.5	0.959517	0.960991	-0.0015
0.7	0.889697	0.890649	-0.0013
0.9	0.855174	0.856172	-0.0011

The fact that the values of the gradient that we obtain are not defined at the nodal points is inconvenient, though, because it does not provide us with a continuous representation of the derivatives over the domain. In order to obtain values of the derivatives with similar accuracy at the nodal points, we evaluate the weighted least squares fit to the derivatives calculated at the Gauss points, using the shape functions. In the above example we express the derivative as a function,

$$d(x) = \sum_{i=1}^{N} N_i(x) d_i \qquad (5.32)$$

where N is the number of nodes and $N_i(x)$ the linear shape functions. The local weighted least squares functional to be minimized can be written as

$$I = \int_{\Omega} \left[\left(\frac{du^h}{dx} \right) - \sum_{j=1}^{N} N_j d_j \right]^2 dx \qquad (5.33)$$

where du^h/dx denotes the derivatives calculated from the finite element solution.

Differentiating I with respect to the degrees of freedom d_i and setting the derivatives equal to zero, we obtain the linear system of equations

$$\sum_{j=1}^{N}\left[\int_{\Omega} N_i N_j dx\right] d_j = \int_{\Omega} N_i \left(\frac{du^h}{dx}\right) dx \qquad (5.34)$$

where the right-hand side integral must be evaluated using the reduced integration Gauss quadrature.

It is usually convenient to make the least squares fit locally. To achieve this, the matrix on the left-hand side is diagonalized and Eq. (5.34) is replaced by

$$\left[\int_{\Omega} N_i dx\right] d_i = \int_{\Omega} N_i \left(\frac{du^h}{dx}\right) dx \qquad (5.35)$$

which yields the values d_i of the derivatives at the nodal points explicitly. Notice that in the case of linear elements with a regular grid it reduces to the average of the derivative over two adjacent elements. The diagonalization process used above is known as mass lumping and will be discussed in more detail in Chapter 7.

Let us now return to Example 5.9 and calculate the derivatives at the nodal points using the above method. The resulting values of the gradient, together with the relative error at the nodal points, are given in Table 5.9.

It can be immediately observed that this procedure is indeed accurate for the interior nodes, where the error remains of the same order as it was at the reduced integration Gauss points. However, it is

Table 5.9 Error in the local weighted least square approximation of the derivatives at the nodal points for Example 5.9

x	Exact	Finite elements	Relative error
0.0	1.31304	1.295135	0.01364
0.2	1.13805	1.146095	-0.00706
0.4	1.00877	1.01378	-0.00501
0.6	0.919904	0.924507	-0.00500
0.8	0.867993	0.872337	0.005000
1.0	0.850918	0.838016	0.01516

inadequate at the boundary nodes, where the error is an order of magnitude larger. As we saw in Chapter 3, the boundary residual equations should be used to calculate the derivatives at those points. In this case, using the boundary residual equations results in the following approximations:

$$u'(0) \cong 1.31333 \qquad \text{relative error} = 0.00022$$

$$u'(1) \cong 0.850892 \qquad \text{relative error} = 0.0005$$

These approximations are an order of magnitude better than in the interior of the region. The importance of the residual equations to calculate boundary fluxes cannot be overstressed.

We should point out that the weighted residuals can also be used to calculate the derivatives in the interior of the computational region. However, in general, they do not offer better accuracy than the use of reduced integration Gauss points, and the extra work is significant.

We will close this chapter with a two-dimensional example to illustrate the use of reduced integration in the evaluation of the gradients of the solution.

Example 5.10

Consider the two-dimensional boundary value problem given in Example 3.9, that is,

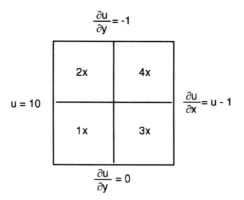

Figure 5.6 Domain and boundary conditions for Example 5.10 showing the position of the points where the gradients given in Table 5.10 are evaluated.

$$\frac{\partial^2 u}{\partial x^2} + \frac{\partial^2 u}{\partial y^2} = 0 \qquad 0 > x, y < 1$$

$$u(0,y) = 10 \qquad x = 0,\, 0 \le y \le 1$$

$$\frac{\partial u}{\partial x} = u - 1 \qquad x = 1,\, 0 \le y \le 1$$

$$\frac{\partial u}{\partial y} = 0 \qquad 0 \le x \le 1,\, y = 0$$

$$\frac{\partial u}{\partial y} = -1 \qquad 0 \le x \le 1,\, y = 1$$

The domain, boundary conditions and reduced integration Gauss points for a uniform mesh of four elements are shown in Fig. 5.6. In Table 5.10 we present the calculated gradients at the reduced integration points and compare with those obtained with a very fine grid of 40 x 40 elements, denoted as the "exact" solution.

Table 5.10: Gradients at the reduced integration points of a four-element mesh for Example 5.10.

Point	u_x^h	Exact u_x	Error	u_y^h	Exact u_y	Error
1	-4.289	-4.318	0.7%	0.073	-0.090	19%
2	-4.712	-4.670	0.9%	-0.350	0.458	23%
3	-4.519	-4.512	0.2%	-0.154	-0.166	7%
4	-4.481	-4-490	0.2%	-0.654	-0.640	2%

The derivatives in the x-direction are very accurate, but we do not observe the same accuracy in the y-direction. The reason for this is that the derivatives in the y direction are 1 and 2 orders of magnitude smaller values than those in the x-direction. We also point out that the errors are maximum at point 2, which lies within the element that contains the corner (0,1) where the gradients vary sharply. Further examples are left as exercises.

5.6 CLOSURE

The use of numerical integration is an integral part of finite element algorithms. It is necessary when we use isoparametric elements to represent irregular geometries. Among the many numerical integration families of algorithms available, Gauss quadratures are particularly important because of their accuracy and further properties that make these especially useful in finite element methodology. In particular, we have used them in this chapter to obtain accurate approximations to the derivatives of the finite element solutions.

The error in finite element approximations has been discussed. We analyzed the effect of the degree of precision of quadrature formulae in the finite element solutions and introduced the concepts of "full" and "reduced" integration, both of which play an important role in the methodology.

At this point we have advanced far enough in the solution of steady state linear problems that any further requirements would become specific to the problem being solved. The references given in the chapter are a good start for those in need of more specialized material.

REFERENCES

Barlow, J. (1976) "Optimal Stress Locations in Finite Element Models," *Int. J. Numer. Methods Eng.*, Vol. 10, pp. 243–251.

Cowper, G. R. (1973) "Gaussian Quadrature Formulas for Triangles," *Int. J. Numer. Methods Eng.*, Vol. 7, pp. 405–408.

Fix, G. J. (1972) "On the Effect of Quadrature Errors in the Finite Element Method," pp. 525-556 in *The Mathematical Foundations of the Finite Element Method with Applications to Differential Equations* (A. K. Aziz, editor). New York: Academic Press.

Fried, I. (1974) "Numerical Integration in the Finite Element Method," *Comput. Struct.*, Vol. 4, pp. 921–932.

Greenberg, M. D. (1978) *Foundations of Applied Mathematics.* Englewood Cliffs, N.J.: Prentice-Hall.

Hammer, P. C., Marlowe, O. P. and Stroud, A. H. (1956) "Numerical Integration over Simplexes and Lower," *Math Table Aids Comput.*, Vol. 10, pp. 130–137.

Herrmann, L. R. (1972) "Interpolation of Finite Element Procedure in Stress Error Minimization," *Proc. Am. Soc. Civ. Eng.*, Vol. 98, pp. 1331–1336.

Irons, B. M. (1966) "Engineering Application of Numerical Integration in Stiffness Method," *AIAA J.*, Vol. 14, pp. 2035–2037.

Isaacson, E. and Keller, H. B. (1966) *Analysis of Numerical Methods*. New York: John Wiley and Sons.

Moan, T. (1973) "On the Local Distribution of Errors by the Finite Element Approximation," in *Theory and Practice in Finite Element Standard Analysis* (Y. Yamada and R. H. Gallagher, editors). Univ. of Tokyo Press.

O'Neil, P. V. (1983) *Advanced Engineering Mathematics*. Belmont, Calif.: Wadsworth Publishing.

Strang, G. and Fix, G. J. (1973) *An Analysis of the Finite Element Method*. Englewood Cliffs, N.J.: Prentice-Hall.

Yakowitz, S. and Szidarovszky, F. (1989) *An Introduction to Numerical Computations*, 2nd Ed. New York: Macmillan.

Zienkiewicz, O. C. (1977) *The Finite Element Method*, 3rd ed. London: McGraw-Hill.

Zienkiewicz, O. C. and Taylor, R. L. (1991) *The Finite Element Method*, Vol. 2, 4th ed. London: McGraw-Hill.

EXERCISES

5.1   Show that if we set n = 2k in Eq. (5.10a), and n = 2k + 1 in Eq. (5.10b), where k is an integer, the error for both quadrature formulae is of the same order.

5.2   Set n = 2 in Eqs. (5.6) through (5.9) and derive Simpson's integration formula.

5.3   Use Eqs. (5.10a) and (5.10b) to show that the trapezoidal rule is third-order accurate and Simpson's rule is fifth-order accurate.

5.4   Find the exact integration error when a quadratic polynomial is integrated using the trapezoidal rule.

5.5   Find the exact integration error when a quadratic polynomial is integrated using Simpson's rule.

5.6   Show that the midpoint rule is a third-order accurate approximation, and find the exact integration error when used to integrate a quadratic polynomial. Compare with the error using the trapezoidal rule.

5.7   Using the integral

$$I = \int_{-1}^{1} \frac{1}{3+\xi} d\xi$$

establish a comparison like the one given in Table 5.1 using Newton-Cotes and Gauss formulae for n = 2,3,...,7. Discuss the results in relation to the error bounds for each quadrature. (The true value is I = Log (2) ≈ 0.693147.)

5.8   (a) Find the open Newton-Cotes formulae corresponding to n = 2, 3, and 4.

(b) Estimate the errors in these formulae, and compare them to those for the closed Newton-Cotes formulae with the same number of points.

5.9   Using separation of variables, find the expressions for the two-dimensional closed Newton-Cotes quadrature formulae corresponding to n = 1, 2, and 3 in each coordinate direction.

5.10   Repeat Exercise 5.9 for the three-dimensional case.

5.11   Write the expressions for the two-dimensional Gauss quadrature formulae corresponding to n = 1, 3, and 4 in each coordinate direction.

5.12   Extend the Gauss quadrature formulae to three dimensions for n = 1, 2, and 3.

5.13   Construct a Gauss quadrature formula for a rectangular element with degree of precision 5 in the x-direction and 3 in the y-direction.

5.14   Verify that the quadrature in Example 5.2 has degree of precision m = 3 by applying it to each of the terms in Pascal's triangle (Fig. 4.3) and showing that the error vanishes for all terms up to row n = 3 but not for those corresponding to row n = 4.

5.15   Show that an approximate integration rule for the quadratic triangle must exhibit degree of precision m = 2, while for the biquadratic quadrilateral it must be m = 4.

5.16   Solve Eq. (5.26) with two and four quadratic elements and using a two Gauss-point and a three Gauss-point integration rule. Check the accuracy and rates of convergence of the solutions in the same way as in Example 5.5.

5.17   (a) Find the full and reduced integration rule for a quadratic triangle using Table 5.3.
         (b) Find the full and reduced integration Gauss quadratures for the element of Fig. 4.24.
         (c) Find the full and reduced integration Gauss quadratures for the bicubic rectangular element.

5.18   Repeat Example 5.5 using two linear elements to interpolate f(x) = e^x and evaluating the derivative at the nodes, at the two full integration Gauss points, and at one reduced integration point.

5.19   Show that if p(x) is a linear polynomial in $-1 \leq x \leq 1$, the constant least squares fit intersects p(x) at x = 0. Repeat for the case when p(x) is a cubic polynomial, and show that the quadratic polynomial q(x), which is the least squares fit to p(x), intersects p(x) at the points corresponding to n = 3 in Table 5.2.

5.20   Repeat the solution of Example 5.9 using two quadratic elements. Evaluate the derivatives at the nodal points using Eq. (5.31), and compare the results at the boundary points with those obtained from the boundary residuals.

5.21   Modify Example 5.10 by changing the boundary conditions along x = 0 to be u = 10 for $0 \leq y \leq 1/2$, and $\partial u / \partial x = 0$ for $1/2 \leq y \leq 1$. Solve using meshes of 5 x 5, 11 x 11, and 21 x 21 nodes. This solution will be much better behaved near the corner (0,1). What happens at the point (0, 0.5)?

5.22   For Exercise 3.33, calculate the heat fluxes along the boundary x = 0.

5.23   Repeat Exercise 5.22 for Exercise 5.21.

5.24   Using Exercise 3.32, calculate the heat flux through the chilling plate and across the plane x = C − F (see Fig. 3.2).

5.25   A pipe is carrying fluid at 120°C and sits half submerged in water at 5°C, as shown in the figure below. The convective heat transfer coefficient with air at 20°C is h = 80 W/m²C, and the conductivity of the material is 50 W/mC. Find the temperature distribution in the pipe cross section.

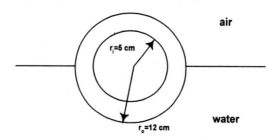

5.26   Consider the Laplace equation $\nabla^2 u = 0$ in the domain $\Omega = \{(x,y)/\ 0 < x < 1, 0 < y < 1\}$ with boundary conditions $u(0,y) = u(x,0) = 0$, $u(1,y) = x$, and $u(x,1) = y$. (Exact solution is $u(x,y) = xy$). Use four bilinear elements to construct the stiffness matrix using
        (a) 2 x 2 Gauss integration (exact)
        (b) 1 x 1 reduced integration
Find each solution and discuss the differences.

5.27   Repeat Exercise 5.26 using a uniform mesh of 16 bilinear elements, and discuss the results.

NONLINEARITY

6.1 OVERVIEW

In the previous chapters we have addressed the solution of linear differential equations, and established a finite element methodology that can be applied to a wide variety of linear problems. However, in many situations the mathematical models lead to nonlinear equations or systems of equations representing the physical system that cannot be reduced to linear approximations without oversimplification. We must therefore extend our method to encompass the solution of nonlinear problems. This represents a formidable challenge. Unlike the linear case where general theories can be developed that apply to all problems governed by the same type of equations, individual nonlinearities require different algorithms; the nonlinearity itself can be of many different types. It is difficult to pretend that we can address all kinds of nonlinearities here. In fact, we will restrict ourselves to a few important techniques that are applicable to a large number of situations. These are nonlinear material properties (coefficients); product nonlinearities like the convective nonlinearity in the Navier-Stokes equations; transcendental functions of the dependent variable such as the exponential dependence in chemical reactions; and nonlinearities occurring on the boundaries, as in the case of radiative heat transfer. We will not address other important cases such as the determination of the position of a free surface or nonlinear constitutive relations, even though these are also important subjects. Such topics are complex and would require lengthy treatment. Our intent is that readers interested in nonlinear topics not covered in this book will still be able to obtain sufficient information and references to be able to undertake them on their own.

6.2 BASIC METHODS FOR NONLINEAR EQUATIONS

There are an abundance of different methods for the numerical solution of nonlinear equations. An introduction and analysis of the basic methods can be found in the work by Isaacson and Keller (1966). A thorough treatment of the solution of systems of nonlinear equations is given by Ortega and Rheinboldt (1970). The so-called quasi-Newton algorithms are favored in optimization; a good survey of these and other methods is given by Gill et al. (1981). In this book we will concentrate on the two methods most commonly used in finite element modeling – the Newton-Raphson method (Bicanic and Johnson, 1979) and the method of direct iteration (Ortega and Rheinboldt, 1970).

6.2.1 The Newton-Raphson Method

Let us first recall Newton's method for a function of one independent variable. If $f(x)$ is a real function of x, we want to find the value x^* such that $f(x^*) = 0$. If we know an approximation x^k to x^*, the linearized Taylor series expansion of f about x^k approximates $f(x)$ in a neighborhood of $x = x^k$:

$$f(x) = f(x^k) + (x - x^k)f'(x^k) + O(x - x^k)^2 \qquad (6.1)$$

Therefore the equation

$$y = f(x^k) + (x - x^k)f'(x^k) \qquad (6.2)$$

represents the line tangent to $f(x)$ at $(x^k, f(x^k))$. Its intersection with the axis defines a point x^{k+1} that will be a better approximation to x^* than x^k, as shown in Fig. 6.1, provided that x^k was already reasonably close to x^* (see Exercise 6.1). We can then define Newton's algorithm as

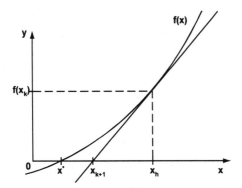

Figure 6.1 One-dimensional Newton's method.

$$x^{k+1} = x^k + \Delta x \tag{6.3}$$

$$f'\left(x^k\right) \Delta x = -f\left(x^k\right) \tag{6.4}$$

where, given x^k, Δx can be readily calculated provided that the derivative $f'(x^k) \neq 0$, in order to find the next approximation x^{k+1} to x^*.

The method can be readily extended to find the roots $\mathbf{x}^*$ of a vector function of a vector variable $\mathbf{F(x)}$ where $\mathbf{F}$ and $\mathbf{x}$ both have dimension n. The generalization of Eqs. (6.3) – 6.4) is accomplished by means of the *tangent matrix* of $\mathbf{F}$, also known as the Jacobian matrix or as the Gateaux derivative of $\mathbf{F}$, and is given by

$$\mathbf{F'(x)} \equiv \left[\frac{\partial f_i}{\partial x_j}\right] = \begin{bmatrix} \dfrac{\partial f_1}{\partial x_1} & \dfrac{\partial f_1}{\partial x_2} & \cdots & \dfrac{\partial f_1}{\partial x_n} \\[2mm] \dfrac{\partial f_2}{\partial x_1} & \dfrac{\partial f_2}{\partial x_2} & \cdots & \dfrac{\partial f_2}{\partial x_n} \\[2mm] \vdots & \vdots & & \vdots \\[2mm] \dfrac{\partial f_n}{\partial x_1} & \dfrac{\partial f_n}{\partial x_2} & \cdots & \dfrac{\partial f_n}{\partial x_n} \end{bmatrix} \tag{6.5}$$

where the components f_i of $\mathbf{F}$ are in general nonlinear functions of the dependent variable $\mathbf{x}$, that is, $f_i = f(x_1, x_2, \ldots x_n)$, $i = 1, 2, \ldots, n$.

The Newton-Raphson method is defined as

$$\mathbf{x}^{k+1} = \mathbf{x}^k + \Delta \mathbf{x} \tag{6.6}$$

$$\mathbf{F}'\left(\mathbf{x}^k\right)\Delta \mathbf{x} = -\mathbf{F}\left(\mathbf{x}^k\right) \tag{6.7}$$

Therefore, in order to obtain $\Delta \mathbf{x}$ at each iteration, we must solve a system of linear algebraic equations with coefficient matrix $\mathbf{F}'(\mathbf{x}^k)$. Let us now illustrate the use of the Newton-Raphson method in the finite element solution to a nonlinear second-order differential equation.

Example 6.1

Consider the one-dimensional steady state Burgers equation

$$-\varepsilon \frac{d^2 u}{\partial x^2} + u \frac{du}{dx} = 0 \tag{6.8}$$

with boundary conditions $u(0) = 1$, $u(1) = 0$, and for simplicity, let $\varepsilon = 1$. Then, the weak form of Eq. (6.8) is

$$\int_0^1 \left(\frac{dw}{dx} \frac{du}{dx} + wu \frac{du}{dx} \right) dx = 0 \tag{6.9}$$

Subdividing the unit interval into four linear elements of equal size and applying Galerkin's method, we obtain (Exercise 6.2) the system of nonlinear equations:

$$\mathbf{F(u)} = \begin{bmatrix} 4u_1 - 4u_2 - \frac{1}{3}u_1^2 + \frac{1}{6}u_1u_2 + \frac{1}{6}u_2^2 \\ -4u_1 + 8u_2 - 4u_3 - \frac{1}{6}u_1^2 - \frac{1}{6}u_1u_2 + \frac{1}{6}u_2u_3 + \frac{1}{6}u_3^2 \\ -4u_2 + 8u_3 - 4u_4 - \frac{1}{6}u_2^2 - \frac{1}{6}u_2u_3 + \frac{1}{6}u_3u_4 + \frac{1}{6}u_4^2 \\ -4u_3 + 8u_4 - 4u_5 - \frac{1}{6}u_3^2 - \frac{1}{6}u_3u_4 + \frac{1}{6}u_4u_5 + \frac{1}{6}u_5^2 \\ -4u_4 + 4u_5 - \frac{1}{6}u_4^2 - \frac{1}{6}u_4u_5 + \frac{1}{3}u_5^2 \end{bmatrix} = \mathbf{0} \quad (6.10)$$

The tangent matrix of $\mathbf{F}$ is readily calculated and yields

$$\mathbf{F'(u)} =$$

$$\begin{bmatrix} 4 - \frac{2}{3}u_1 + \frac{1}{6}u_2 & -4 + \frac{1}{6}u_1 + \frac{1}{3}u_2 & 0 & 0 & 0 \\ -4 - \frac{1}{3}u_1 - \frac{1}{6}u_2 & 8 - \frac{1}{6}u_1 + \frac{1}{6}u_3 & -4 + \frac{1}{6}u_2 + \frac{1}{3}u_3 & 0 & 0 \\ 0 & -4 - \frac{1}{3}u_2 - \frac{1}{6}u_3 & 8 - \frac{1}{6}u_2 + \frac{1}{6}u_4 & -4 + \frac{1}{6}u_3 + \frac{1}{3}u_4 & 0 \\ 0 & 0 & -4 - \frac{1}{3}u_3 - \frac{1}{6}u_4 & 8 - \frac{1}{6}u_3 + \frac{1}{6}u_5 & -4 + \frac{1}{6}u_4 + \frac{1}{3}u_5 \\ 0 & 0 & 0 & -4 - \frac{1}{3}u_4 - \frac{1}{6}u_5 & 4 - \frac{1}{6}u_4 + \frac{2}{3}u_5 \end{bmatrix}$$

$$(6.11)$$

Notice that we can also produce $\mathbf{F_e'(u)}$ at the element level and assemble to obtain Eq. (6.11). This is left to Excercise 6.5. Therefore it suffices to obtain the tangent matrix at the element level.

We now apply the boundary conditions. Notice that the increments $\Delta\mathbf{x}$ must always satisfy *homogeneous Dirchlet conditions* because u^k must satisfy the Dirichlet conditions for all k. Hence Eq. (6.7) can be readily applied using the reduced matrix

$$\mathbf{f'} = \begin{bmatrix} 8 - \dfrac{1}{6} + \dfrac{1}{6}u_3 & -4 + \dfrac{1}{6}u_2 + \dfrac{1}{3}u_3 & 0 \\[2mm] -4 - \dfrac{1}{3}u_2 - \dfrac{1}{6}u_3 & 8 - \dfrac{1}{6}u_2 + \dfrac{1}{6}u_4 & -4 + \dfrac{1}{6}u_3 + \dfrac{1}{3}u_4 \\[2mm] 0 & -4 - \dfrac{1}{3}u_3 - \dfrac{1}{6}u_4 & 8 - \dfrac{1}{6}u_3 \end{bmatrix} \quad (6.12)$$

where the rows and columns corresponding to the degrees of freedom u_1 and u_5 have been eliminated and their values $u_1 = 1$ and $u_5 = 0$ have been replaced. Therefore, using Eqs. (6.10) and (6.12), Eq. (6.7) takes the form

$$\begin{bmatrix} \dfrac{47}{6} + \dfrac{1}{6}u_3^k & -4 + \dfrac{1}{6}u_2^k + \dfrac{1}{3}u_3^k & 0 \\[2mm] -4 - \dfrac{1}{3}u_2^k - \dfrac{1}{6}u_3^k & 8 - \dfrac{1}{6}u_2^k + \dfrac{1}{6}u_4^k & -4 + \dfrac{1}{6}u_3^k + \dfrac{1}{3}u_4^k \\[2mm] 0 & -4 - \dfrac{1}{3}u_3^k - \dfrac{1}{6}u_4^k & 8 - \dfrac{1}{6}u_3^k \end{bmatrix} \begin{bmatrix} \Delta u_2 \\[2mm] \Delta u_3 \\[2mm] \Delta u_4 \end{bmatrix}$$

$$= - \begin{bmatrix} -\dfrac{25}{6} + \dfrac{47}{6}u_2^k - 4u_3^k + \dfrac{1}{6}u_2^k u_3^k + \dfrac{1}{6}\left(u_3^k\right)^2 \\[2mm] -4u_2^k + 8u_3^k - 4u_4^k - \dfrac{1}{6}\left(u_2^k\right)^2 - \dfrac{1}{6}u_2^k u_3^k + \dfrac{1}{6}u_3^k u_4^k + \dfrac{1}{6}\left(u_4^k\right)^2 \\[2mm] -4u_3^k + 8u_4^k - \dfrac{1}{6}\left(u_3^k\right)^2 - \dfrac{1}{6}u_3^k u_4^k \end{bmatrix} \quad (6.13)$$

Let us stop here for a moment; we will come back to this example after some discussion. ∎

REMARKS

1. Notice that the coefficient matrix in Eq. (6.13) is nonsymmetric, and hence the solution of this linear system of equations will be computationally more demanding both in terms of operations and storage requirements. However, the lack of symmetry of the matrix is not

due to the fact that the problem is nonlinear but because the differential operator contains a first derivative in the second term. This term represents convection in the physical system, and its presence causes additional computational difficulties when it is dominant. We will defer the discussion of such difficulties and the methods to deal with them until Chapter 8, where we will address the numerical modeling of convection in detail.

2. The computational work required to obtain a converged solution to Eq. (6.13) has increased enormously when compared to that required to solve a linear problem. At every iteration the coefficient matrix in Eq. (6.13) must be reconstructed and inverted. Furthermore, the right-hand side must be reevaluated at each iteration, which requires a relatively large number of operations.

3. We cannot estimate a priori how many iterations will be required to obtain a converged solution. This will depend on the criteria used to determine when the solution is converged and how much accuracy we demand to the solution.

There are many ways in which we can decide when to terminate the iteration; in some problems this can only be determined by the requirements on the expected solution. A most convenient way that suffices in most practical cases consists in requiring some measure of the right-hand side vector $\mathbf{F}(\mathbf{x})$ to be smaller than a specified tolerance, i.e., we consider the iteration defined by Eqs. (6.6) and (6.7) to be converged when

$$\left\| \mathbf{F}\left(\mathbf{x}^k\right) \right\| < \delta \tag{6.14}$$

where δ is a user-specified parameter and the norm that is most commonly used is the Euclidean norm, i.e.,

$$\left\| \mathbf{F}\left(\mathbf{x}^k\right) \right\| = \left(\sum_{i=1}^{n} \left[F_i\left(\mathbf{x}^k\right) \right]^2 \right)^{\frac{1}{2}} \tag{6.15}$$

This norm is convenient because the components $F_i(\mathbf{x}^k)$ must be calculated as part of the solution process at each iteration. Hence evaluating Eq. (6.15) adds very little overhead to the calculation. For certain applications the L^2 norm is sometimes preferred, but we will use the Euclidean norm defined in Eq. (6.15). Let us now return to our example.

Example 6.1 (Continued)

We will now carry out the iteration on Eq. (6.13) starting with the initial guess $u_2^0 = u_3^0 = u_4^0 = 0$ and setting $\delta = 10^{-6}$ in Eq. (6.14). The iterative procedure yields the results shown in Table 6.1.

Table 6.1 Iterative solution of Burger's equation, Eq. (6.8), using four linear elements

Iteration number	Numerical solution			
	u_2	u_3	u_4	$\left\| \mathbf{F}\left(\mathbf{u}^k\right) \right\|$
0	0.0	0.0	0.0	0.41667E+01
1	0.80645	0.53763	0.26882	0.20153E+00
2	0.80496	0.56709	0.29368	0.37907E-03
3	0.80492	0.56704	0.29369	0.39817E-06
Exact	0.80531	0.56763	0.29412	

It only took three iterations to obtain convergence, and as we can see, the solution obtained is very accurate (at the nodal points). This is, of course, not always the case. As an exercise and to illustrate how the nonlinear iteration depends on the initial guess, the reader should try different initial conditions in this example (Exercise 6.6). ∎

There are many variations of this method designed to accelerate convergence or to be able to apply it in problems where the present form is not applicable. The reader interested in a deeper treatment should consult the text by Ortega and Rheinboldt (1970).

In some cases the above approach is not convenient. In particular, if transcendental functions of the dependent variable appear in the nonlinear operator, integration of the weighted residuals equations may be difficult to perform in order to find the vector $\mathbf{F}(\mathbf{x})$, as in Eq. (6.10). For example, if a reactive term ae^u is added to Eq. (6.8) and we now attempt the solution of the equation

$$-\varepsilon \frac{d^2 u}{dx^2} + u \frac{du}{dx} + ae^u = 0 \qquad (6.16)$$

the Galerkin weighted residuals expression becomes

$$\int_0^1 \left[\varepsilon \frac{dN_i}{dx} \left(\sum_j \frac{dN_j}{dx} u_j \right) + N_i \left(\sum_k N_k u_k \right) \left(\sum_j \frac{dN_j}{dx} u_j \right) \right.$$

$$\left. + aN_i e^{\sum\limits_j N_j u_j} \right] dx = 0 \qquad (6.17)$$

and requires the evaluation of integrals of the form

$$\int_0^1 N_i(x) e^{\sum\limits_j N_j(x)u_j} dx \qquad (6.18)$$

This is not possible if **u** is unknown. Therefore, in these cases, we must linearize the differential operator directly. Denoting the nonlinear differential operator by D(u), we seek to obtain the linear part in Δu, denoted by L(u), of the operator. We can then write

$$D(u + \Delta u) - D(u) = \left[L(u) \right] \Delta u + 0 \left[(\Delta u)^2 \right] \qquad (6.19)$$

Hence the linearized operator L(u) should correspond to the tangent operator of D(u) and, in discretized form, corresponds to the tangent matrix **F'(u)** in Eq. (6.7).

Example 6.2

Let us construct the linear part of the nonlinear operator D(u) of Eq. (6.16) setting $\varepsilon = a = 1$ for simplicity:

$$D(u+\Delta u)-D(u)=\left[-\frac{d^2(u+\Delta u)}{dx^2}+(u+\Delta u)\frac{d(u+\Delta u)}{dx}+e^{u+\Delta u}\right]$$

$$-\left[-\frac{d^2u}{dx^2}+u\frac{du}{dx}+e^u\right]$$

Expanding the first term and simplifying, we have

$$D(u+\Delta u)-D(u)$$

$$=-\frac{d^2\Delta u}{dx^2}+u\frac{d\Delta u}{dx}+\Delta u\frac{du}{dx}+\Delta u\frac{d\Delta u}{dx}+\left(e^{u+\Delta u}-e^u\right)$$

The term $\Delta u\ d\Delta u/dx$ is second order in Δu and is neglected because it should be small. The exponential term is expanded in Taylor series to give

$$e^{u+\Delta u}-e^u=e^u\left(\Delta u+\frac{1}{2}(\Delta u)^2+...\right)\cong e^u\Delta u$$

Therefore the linear part of the operator is

$$L(u)=-\frac{d^2}{dx^2}+u\frac{d}{dx}+\frac{du}{dx}+e^u \qquad (6.20)$$

and the Newton-Raphson method applied to Eq. (6.16) yields the iterative algorithm

$$\left[L(u^k)\right]\Delta u=-D(u^k) \qquad (6.21)$$

$$u^{k+1}=u^k+\Delta u \qquad (6.22)$$

Equation (6.21) can now be discretized using the Galerkin method to directly yield a system of linear algebraic equations to be solved at each iteration. If no transcendental functions of the dependent variable appear in the nonlinear operator, the final system of equations obtained by this method will be identical to that obtained by the method used in Example 6.1. We leave it to Exercise 6.10 to show that this is true for Eq. (6.8). ∎

This form of Newton's method is also known in the literature as the method of *quasi-linearization*. As we saw in Example 6.1, the Newton-Raphson method may converge very fast under the right conditions, and this is the main reason for its widespread use. If the initial iterate $x^{(0)}$ is sufficiently close to the solution x^* that satisfies $F(x^*) = 0$ and the tangent matrix $F'(x^*)$ is nonsingular, then the method is at least of second-order in the sense that the error at any iterate is proportional to the square of the previous error, i.e.,

$$\left\| x^k - x^* \right\|_\infty \le M \left\| x^{k-1} - x^* \right\|_\infty^2 \qquad (6.23)$$

where M is a constant, and the infinity norm is defined by

$$\left\| x \right\|_\infty \equiv \max_{i=1,\ldots,n} \left| x_i \right| \qquad (6.24)$$

with n the dimension of the vector space.

In order to converge, the Newton-Raphson method does not require that $F'(x^*)$ be nonsingular. However, if $F'(x^*)$ is singular, convergence is only of first-order, and in the general case, the error satisfies the relation

$$\left\| x^k - x^* \right\|_\infty \le \frac{M^1}{2^k} \qquad (6.25)$$

where M^1 is a constant. Moreover, sufficient conditions have been developed that can be checked without knowledge of x^* and, if satisfied by $x^{(0)}$, guarantee the convergence of the Newton-Raphson method. The proofs are rather involved and not given here. The interested reader should consult the books by Isaacson and Keller (1966) or Ortega and Rheinboldt (1970).

6.2.2 Direct Iteration Methods

Although the Newton-Raphson method is the most widely used nonlinear iteration method in finite element analysis, there are cases in which a *direct* iteration method is more desirable. It consists of writing the iterative method in the form

$$\mathbf{x}^k = \mathbf{G}\left(\mathbf{x}^k\right) \tag{6.26}$$

Hence, if we seek the solution $\mathbf{x}*$ to $\mathbf{F}(\mathbf{x}*) = 0$, we solve the slightly modified problem

$$\mathbf{x}* - \mathbf{G}(\mathbf{x}*) = 0 \tag{6.27}$$

using a suitable function $\mathbf{G}(\mathbf{x})$. Therefore the method is not uniquely determined, as was the case of the Newton-Raphson scheme, since there are usually many functions of $\mathbf{G}(\mathbf{x})$ that can be used. It can be shown that the method will converge if

$$\left\|\mathbf{G}'(\mathbf{x})\right\|_\infty < \frac{1}{n}$$

for all vectors $\mathbf{x}$ such that

$$\left\|\mathbf{x} - \mathbf{x}*\right\|_\infty \le \left\|\mathbf{x}^{(0)} - \mathbf{x}*\right\|_\infty$$

where n is the dimension of the vector space. Therefore convergence of a direct iteration is *global*: if the method converges for some initial guess $\mathbf{x}^{(0)}$, it converges for *every* initial guess within the radius $\left\|\mathbf{x}^{(0)} - \mathbf{x}*\right\|_\infty$.

However, convergence is only of first order, that is, at each iteration we have

$$\left\|\mathbf{x}^k - \mathbf{x}*\right\|_\infty \le M\left\|\mathbf{x}^{k-1} - \mathbf{x}*\right\|_\infty \tag{6.28}$$

where M is a constant.

In most finite element applications it is not possible to define an explicit iteration of the form of Eq. (6.26), since discretization of the differential equations leads to systems of equations of the form

$$\mathbf{A}(\mathbf{x})\mathbf{x} = \mathbf{B}(\mathbf{x}) \tag{6.29}$$

If $\mathbf{A}$ is invertible, we can define the iteration

$$\mathbf{x}^{k+1} = \left[\mathbf{A}(\mathbf{x}^{k})\right]^{-1}\mathbf{B}(\mathbf{x}^{k}) \tag{6.30}$$

We will use Burgers equation, Eq. (6.8), to illustrate the concepts.

Example 6.3

Let us rewrite the differential equation, Eq. (6.8), in the form of Eq. (6.29), that is,

$$L(u)u = 0 \tag{6.31}$$

where L is a linearized differential operator. We must now choose a particular form of L that can be used to define a nonlinear iteration that will converge to the solution of Eq. (6.8). One possibility is to set

$$L_1\left(u^k\right) = -\varepsilon\frac{d^2}{dx^2} + u^k\frac{d}{dx}$$

and at each iteration solve the linear differential equation

$$-\varepsilon\frac{d^2u^{k+1}}{dx^2} + u^k\frac{du^{k+1}}{dx} = 0 \tag{6.32}$$

Clearly, if the iteration converges, it will be to the solution of the Burgers equation.

Another possibility is to define

$$L_2\left(u^k\right) = -\varepsilon\frac{d^2}{dx^2} + \frac{du^k}{dx}$$

to find the next iterate u^{k+1} through the solution of the linear differential equation

$$-\varepsilon \frac{d^2 u^{k+1}}{dx^2} + \left(\frac{d u^k}{dx}\right) u^{k+1} = 0 \tag{6.33}$$

Yet another possibility is to treat the whole convection term explicitly, i.e., set

$$L_3\left(u^k\right) = -\varepsilon \frac{d^2}{dx^2}$$

and define the iteration

$$-\varepsilon \frac{d^2 u^{k+1}}{dx^2} = -u^k \frac{d u^k}{dx} \tag{6.34}$$

Each of these algorithms will produce a different discretized system. Let us take Eq. (6.32) and use the Galerkin method to discretize the problem domain using four linear elements and $\varepsilon = 1$ as in Example 6.1. The resulting system of equations is (see Exercise 6.14)

$$\begin{bmatrix} 4-\frac{1}{3}u_1^k-\frac{1}{6}u_2^k & -4+\frac{1}{3}u_1^k+\frac{1}{6}u_2^k & 0 & 0 & 0 \\ 4-\frac{1}{6}u_1^k-\frac{1}{3}u_2^k & 8+\frac{1}{6}(u_1^k-u_3^k) & -4+\frac{1}{3}u_2^k-\frac{1}{6}u_3^k & 0 & 0 \\ 0 & -4-\frac{1}{6}u_2^k-\frac{1}{3}u_3^k & 8+\frac{1}{6}(u_2^k-u_4^k) & -4+\frac{1}{3}u_3^k+\frac{1}{6}u_4^k & 0 \\ 0 & 0 & -4-\frac{1}{6}u_3^k-\frac{1}{3}u_4^k & 8+\frac{1}{6}(u_3^k-u_5^k) & -4+\frac{1}{3}u_4^k+\frac{1}{6}u_5^k \\ 0 & 0 & 0 & -4-\frac{1}{6}u_4^k-\frac{1}{3}u_5^k & 4+\frac{1}{3}u_4^k+\frac{1}{6}u_5^k \end{bmatrix} \begin{bmatrix} u_1^{k+1} \\ u_2^{k+1} \\ u_3^{k+1} \\ u_4^{k+1} \\ u_5^{k+1} \end{bmatrix} = \begin{bmatrix} 0 \\ 0 \\ 0 \\ 0 \\ 0 \end{bmatrix}$$

$$\tag{6.35}$$

and applying the boundary conditions $u_1 = 1$, $u_5 = 0$, we get

$$
\begin{bmatrix}
\dfrac{49}{6} - u_3^k & -4 + \dfrac{1}{3}u_2^k + \dfrac{1}{6}u_3^k & 0 \\[2ex]
-4 - \dfrac{1}{6}u_2^k - \dfrac{1}{3}u_3^k & 8 + \dfrac{1}{6}(u_2^k - u_4^k) & -4 + \dfrac{1}{3}u_3^k + \dfrac{1}{6}u_4^k \\[2ex]
0 & 4 - \dfrac{1}{6}u_3^k - \dfrac{1}{3}u_4^k & 8 + \dfrac{1}{6}u_3^k
\end{bmatrix}
\begin{bmatrix} u_2^{k+1} \\[2ex] u_3^{k+1} \\[2ex] u_4^{k+1} \end{bmatrix}
$$

$$
=
\begin{bmatrix} \dfrac{25}{6} + \dfrac{1}{3}u_2^k \\[2ex] 0 \\[2ex] 0 \end{bmatrix}
$$

(6.36)

The solution is left to Exercise (6.16). ∎

Notice that the stiffness matrix in Eq. (6.36) is different from Eq. (6.13) generated by Newton's method. Furthermore, the iteration defined by Eq. (6.36) yields the vector $\mathbf{u}^{k+1}$ directly, while the Newton-Raphson method calculates only an increment $\Delta\mathbf{u}$ to the previous approximations. However, both methods give similar solutions to the problem as will be shown in Example 6.4.

The question of which of the direct iterations methods (6.32), (6.33), or (6.34) will be best is not easy to answer. This is because their convergence characteristics depend on the inverse operator, as shown in Eq. (6.30). As a general rule, methods that are more strongly implicit will have better convergence properties – this criteria would favor Eqs. (6.32) and (6.33) over Eq. (6.34).

It is usually best to maintain the highest derivatives of every nonlinear term in the equation implicit, which in this case favors Eq. (6.32) over (6.33). It also happens that some methods are always unstable. Simple examples involving scalar equations are given by Isaacson and Keller (1966) and Greenberg (1978); see also Exercise 6.15.

Example 6.4

The solution of the equation

$$-0.1\frac{d^2u}{dx^2} + u\frac{du}{dx} = 0 \qquad u(0) = 1 \quad u(1) = 0$$

using 10 linear elements of size h = 0.1 was performed using both Newton's method and the direct iteration defined by Eq. (6.32). The iterations were stopped when the relative difference between two iterates was less than 1% in the Euclidean norm, i.e., when

$$\left(\frac{\sum_{i=1}^{11}\left(u_i^k - u_i^{k-1}\right)^2}{\sum_{i=1}^{11}\left(u_i^k\right)^2}\right)^{\frac{1}{2}} < 0.01$$

The results (at five nodes, to be more concise) are shown in Table 6.2 together with the exact solution. The approximations are practically the same; however, the direct iteration method required six iterations to converge, compared to only four for Newton-Raphson's method. This difference in the number of iterates can become significantly larger, depending on the problem. ∎

Table 6.2 Solution to Burgers equation for ε = 0.1 using Newton's method and Picard iteration

x	Picard	Newton	Exact
0.2	1.000	1.000	0.999
0.4	0.997	0.997	0.995
0.6	0.969	0.970	0.964
0.8	0.766	0.767	0.762
0.9	0.460	0.463	0.462

In the next section, we present several examples to illustrate the use of the methods discussed here.

6.3 NONLINEAR EXAMPLES

In this book a large effort is devoted to the solution of the nonlinear Navier-Stokes equations that will be discussed in detail in subsequent chapters. Here we will present some simple examples of application of the Newton-Raphson and Picard iteration methods to heat transfer and the stationary Navier-Stokes equations for slow flows.

6.3.1 Heat Transfer with Temperature–Dependent Conductivity

In many applications the physical properties are not constant but are rather a function of the temperature and/or various species concentrations, or even the state of stress of the material. The dependence may be very strong, as in the case of the viscosity of some lubricants, which can vary several orders of magnitude over the range of working temperatures. Because these properties can only be determined by measurement, they are usually available in the form of tables, where the values between given points can be obtained by interpolation or through a curve fit. In these situations the Picard iteration method is usually preferred to the Newton-Raphson method. The variable coefficients are evaluated at the latest iteration and used to obtain the next approximation.

Example 6.5

Tungsten has a variable thermal conductivity that ranges from 1.1 W/(cm °C) at 20°C to 1.3 W/(cm °C) at 100°C (Rohsenow and Choi, 1961). If we assume a linear variation, we can write

$$K(T) = 1.05 + 0.0025T$$

To understand how this variation in the conductivity affects the physical process, we will examine the solution of a simple conduction problem. The governing equations for one-dimensional conduction with variable thermal conductivity can be written for the problem in the following form:

$$-\frac{d}{dx}\left[K(T)\frac{dT}{dx}\right] = 0 \qquad 0 < x < 8 \text{ cm}$$

$$T(0) = 20^\circ C$$

$$-K\frac{dT}{dx}\bigg|_{x=8} = -12 \text{ W} / \text{cm}^2$$

The analytical solution is given by

$$T(x) = 400\left(-1.05 + \sqrt{1.21 + 0.6x}\right)$$

A direct iteration is set up using the linearized operator

$$-\frac{d}{dx}\left[\left(1.05 + 0.0025T^k\right)\frac{dT^{k+1}}{dx}\right] = 0$$

We leave it to Exercise 6.17 to determine an appropriate mesh spacing and to obtain a solution exact to four significant digits. The solution is shown in Table 6.3 together with solutions obtained using constant values of the conductivity, which are solved exactly by the finite element method.

Table 6.3 Solution to the nonlinear conduction equation and linear approximations using K = 1.1, 1.2, and 1.3

x (cm)	K = 1.1	K = 1.2	K = 1.3	K = 1.05 + 0.0025T
2	41.82	40.00	38.46	41.30
4	63.64	60.00	56.92	61.66
6	85.45	80.00	75.38	81.20
8	107.27	100.00	93.85	100.00

We note that choosing the right average constant conductivity gives a very good approximation to the solution of the nonlinear equation. However, keep in mind that this is still a very simplified problem and that we may not know a good value for the average conductivity a priori.

On the other hand, the other two constant values chosen lead to departures of as much as 8% from the solution. ∎

Direct iteration methods may converge very slowly when the coefficients have a strong variation, and the convergence is often not monotonic, but the iterates oscillate around the desired solution. To accelerate convergence in these cases, we may use *underrelaxation*. That is, we consider the calculated solution $\overline{\mathbf{X}}^{k+1}$ at iterate k+1 as an intermediate value, and define $\mathbf{X}^{k+1}$ from

$$\mathbf{X}^{k+1} = \Theta\mathbf{X}^k + (1-\Theta)\overline{\mathbf{X}}^{k+1} \tag{6.37}$$

where Θ is a number between zero and 1 used to relax the solution. This can be determined through numerical experiments to obtain an optimal convergence rate.

An example involving a complex application of the direct iteration method in the context of plastic flow of metals, for which underrelaxation was used, is presented by Zienkiewicz et al. (1981). We will illustrate the use of underrelaxation through a simple example involving one algebraic equation.

Example 6.6

Let us find the root of the function

$$f(x) = xe^x - 1 = 0$$

which is at x* = 0.56714329. We will stop when five significant digits are exact, i.e., when $|x^* - x^k| < 5 \times 10^{-6}$. An application of Newton's method starting with $x^0 = 1.0$ converges monotonically in five iterations, while the direct iteration

$$x^{k+1} = e^{-k}$$

needs 19 iterates and the error oscillates. However, if we use an underrelaxation parameter $\Theta = 1/2$, the number of iterations drops to nine, and the convergence is monotonic. The behavior of the error is shown in Fig. 6.2. Let us now turn to an application of the Newton-Raphson method. ∎

6.3.2 Stationary Navier-Stokes Equations

We will now take our first look at discretizing the Navier-Stokes equations for laminar incompressible flow and apply the Newton-Raphson method to a nonlinear system of equations. We will restrict ourselves to describing how the nonlinearity is treated and assume low Reynolds number flows to avoid having to deal with convective instabilities, which are addressed in Chapter 8.

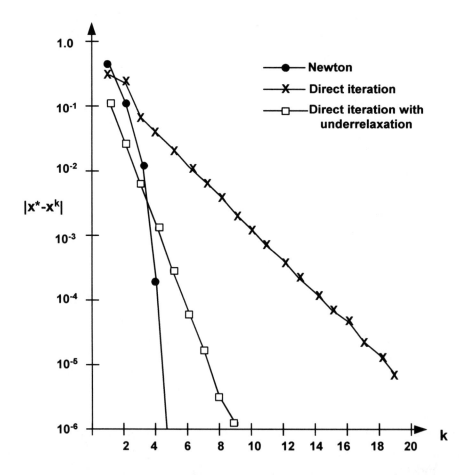

Figure 6.2 Convergence history for Example 6.6.

Assume two-dimensional flow and rewrite the continuity equation, Eq. (2.6), and the momentum equations, Eq. (2.10), in nondimensional form as

$$D_1(u,v,p) \equiv Re\left(u\frac{\partial u}{\partial x} + v\frac{\partial u}{\partial y}\right) + \frac{\partial p}{\partial x} - \left(\frac{\partial^2 u}{\partial x^2} + \frac{\partial^2 u}{\partial y^2}\right) = 0 \qquad (6.38)$$

$$D_2(u,v,p) \equiv Re\left(u\frac{\partial v}{\partial x} + v\frac{\partial v}{\partial y}\right) + \frac{\partial p}{\partial y} - \left(\frac{\partial^2 v}{\partial x^2} + \frac{\partial^2 v}{\partial y^2}\right) = 0 \qquad (6.39)$$

$$D_3(u,v,p) \equiv \frac{\partial u}{\partial x} + \frac{\partial v}{\partial y} = 0 \qquad (6.40)$$

where the Reynolds number Re is defined as

$$Re \equiv \frac{UL}{\nu}$$

Here U is a reference velocity, L is a reference length, and ν is the kinematic viscosity. To obtain Eqs. (6.38) and (6.39), we use a reference pressure $p_o = \mu U/L$ with μ the dynamic viscosity. To find the tangent operators in the momentum equations, we proceed as in Example 6.2 (Exercise 6.20) to obtain

$$\left[L_1\left(u^k, v^k, p^k\right)\right]\begin{bmatrix}\Delta u \\ \Delta v \\ \Delta p\end{bmatrix} = \left[Re\left(u^k\frac{\partial}{\partial x} + \frac{\partial u^k}{\partial x} + v^k\frac{\partial}{\partial y}\right) - \frac{\partial^2}{\partial x^2} - \frac{\partial^2}{\partial y^2}\right]\Delta u$$

$$+ \left[Re\frac{\partial u^k}{\partial y}\right]\Delta v + \left[\frac{\partial}{\partial x}\right]\Delta p = 0$$

$$(6.41)$$

$$\left[L_2\left(u^k, v^k, p^k\right)\right]\begin{bmatrix}\Delta u \\ \Delta v \\ \Delta p\end{bmatrix} \equiv \left[Re\left(u^k\frac{\partial}{\partial x} + v^k\frac{\partial}{\partial y} + \frac{\partial v^k}{\partial y}\right) - \frac{\partial^2}{\partial x^2} - \frac{\partial^2}{\partial y^2}\right]\Delta v$$

$$+ \left[Re\frac{\partial v^k}{\partial x}\right]\Delta u + \left[\frac{\partial}{\partial y}\right]\Delta p = 0$$

$$(6.42)$$

$$\left[L_3\left(u^k, v^k, p^k\right)\right]\begin{bmatrix}\Delta u \\ \Delta v \\ \Delta p\end{bmatrix} \equiv \left[\frac{\partial}{\partial x}\right]\Delta u + \left[\frac{\partial}{\partial y}\right]\Delta v = 0 \qquad (6.43)$$

The Newton-Raphson iteration can then be written as

$$\begin{bmatrix} Re\left(u^k\frac{\partial}{\partial x} + v^k\frac{\partial}{\partial y} + \frac{\partial u^k}{\partial x}\right) - \frac{\partial^2}{\partial x^2} - \frac{\partial^2}{\partial y^2} & Re\frac{\partial u^k}{\partial y} & \frac{\partial}{\partial x} \\ Re\frac{\partial v^k}{\partial x} & Re\left(u^k\frac{\partial}{\partial x} + v^k\frac{\partial}{\partial y} + \frac{\partial v^k}{\partial y}\right) - \frac{\partial^2}{\partial x^2} - \frac{\partial^2}{\partial y^2} & \frac{\partial}{\partial y} \\ \frac{\partial}{\partial x} & \frac{\partial}{\partial y} & 0 \end{bmatrix}$$

$$
\bullet \begin{bmatrix} \Delta u \\ \\ \\ \Delta v \\ \\ \\ \\ \Delta p \end{bmatrix} = - \begin{bmatrix} D_1\!\left(u^k, v^k, p^k\right) \\ \\ \\ D_2\!\left(u^k, v^k, p^k\right) \\ \\ \\ \\ D_3\!\left(u^k, v^k, p^k\right) \end{bmatrix}
$$

$$(6.44)$$

Let us now discretize the domain using bilinear isoparametric elements for the velocity components, i.e., over an element Ω^e, we have

$$
u^{k+1}(x,y) = \sum_{j=1}^{4} N_j(x,y)\,u_j^{k+1} = \sum_{j=1}^{4} N_j(x,y)\,u_j^{k} + \sum_{i=1}^{4} N_j(x,y)\,\Delta u_j
$$

$$(6.45)$$

$$
v^{k+1}(x,y) = \sum_{j=1}^{4} N_j(x,y)\,v_j^{k+1} = \sum_{j=1}^{4} N_j(x,y)\,v_j^{k} + \sum_{i=1}^{4} N_j(x,y)\,\Delta v_j
$$

For reasons that we explain in Chapter 9, the pressure is approximated by a piecewise constant function in such a way that it is constant over each element. Hence

$$
p^{k+1} = p^{k} + \Delta p = \text{const.} \quad (x,y) \in \Omega^e
$$

$$(6.46)$$

We want to obtain the Galerkin weighted residuals formulation for Eq. (6.44). To do this, let us first rewrite Eq. (6.44) in a more condensed form as

$$\mathbf{L}^k \mathbf{a} = -\mathbf{D}^k \tag{6.47}$$

where

$$
\mathbf{L}^k =
\begin{bmatrix}
\operatorname{Re}\left(u^k \dfrac{\partial}{\partial x} + v^k \dfrac{\partial}{\partial y} + \dfrac{\partial u^k}{\partial x} \right) & \operatorname{Re}\dfrac{\partial u^k}{\partial y} & \dfrac{\partial}{\partial x} \\[2mm]
-\dfrac{\partial^2}{\partial x^2} - \dfrac{\partial^2}{\partial y^2} & & \\[4mm]
\operatorname{Re}\dfrac{\partial v^k}{\partial x} & \operatorname{Re}\left(u^k \dfrac{\partial}{\partial x} + v^k \dfrac{\partial}{\partial y} + \dfrac{\partial v^k}{\partial y} \right) & \dfrac{\partial}{\partial y} \\[2mm]
 & -\dfrac{\partial^2}{\partial x^2} - \dfrac{\partial^2}{\partial y^2} & \\[4mm]
\dfrac{\partial}{\partial x} & \dfrac{\partial}{\partial y} & 0
\end{bmatrix}
$$

$$
\mathbf{a} =
\begin{bmatrix}
\Delta u \\
\Delta v \\
\Delta p
\end{bmatrix}
\qquad
\mathbf{D}^k =
\begin{bmatrix}
\operatorname{Re}\left(u^k \dfrac{\partial u^k}{\partial x} + v^k \dfrac{\partial u^k}{\partial y} \right) + \dfrac{\partial p^k}{\partial x} - \dfrac{\partial^2 u^k}{\partial x^2} - \dfrac{\partial^2 u^k}{\partial y^2} \\[3mm]
\operatorname{Re}\left(u^k \dfrac{\partial v^k}{\partial x} + v^k \dfrac{\partial v^k}{\partial y} \right) + \dfrac{\partial p^k}{\partial y} - \dfrac{\partial^2 v^k}{\partial x^2} - \dfrac{\partial^2 v^k}{\partial y^2} \\[3mm]
\dfrac{\partial u^k}{\partial x} + \dfrac{\partial v^k}{\partial y}
\end{bmatrix}
$$

In the Galerkin formulation, we must choose the weighting functions equal to the shape functions, which are N_i for the velocities u and v, and 1 for the pressure p. Hence we have

$$\mathbf{W}_i = \begin{bmatrix} N_i \\ N_i \\ 1 \end{bmatrix}$$

and we can write

$$\left[\int_{\Omega^e} \mathbf{W}_i^T \mathbf{L}^k d\Omega\right]\mathbf{a} = -\int_{\Omega^e} \mathbf{W}_i^T \mathbf{D}^k d\Omega \qquad (6.48)$$

Because the pressure is assumed to be piecewise constant, we must apply Gauss's theorem to the pressure gradient term as well as the second-order terms in Eq. (6.48). The details are left to Exercise (6.21). The left- and right-hand sides of Eq. (6.48) take the form

$$\left[\int_{\Omega^e} \mathbf{W}_i^T \mathbf{L}^k d\Omega\right]\mathbf{a} =$$

$$\int_{\Omega_e} \begin{bmatrix} \mathrm{Re}N_i\left(u^k\dfrac{\partial}{\partial x}+\dfrac{\partial u^k}{\partial x}+v^k\dfrac{\partial}{\partial y}\right) \\ +\dfrac{\partial N_i}{\partial x}\dfrac{\partial}{\partial x}+\dfrac{\partial N_i}{\partial y}\dfrac{\partial}{\partial y} \end{bmatrix} \quad \mathrm{Re}N_i\dfrac{\partial u^k}{\partial y} \quad -\dfrac{\partial N_i}{\partial x} \\ \mathrm{Re}N_i\dfrac{\partial v^k}{\partial x} \quad \begin{matrix}\mathrm{Re}N_i\left(u^k\dfrac{\partial}{\partial x}+v^k\dfrac{\partial}{\partial y}+\dfrac{\partial v^k}{\partial y}\right)\\ +\dfrac{\partial N_i}{\partial x}\dfrac{\partial}{\partial x}+\dfrac{\partial N_i}{\partial y}\dfrac{\partial}{\partial y}\end{matrix} \quad -\dfrac{\partial N_i}{\partial y} \\ \dfrac{\partial}{\partial x} \quad \dfrac{\partial}{\partial y} \quad 0 \end{bmatrix} d\Omega\,\mathbf{a}$$

$$
-\int_{\Gamma^e}
\begin{bmatrix}
N_i\left[\left(-p+\dfrac{\partial}{\partial x}\right)n_x + \dfrac{\partial}{\partial y}n_y\right] \\[2em]
N_i\left[\dfrac{\partial}{\partial x}n_x + \left(-p+\dfrac{\partial}{\partial y}\right)n_y\right] \\[2em]
0
\end{bmatrix} d\Gamma
$$

$$(6.49)$$

and

$$
\left[\int_{\Omega^e} W_i^T D^k d\Omega\right]
$$

$$
= \int_{\Omega^e}
\begin{bmatrix}
\operatorname{Re}N_i\left(u^k\dfrac{\partial u^k}{\partial x} + v^k\dfrac{\partial u^k}{\partial y}\right) - \dfrac{\partial N_i}{\partial x}p^k + \dfrac{\partial N_i}{\partial x}\dfrac{\partial u^k}{\partial x} + \dfrac{\partial N_i}{\partial y}\dfrac{\partial u^k}{\partial y} \\[2em]
\operatorname{Re}N_i\left(u^k\dfrac{\partial v^k}{\partial x} + v^k\dfrac{\partial v^k}{\partial y}\right) - \dfrac{\partial N_i}{\partial y}p^k + \dfrac{\partial N_i}{\partial x}\dfrac{\partial v^k}{\partial x} + \dfrac{\partial N_i}{\partial y}\dfrac{\partial v^k}{\partial y} \\[2em]
\dfrac{\partial u^k}{\partial x} + \dfrac{\partial v^k}{\partial y}
\end{bmatrix} d\Omega
$$

$$
-\int_{\Gamma^e}
\begin{bmatrix}
N_i\left[\left(-p+\dfrac{\partial u^k}{\partial x}\right)n_x + \dfrac{\partial u^k}{\partial y}n_y\right] \\[2em]
N_i\left[\dfrac{\partial v^k}{\partial x}n_x + \left(-p+\dfrac{\partial v^k}{\partial y}\right)n_y\right] \\[2em]
0
\end{bmatrix} d\Gamma
$$

$$(6.50)$$

where n_x and n_y are the components of the unit normal to the boundary Γ, as before.

Finally, substituting Eqs. (6.45) and (6.46), we obtain the 9 x 9 system of element equations

$$\left[\mathbf{K}\left(\mathbf{x}^k\right)\right]\Delta\mathbf{x} = -\mathbf{F}\left(\mathbf{x}^k\right) \tag{6.51}$$

where

$$\mathbf{x}^k = \begin{bmatrix} u_1^k \\ v_1^k \\ u_2^k \\ v_2^k \\ u_3^k \\ v_3^k \\ u_4^k \\ v_4^k \\ p^k \end{bmatrix} \qquad \Delta\mathbf{x} = \begin{bmatrix} \Delta u_1 \\ \Delta v_1 \\ \Delta u_2 \\ \Delta v_2 \\ \Delta u_3 \\ \Delta v_3 \\ \Delta u_4 \\ \Delta v_4 \\ \Delta p \end{bmatrix}$$

The element stiffness matrix $\mathbf{K}(\mathbf{x}^k)$ is defined by

$$k_{ij} = \int_{\Omega^e} \left\{ \mathrm{Re}\, N_i \left[\left(\sum_{\ell=1}^{4} N_\ell u_\ell^k \right) \frac{\partial N_j}{\partial x} + \left(\sum_{\ell=1}^{4} \frac{\partial N_\ell}{\partial x} u_\ell^k \right) N_j \right. \right.$$

$$\left. \left. + \left(\sum_{\ell=1}^{4} N_\ell v_\ell^k \right) \frac{\partial N_j}{\partial y} \right] + \frac{\partial N_i}{\partial x}\frac{\partial N_j}{\partial x} + \frac{\partial N_i}{\partial y}\frac{\partial N_j}{\partial y} \right\} d\Omega$$

$$-\int_{\Gamma^e} N_i \left[\frac{\partial N_j}{\partial x} n_x + \frac{\partial N_j}{\partial y} n_y \right] d\Gamma \qquad \begin{cases} i = 2m-1 \\ j = 2n-1 \end{cases} \quad m,n = 1,2,3,4 \tag{6.52}$$

$$k_{ij} = \int_{\Omega^e} \text{Re}\left(\sum_{\ell=1}^{4} \frac{\partial N_\ell}{\partial y} u_\ell^k \right) N_i N_j d\Omega \quad \begin{cases} i = 2m-1 \\ \\ j = 2n \end{cases} \quad m,n = 1,2,3,4 \quad (6.53)$$

$$k_{ij} = -\int_{\Omega^e} \frac{\partial N_i}{\partial x} d\Omega + \int_{\Gamma^e} N_i d\Gamma \quad \begin{cases} i = 2m-1 \\ \\ j = 9 \end{cases} \quad m = 1,2,3,4 \quad (6.54)$$

$$k_{ij} = \int_{\Omega^e} \text{Re}\left(\sum_{\ell=1}^{4} \frac{\partial N_\ell}{\partial x} v_\ell^k \right) N_i N_j d\Omega \quad \begin{cases} i = 2m \\ \\ j = 2n-1 \end{cases} \quad m,n = 1,2,3,4 \quad (6.55)$$

$$k_{ij} = \int_{\Omega^e} \left\{ \text{Re} N_i \left[\left(\sum_{\ell=1}^{4} N_\ell u_\ell^k \right) \frac{\partial N_j}{\partial x} + \left(\sum_{\ell=1}^{4} N_\ell v_\ell^k \right) \frac{\partial N_j}{\partial y} \right. \right.$$

$$\left. \left. + \left(\sum_{\ell=1}^{4} \frac{\partial N_\ell}{\partial y} v_\ell^k \right) N_j \right] + \frac{\partial N_i}{\partial x} \frac{\partial N_j}{\partial x} + \frac{\partial N_i}{\partial y} \frac{\partial N_j}{\partial y} \right\} d\Omega$$

$$(6.56)$$

$$-\int_{\Gamma^e} N_i \left[\frac{\partial N_j}{\partial x} n_x + \frac{\partial N_j}{\partial y} n_y \right] d\Gamma \quad \begin{cases} i = 2m \\ \\ j = 2n \end{cases} \quad m,n = 1,2,3,4$$

$$k_{ij} = -\int_{\Omega^e} \frac{\partial N_i}{\partial y} d\Omega + \int_{\Gamma^e} N_i d\Gamma \quad \begin{cases} i = 2m \\ \\ j = 9 \end{cases} \quad m = 1,2,3,4 \quad 6.57)$$

$$k_{ij} = \int_{\Omega^e} \frac{\partial N_j}{\partial x} d\Omega \quad \begin{cases} i = 9 \\ \\ j = 2n-1 \end{cases} \quad n = 1,2,3,4 \quad (6.58)$$

$$k_{ij} = \int_{\Omega^e} \frac{\partial N_j}{\partial y} d\Omega \qquad \begin{cases} i = 9 \\ \\ j = 2n \end{cases} \quad n = 1,2,3,4 \qquad (6.59)$$

$$k_{ij} = 0 \qquad\qquad i = j = 9 \qquad\qquad (6.60)$$

and the right-hand side vector $\mathbf{F}(\mathbf{x}^k)$ is calculated from

$$f_{i=} \int_{\Omega^e} \left\{ \operatorname{Re} N_i \left[\left(\sum_{\ell=1}^4 N_\ell u_\ell^k \right) \left(\sum_{\ell=1}^4 \frac{\partial N_\ell}{\partial x} u_\ell^k \right) + \left(\sum_{\ell=1}^4 N_\ell v_\ell^k \right) \left(\sum_{\ell=1}^4 \frac{\partial N_\ell}{\partial y} u_\ell^k \right) \right] \right.$$

$$+ \frac{\partial N_i}{\partial x} \left(\sum_{\ell=1}^4 \frac{\partial N_\ell}{\partial x} u_\ell^k \right) + \frac{\partial N_i}{\partial y} \left(\sum_{\ell=1}^4 \frac{\partial N_\ell}{\partial y} u_\ell^k \right) - \frac{\partial N_i}{\partial x} p^k \left. \right\} d\Omega$$

$$- \int_\Gamma N_i \left[\left(-p^k + \sum_{\ell=1}^4 \frac{\partial N_\ell}{\partial x} u_\ell^k \right) n_x + \left(\sum_{\ell=1}^4 \frac{\partial N_\ell}{\partial y} u_\ell^k \right) n_y \right] d\Gamma$$

$$i = 2m - 1 \qquad\qquad m = 1,2,3,4$$

$$(6.61)$$

$$f_{i=} \int_{\Omega^e} \left\{ \operatorname{Re} N_i \left[\left(\sum_{\ell=1}^4 N_\ell u_\ell^k \right) \left(\sum_{\ell=1}^4 \frac{\partial N_\ell}{\partial x} v_\ell^k \right) + \left(\sum_{\ell=1}^4 N_\ell v_\ell^k \right) \left(\sum_{\ell=1}^4 \frac{\partial N_\ell}{\partial y} v_\ell^k \right) \right] \right.$$

$$+ \frac{\partial N_i}{\partial x} \left(\sum_{\ell=1}^4 \frac{\partial N_\ell}{\partial x} v_\ell^k \right) + \frac{\partial N_i}{\partial y} \left(\sum_{\ell=1}^4 \frac{\partial N_\ell}{\partial y} v_\ell^k \right) - \frac{\partial N_i}{\partial y} p^k \left. \right\} d\Omega$$

$$- \int_{\Gamma^e} N_i \left[\left(\sum_{\ell=1}^4 \frac{\partial N_\ell}{\partial x} v_\ell^k \right) n_x + \left(-p^k + \sum_{\ell=1}^4 \frac{\partial N_\ell}{\partial y} v_\ell^k \right) n_y \right] d\Gamma$$

$$i = 2m \qquad\qquad m = 1,2,3,4$$

$$(6.62)$$

$$f_{i} = \int_{\Omega^e} \left(\sum_{\ell=1}^{4} \frac{\partial N_\ell}{\partial x} u_\ell^k + \sum_{\ell=1}^{4} \frac{\partial N_\ell}{\partial y} v_\ell^k \right) d\Omega \qquad i = 9 \qquad (6.63)$$

REMARKS

1. In the above equations the line integrals have been carried in every term for completion; however, they will vanish most of the time. This is true for every side of an element that is not on the boundary Γ of Ω. Moreover, the line integrals will also vanish along those boundaries where Dirichlet conditions are imposed. Only on occasions in that very specific boundary conditions are imposed at boundaries that are open, in the sense that they allow for flow in or out of the region, will the need to evaluate these line integrals arise. The subject of which are the most appropriate boundary conditions to apply in such *open* boundaries is discussed in Chapter 9.

2. The expressions given in Eqs. (6.52) through (6.63) are algebraically involved. However, the integrations can be readily performed, since they only involve shape functions and known values of the dependent parameters at the k-th iteration. We wrote them for the specific case of the four-noded bilinear isoparametric element, but it should be clear that the same expressions are valid for other elements. We must be careful to rearrange the degrees of freedom in the vector $\mathbf{x}^k$ and keep in mind that the shape functions for the pressure p may no longer be constant.

3. To obtain the system of Eq. (6.51), we used the shape functions for the pressure as weighting functions on the continuity equation in order to obtain additional equations for the pressure. We will discuss the reasons and implications later. As a consequence, the element stiffness matrix $\mathbf{K}(\mathbf{x}^k)$ contains a zero in the last diagonal element. Furthermore, the matrix is not symmetric. After assembly, the zero will remain in the diagonal, and the lack of symmetry will also persist. Therefore we must make sure that an appropriate algorithm for the solution of linear systems is used that can deal with these special situations.

A word of warning is that the Galerkin method just described to solve the Navier-Stokes equations is only effective for low values of Re. When Re is increased, the solution will exhibit oscillations in the velocity field that eventually will prevent the method from converging.

To eliminate this problem, special Petrov-Galerkin techniques are introduced in Chapter 8.

We will now look at a simple example where Eq. (6.51) can be constructed by hand.

Example 6.7

Consider a plane Couette flow between two parallel plates. Assume that the bottom plate is fixed and the upper plate, which is at a distance $L = 2$, is moving in the x-direction with velocity $U = 2$, as depicted in Fig. 6.3. The Navier-Stokes equations can be easily solved for this case, and the solution is

$$u(x,y) = 2y$$

$$v(x,y) = 0$$

$$p(x,y) = \text{const.}$$

The solution is independent of Re so we can set Re $= 1$. We will establish the finite element equations over two bilinear elements of size $\Delta x = \Delta y = 1$, as shown in Fig. 6.3. The boundary conditions are the exact solution at nodes 1,2,3,5, and 6. Hence we must calculate the velocities u_4 and v_4 and the pressures p_1 and p_2. Along the side $x = 1$, we assume the natural boundary conditions

$$-p + \frac{\partial u}{\partial x} = 0 \qquad \frac{\partial v}{\partial x} = 0$$

so that the line integrals in Eq. (6.50) vanish.

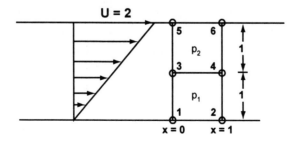

Figure 6.3 Plane Couette flow with two-element discretization.

For the first element, containing nodes 1,2,3,4, after applying the boundary conditions, the velocity components are

$$u(x,y) = N_3 + N_4 u_4 = y(1-x) + xy\,u_4 \tag{6.64}$$

$$v(x,y) = N_4 v_4 = xy\,v_4 \tag{6.65}$$

and the pressure p_1 is constant. The 3 x 3 element system corresponding to Eq. (6.51) can be obtained directly from Eqs. (6.52) through (6.63) by substituting $N_i = N_4(x,y)$ and the expressions for u and v given in Eqs. (6.64) and (6.65) and performing the integrations. For example, we have

$$k_{44} = \int_0^1\int_0^1 \left\{ N_4\left[(N_3 + N_4 u_4^k)\frac{\partial N_4}{\partial x} + \left(\frac{\partial N_3}{\partial x} + \frac{\partial N_4}{\partial x}u_4^k\right)N_4 \right. \right.$$

$$\left. \left. + (N_4 v_4^k)\frac{\partial N_4}{\partial y} \right] + \left(\frac{\partial N_4}{\partial x}\right)^2 + \left(\frac{\partial N_4}{\partial y}\right)^2 \right\} dx\,dy$$

$$= \int_0^1\int_0^1 \left\{ xy\left[(y - xy + xy\,u_4^k)y + (-y + y\,u_4^k)xy + (xy\,v_4^k)x \right] \right.$$

$$\left. + (y^2) + (x^2) \right\} dx\,dy = \frac{u_4^k}{6} + \frac{v_4^k}{12} + \frac{15}{24}, \text{ etc.}$$

The resulting system of linear element equations is

$$
\begin{bmatrix}
\dfrac{u_4^k}{6} + \dfrac{u_4^k}{12} + \dfrac{15}{24} & \dfrac{u_4^k}{12} + \dfrac{1}{36} & -\dfrac{1}{2} \\[2mm]
\dfrac{v_4^k}{12} & \dfrac{u_4^k}{12} + \dfrac{v_4^k}{6} + \dfrac{17}{24} & -\dfrac{1}{2} \\[2mm]
\dfrac{1}{2} & \dfrac{1}{2} & 0
\end{bmatrix}
\begin{bmatrix}
\Delta u_4 \\[2mm]
\Delta v_4 \\[2mm]
\Delta p_1
\end{bmatrix}
$$

$$
= - \begin{bmatrix} \dfrac{\left(u_4^k\right)^2}{12} + \dfrac{15 u_4^k}{24} + \dfrac{v_4^k}{36} + \dfrac{u_4^k v_4^k}{12} - \dfrac{5}{24} - \dfrac{p_1^k}{2} \\[12pt] \dfrac{\left(v_4^k\right)^2}{12} + \dfrac{u_4^k v_4^k}{12} + \dfrac{17 v_4^k}{24} - \dfrac{p_1^k}{2} \\[12pt] \dfrac{1}{2}\left(u_4^k + v_4^k - 1\right) \end{bmatrix}
$$

A similar system is obtained for the second element, Exercise (6.22), and after assembling, the final system of equations is

$$
\begin{bmatrix} \dfrac{u_4^k}{3} + \dfrac{4}{3} & \dfrac{2}{9} & -\dfrac{1}{2} & -\dfrac{1}{2} \\[10pt] \dfrac{v_4^k}{6} & \dfrac{u_4^k}{6} + \dfrac{3}{2} & -\dfrac{1}{2} & \dfrac{1}{2} \\[10pt] \dfrac{1}{2} & \dfrac{1}{2} & 0 & 0 \\[10pt] \dfrac{1}{2} & -\dfrac{1}{2} & 0 & 0 \end{bmatrix} \begin{bmatrix} \Delta u_4 \\[10pt] \Delta v_4 \\[10pt] \Delta p_1 \\[10pt] \Delta p_2 \end{bmatrix}
$$

$$
= \begin{bmatrix} \dfrac{\left(u_4^k\right)^2}{6} + \dfrac{4 u_4^k}{3} + \dfrac{2 v_4^k}{9} - \dfrac{3}{2} - \dfrac{p_1^k}{2} - \dfrac{p_2^k}{2} \\[12pt] \dfrac{\left(v_4^k\right)^2}{6} + \dfrac{3 v_4^k}{2} + \dfrac{u_4^k v_4^k}{6} - \dfrac{p_1^k}{2} - \dfrac{p_2^k}{2} \\[12pt] \dfrac{1}{2}\left(u_4^k + v_4^k - 1\right) \\[12pt] \dfrac{1}{2}\left(u_4^k - v_4^k - 1\right) \end{bmatrix}
$$

(6.66)

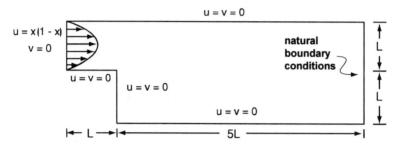

Figure 6.4 Domain and boundary conditions for flow over a backward–facing step.

It can be easily shown that the nonlinear iteration converges to the exact solution $u_4 = 1$, $v_4 = p_1 = p_2 = 0$. ∎

We close this section with an example of a slightly more complicated flow over a backward facing step at Re = 10. The geometry and boundary conditions are shown in Fig. 6.4. At the open downstream boundary, we apply a natural boundary condition similar to that used in Example 6.7. We use a uniform mesh of 10 x 30 bilinear elements of size $\Delta x = \Delta y = 0.2L$, and the calculation is stopped when Eq. (6.14) is satisfied for $\delta = 10^{-5}$.

The results of the calculation are shown in Fig. 6.5 in the form of stream function contours. We have restricted ourselves to a very low value of Re because at higher values of Re the Galerkin formulation becomes unstable and must be modified. The stabilized methods are commonly known as Petrov-Galerkin formulations and are developed in Chapters 8 and 9. The stream function is obtained from the calculated velocity field directly from the definition, that is,

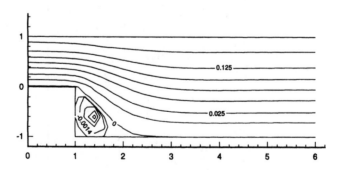

Figure 6.5 Streamlines for flow over a backward-facing step at Re = 10.

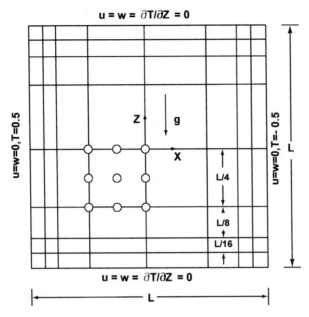

Figure 6.6 Domain, finite element mesh, and temperature boundary conditions for natural convection in a square enclosure.

$$\varphi(x,y) = \int_{y_0}^{y} u\,d\,y - \int_{x_0}^{x} v\,d\,x$$

where (x_0, y_0) is a starting reference point.

6.3.3 Steady State Natural Convection

We close this chapter with an example of a combined application of the Newton-Raphson and the direct iteration methods. Let us look at the calculation of natural convection in a differentially heated square enclosure in two dimensions, as it was first presented by Marshall et al. (1978). This time, nine-noded biquadratic elements are used to approximate the dependent variables of velocity and temperature as shown in Fig. 6.6. We will not discuss the approximation to the pressure at this time.

We assume that no mass exchange can take place between the enclosure and its surroundings. Therefore the only driving forces in this system are the buoyancy forces induced by the temperature difference in the two vertical sides. We will use the Boussinesq approximation and assume that the density is constant except in the body force term, where it obeys the state equation

$$\rho = \rho_0\left(1 + \beta\left(T^* - T_0\right)\right) \tag{6.67}$$

with ρ_0 the reference density, T_0 the reference temperature, and β the coefficient of thermal expansion. The coupled Navier-Stokes equations and energy equation for a single-phase laminar incompressible fluid are nondimensionalized and written in the form

$$u\frac{\partial u}{\partial x} + w\frac{\partial u}{\partial z} = -\frac{\partial p}{\partial x} + P_r\left(\frac{\partial^2 u}{\partial x^2} + \frac{\partial^2 u}{\partial z^2}\right) \tag{6.68}$$

$$u\frac{\partial w}{\partial x} + w\frac{\partial w}{\partial z} = -\frac{\partial p}{\partial z} + P_r\left(\frac{\partial^2 w}{\partial x^2} + \frac{\partial^2 w}{\partial z^2}\right) + P_r RaT \tag{6.69}$$

$$\frac{\partial u}{\partial x} + \frac{\partial w}{\partial z} = 0 \tag{6.70}$$

$$u\frac{\partial T}{\partial x} + w\frac{\partial T}{\partial z} = \frac{\partial^2 T}{\partial x^2} + \frac{\partial^2 T}{\partial z^2} \tag{6.71}$$

The nondimensional parameters are the Prandtl number Pr and the Rayleigh number Ra, defined as

$$Pr = \frac{\nu}{D_T} \tag{6.72}$$

and

$$Ra = \frac{\beta g \Delta T^* L^3}{\nu D_T} \tag{6.73}$$

Here g is gravity, ΔT^* is a reference temperature difference, and D_T is the thermal diffusivity. The nondimensionalization is accomplished using

$$X = \frac{x^*}{L} \quad Z = \frac{z^*}{L} \quad u = \frac{u^* L}{D_T,} \quad v = \frac{v^* L}{D_T}$$

$$T = \frac{T^* - \frac{\Delta T^*}{2}}{\Delta T^*} \qquad p = \frac{p^* - \rho_0 g z}{\rho_0 \frac{D_T^2}{L^2}}$$

The asterisk denotes a dimensional quantity. The analysis of this kind of flow can be found in the books by Jaluria (1980) and Gebhart et al. (1988).

The boundary conditions associated with Eqs. (6.68) through (6.71) are no slip on all walls, insulated horizontal walls, and prescribed temperatures along the vertical walls, i.e.,

$$u = w = 0 \qquad x \in \Gamma$$

$$T = .5 \qquad x = -\frac{L}{2}$$

$$T = -.5 \qquad x = \frac{L}{2}$$

$$\frac{\partial T}{\partial y} = 0 \qquad y = \pm \frac{L}{2}$$

The equations may be solved using a combination of a direct iteration and the Newton-Raphson iteration developed in Chapter 5. To do this, we define the linearized operator for the temperature

$$\left[L_T \left(u^k, v^k, T^k \right) \right] [T]$$

$$\equiv \left[u^k \frac{\partial}{\partial x} + w^k \frac{\partial}{\partial z} - \frac{\partial^2}{\partial x^2} - \frac{\partial^2}{\partial z^2} \right] T = 0 \qquad (6.74)$$

The operators $\mathbf{D}_1, \mathbf{D}_2, \mathbf{D}_3, \mathbf{L}_1, \mathbf{L}_2$, and $\mathbf{L}_3$ defined in Section 6.3.2 must also be modified slightly and become

$$\mathbf{D}_1(u,w,p) \equiv u\frac{\partial u}{\partial x} + w\frac{\partial u}{\partial z} + \frac{\partial p}{\partial x} - Pr\left(\frac{\partial^2 u}{\partial x^2} + \frac{\partial^2 u}{\partial z^2}\right) = 0 \quad (6.75)$$

$$\mathbf{D}_2(u,w,p,T) \equiv u\frac{\partial v}{\partial x} + w\frac{\partial v}{\partial z} + \frac{\partial p}{\partial z}$$

$$- Pr\left(\frac{\partial^2 w}{\partial x^2} + \frac{\partial^2 w}{\partial z^2}\right) - Pr\ Ra\ T = 0 \qquad (6.76)$$

$$\mathbf{D}_3(u,w) \equiv \frac{\partial u}{\partial x} + \frac{\partial w}{\partial z} = 0 \qquad (6.77)$$

$$\left[\mathbf{L}_1\left(u^k,w^k,p^k\right)\right]\begin{bmatrix}\Delta u \\ \Delta w \\ \Delta p\end{bmatrix} \equiv$$

$$\left[u^k\frac{\partial}{\partial x} + \frac{\partial u^k}{\partial x} + w^k\frac{\partial}{\partial z} - Pr\left(\frac{\partial^2}{\partial x^2} + \frac{\partial^2}{\partial z^2}\right)\right]\Delta u$$

$$+ \left[\frac{\partial u^k}{\partial z}\right]\Delta u + \left[\frac{\partial}{\partial x}\right]\Delta p = 0 \qquad (6.78)$$

$$\left[\mathbf{L}_2\left(u^k,w^k,p^k,T^k\right)\right]\begin{bmatrix}\Delta u \\ \Delta w \\ \Delta p\end{bmatrix} \equiv \left[\frac{\partial w^k}{\partial x}\right]\Delta u$$

$$+ \left[u^k\frac{\partial}{\partial x} + w^k\frac{\partial}{\partial z} + \frac{\partial w^k}{\partial z} - P_r\left(\frac{\partial^2}{\partial x^2} + \frac{\partial^2}{\partial z^2}\right)\right]\Delta v$$

$$+ \left[\frac{\partial}{\partial y}\right]\Delta p - Pr\ Ra\ T^k = 0 \qquad (6.79)$$

and

$$\left[L_3\left(u^k, w^k\right)\right]\begin{bmatrix} \Delta u \\ \Delta w \end{bmatrix} \equiv \left[\frac{\partial}{\partial x}\right]\Delta u + \left[\frac{\partial}{\partial z}\right]\Delta w \qquad (6.80)$$

The solution algorithms can now be established as follows:

1. Our known initial guesses are $k = 0$, u^0, w^0, p^0, and T^0.

2. Perform a Newton-Raphson iteration with Eqs. (6.75) through (6.80) with T^k fixed to calculate u^{k+1}, w^{k+1}, and p^{k+1}.

3. Solve for T^{k+1} from Eq. (6.74) using the values of u^{k+1} and w^{k+1} from step 2.

4. Check for convergence in all variables. If satisfied, stop; otherwise, set $k = k+1$, and go back to step 2.

REMARKS

1. Clearly, we can apply the Newton-Raphson method to the fully coupled system of equations, which is left as Exercise 6.24. The question is, "What do we gain by performing this seemingly more complex double iteration of which the Newton-Raphson iteration explained in the pervious section is a subiteration?"

The answer lies in the size of the problem to be solved. For the mesh shown in Fig. 6.6 the algorithm above requires the solution of a banded matrix with a bandwidth of 65 and storage requirement of 14,625 words for the temperature, plus a matrix with bandwidth 130 and 58,500 words of storage for the flow calculation. A total memory requirement of 73,125 words is thus necessary.[*] On the other hand, the fully coupled Newton-Raphson method yields a bandwidth of 195 and requires the storage of 131,625 words. This last value is 80% larger than the storage required by the double-iteration scheme. Moreover, a direct method will require significantly more CPU time to solve the large system than the two smaller ones together. Of course, our present scheme may need more iterations to converge and still be slower than a full Newton-Raphson iteration. Comparisons are very difficult to make. However, our present method offers a clear advantage if the amount of storage is a concern.

---

[*] These values assume that we solve for velocity and temperature degrees of freedom only; the pressure has not been included. We will see in Chapter 10 how we can achieve this using the penalty function approximation.

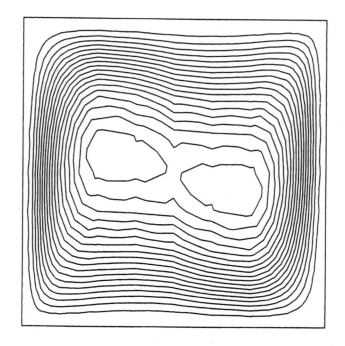

Figure 6.7 Streamlines for natural convection in a square enclosure at Ra = 10^5 and Pr = 1.

In general, the solutions of nonlinear problems will always lead to many alternative approaches. Choices will invariably have to be made as to the type of algorithm to be used and how to implement it. There is certainly *no best algorithm* that will always lead us to the right solution in the most efficient way. The specific approach to the solution of a problem will always depend on the particular problem itself and the analyst's personal expertise, time availability, experience, and preferences.

2. The direct solution of steady state problems involving the Navier-Stokes equations can be difficult in the sense that convergence may be elusive, especially at high Re or Ra. It is always advisable to approach the desired solutions in steps, starting from low values of the parameters and using the solutions at those values as the initial guess to reach higher ones. As an example, in Fig. 6.7 we show the streamlines for the solution of our natural convection problem at Ra = 10^5 and Pr = 1. The solution was reached in seven iterations, starting from

homogeneous initial conditions. This solution was then used as the starting point to reach Ra = 10^6 and 10^7.

Physical systems exhibiting natural convection are often unstable and do not reach steady state. For this reason, it may be more desirable to solve the time-dependent equations to steady state. If the steady state solutions are unstable this will be captured by the numerical algorithm. We discuss this in more detail in Chapter 10.

3. The heat transfer across the enclosure can be readily calculated as part of the solution using the boundary residual equations as explained in section 3.7.2. The quantity we are most interested in is the average Nusselt number, $\overline{Nu}$, which measures the net energy transfer across the enclosure. We define this as

$$\overline{Nu} = -\int_{-L/2}^{L/2} \frac{\partial T}{\partial x}(L/2, y) dy \qquad (6.81)$$

The calculated value in this case is $\overline{Nu}$ = 4.6. In Fig. 6.8 we show the local Nusselt number distribution calculated along the right-hand wall x = L/2.

6.4 CLOSURE

In this chapter we reviewed the basic iterative methods for the solution of nonlinear equations as they have been usually applied in finite element discretizations. The Newton-Raphson method is normally preferred to a direct iteration because of its faster convergence rate and the fact that, given the nonlinear operator, the linearized tangent operator is always uniquely defined. However, there are instances in which a direct iteration can provide a better way to solve a problem.

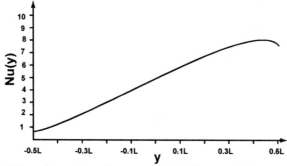

Figure 6.8 Local Nusselt number along vertical wall x = L/2 for Ra = 10^5 and Pr = 1.

We have had our first excursion into the solution of the Navier-Stokes equations in the time-independent case. We saw there were several different ways to approach the solution of the sample problems, with each alternate method offering some kind of advantage or improvement. This will always be the case in the modeling of physical systems. There will be choices as to the mathematical model to be used, the numerical formulation that may be considered most appropriate, the type of nonlinear iteration or time marching scheme to be used, the method to solve the linear equations, etc.

Throughout the rest of the book, we will choose methods that we have tested and that are widely used by a high percentage of finite element practitioners, but we do not claim that these are always the best or most convenient methods to use. Our objective is to provide the reader with good basic algorithms, not to attempt to discuss all of them. We will now move on to the solution of time-dependent problems where even more possibilities and alternatives will open up.

REFERENCES

Bicanic, N. and Johnson, K. W. (1979) "Who Was Raphson?," *Int. J. Numer. Methods Eng.*, Vol. 14, pp. 148–152.

Isaacson, E. and Keller, H. B. (1966) *Analysis of Numerical Method,* New York: John Wiley & Sons.

Gebhart, B., Jaluria, Y., Mahajan, R. L. and Sammakia, B. (1988) *Buoyancy Induced Flows and Transport.* Washington, D.C.: Hemisphere (Taylor and Francis).

Gill, P. E., Murray, W. and Wright, M. H. (1981) *Practical Optimization.* New York: Academic Press.

Jaluria, Y. (1980) *Natural Convection Heat and Mass Transfer.* Oxford: Pergamon Press.

Marshall, R. S., Heinrich, J. C. and Zienkiewicz, O. C. (1978) "Natural Convection in a Square Enclosure by a Finite Element, Penalty Function Method Using Primitive Fluid Variables," *Numer. Heat Transfer*, Vol. 1, pp. 315–330.

Ortega, J. M. and Rheinboldt, W. C. (1970) *Iterative Solution of Nonlinear Equations in Several Variables.* New York: Academic Press.

Rohsenow, W. R. and Choi, H. Y. (1961) *Heat, Mass and Momentum Transfer.* Englewood Cliffs, N.J.: Prentice-Hall.

Zienkiewicz, O. C., Onate, E. and Heinrich, J. C. (1981) "A Generic Formulation for Coupled Thermal Flow of Metals Using Finite Elements," *Int. J. Numer. Methods Eng.*, Vol. 17, pp. 1497–1514.

EXERCISES

6.1   Consider the function $f(x) = \sin x$. We want to obtain the root $x^* = 0$ using Newton's method. Set up the iteration, starting with

(a) $x_0 = -1.262627$
(b) $x_0 = -.5$
(c) $x_0 = -2.0$

Perform at least three iterations in cases (a) and (b) and six in case (c). Interpret the results geometrically using graphs. This example illustrates the local convergence properties of Newton's algorithms.

6.2   Replace $u = \sum\limits_{i=1}^{5} N_i(x)u_i$ and $w_i = N_i$ in Eq. (6.9) to find Eq. (6.10).

6.3   Consider the Blasius equation

$$\frac{d^3 f}{dx^3} + 2f\frac{d^2 f}{dx^2} = 0$$

with boundary conditions $f(0) = f'(0) = 0$, $f'(10) = 1^*$. Introduce the auxiliary variable $g(x) = f'(x)$, and find the tangent matrix for the Newton-Raphson iteration using four linear elements.

6.4   Find the tangent matrix in Eq. (6.11), applying Eq. (6.5) to the vector function **F** given in Eq. (6.10).

6.5   In Example 6.1, show that the element contribution $F_e(u)$ to $F(u)$ is given by

---

* The third boundary condition is actually $\lim\limits_{x \to \infty} f'(x) = 1$.

$$F_e(u) = \begin{bmatrix} 4(u_1 - u_2) - \dfrac{1}{3}u_1^2 + \dfrac{1}{6}u_1 u_2 + \dfrac{1}{6}u_2^2 \\ 4(u_2 - u_1) - \dfrac{1}{6}u_1^2 - \dfrac{1}{6}u_1 u_2 + \dfrac{1}{3}u_2^2 \end{bmatrix}$$

Using $F_e(u)$, calculate the element tangent matrix and assemble to obtain Eq. (6.11).

6.6   Solve the system of equations in Eq. (6.13) iteratively trying different initial guesses u^0 such as $u_i^0 = -100, -10, 10,$ and 100 for $i = 2,3,4$. You will find that the number of iterations needed for convergence may vary, depending on the initial condition but, in general, convergence will always occur – which means the Newton-Raphson method is quite robust.

6.7   Consider the nonlinear equation

$$\frac{d^2 u}{dx^2} + \left(\frac{du}{dx}\right)^2 + 1 = 0 \qquad u(0) = u(1) = 0$$

(a) Using linear elements, find the element vector $F_e(u)$ and the element tangent matrix $F_e'(u)$.
(b) Using 4, 8, and 16 elements, solve the equation using the Newton-Raphson method. Determine the rate of convergence comparing with the exact solution

$$u(x) = \ln\left\{\cos(x - 1/2) / \cos\,(1/2)\right\}$$

6.8   Repeat Exercise 6.7 for the equation $u'' = e^u$ with general solution
$$u(x) = \ln\left\{1/2\ c_1^2 / \cos^2\left[c_1(x + c_2)/2\right]\right\}$$

6.9   Find the linear part of the nonlinear operator $D(u) = u'' - \sin u$.

6.10 Find the linear part of the nonlinear operator of Eq. (6.8). Use it to define an iterative scheme of the form of Eqs. (6.21) and (6.22) and discretize using the Galerkin method with four linear elements. Show that the element stiffness matrices are the same as those obtained in Exercise 6.6 and that the global system of equations is identical to Eq. (6.13).

6.11 Linearize the Sine-Gordon wave equation

$$\frac{\partial^2 u}{\partial x^2} + \frac{\partial^2 u}{\partial y^2} - \frac{\partial^2 u}{\partial t^2} = F(x,y)\sin u$$

This equation is important because it accepts "soliton" solutions.

6.12 The first-order nonlinear equation

$$\left(\frac{du}{dx}\right)^2 + u^2 = 1$$

has the solution $u(x) = \sin(x + c)$, where c is the integration constant. Find the linearized operator.

6.13 Find the element stiffness matrices and element right-hand side vector when the Newton-Raphson method is applied to the boundary layer equations:

$$\text{(a)} \qquad \varepsilon\frac{d^2 u}{dx^2} - \left(\frac{du}{dx}\right)^2 + e^u = 0$$

$$\text{(b)} \qquad \varepsilon\frac{d^2 u}{dx^2} - \left(\frac{du}{dx}\right)^2 + u^2 = 0$$

6.14 Using the Galerkin formulation of Eq. (6.32) with linear elements and $h = 1/4$, show that the contribution of the element stiffness matrix of the convective term is

$$\frac{1}{6}\begin{bmatrix} -2u_1 - u_2 & 2u_1 + u_2 \\ -u_1 - 2u_2 & u_1 + 2u_2 \end{bmatrix}$$

and assemble to obtain Eq. (6.35).

6.15   To find the roots of $f(x) = x^2 - x - 12$, consider the three iterations defined by

(a) $x = x^2 - 12$ (b) $x = \dfrac{12}{x-1}$ (c) $x = \sqrt{x + 12}$

Show that (a) always diverges, while (b) and (c) always converge to only one of the roots each.

6.16   Find the solution to the iteration defined by Eq. (6.36). Stop the iteration when the relative differences between two consecutive iterations are less than 0.01 as in Example 6.3. Compare the solution to that of Table 6.1, and compare the number of iterates needed to that of Newton's method.

6.17   Can you find an optimal mesh size such that the solution to the equation in Example 6.5 can be obtained at the minimum CPU cost?

6.18   Repeat Table 6.3 for the equation

$$-\frac{d}{dx}\left[(1.05 + 0.0025\,T)\frac{dT}{dx}\right] = 2$$

$$T(0) = 20, \quad -k\frac{dT}{dx}\bigg|_{x=8} = -4$$

6.19   Apply the Newton-Raphson method and the direct iteration method to the function $f(x)$ in Example 6.5 starting with $x^0 = 2$, and plot the error in the iterations in a figure similar to Fig. 6.2.

6.20   Follow the steps in Example 6.2 using

$$D(u + \Delta u, v + \Delta v, p + \Delta p) - D(u, v, p)$$

to obtain the linearized operators of Eqs. (6.41), (6.42), and (6.43).

6.21   Apply Gauss's theorem to the second-order and pressure gradient terms of the weighted residual form of Eqs. (6.38), (6.39), (6.41), and (6.42) to obtain Eqs. (6.49) and (6.50).

6.22   Compute the element equations for the second element in Example 6.7, and solve the system of Eqs. (6.66) starting with $u_4^\circ = v_4^\circ = p_1^\circ = p_2^\circ = 0$.

6.23   Use the enclosed computer program to reproduce the results of Fig. 6.5 for flow over a backward-facing step.

6.24   Formulate the fully coupled Newton-Raphson nonlinear iteration for steady state natural convection in Section 6.3.3, and write it in the same form as Eq. (6.44).

6.25   Use the enclosed computer program to reproduce the results of natural convection in a square enclosure shown in Fig. 6.7. A 10 x 10 bilinear element mesh should suffice.

TIME DEPENDENCE

7.1 OVERVIEW

Up to this point, we have only considered equations whose independent variables are spatial variables in one or two (and three) dimensions and are independent of time. Therefore, their solutions require only discretization in space, which is accomplished using the techniques discussed so far. If the solutions change with time, the spatial discretization is no longer sufficient, and we must follow the evolution of the solutions in the time dimension. We will now address the basic techniques used to model time-dependent problems that are parabolic or first-order hyperbolic using finite elements, beginning with the time-dependent heat diffusion equation and working our way up to the nonlinear Navier-Stokes equations in subsequent chapters. We will also touch briefly on second-order hyperbolic equations. Implicit and explicit schemes will be discussed in terms of accuracy and stability properties. The concepts of *mass* matrix and *lumped mass* matrix are introduced. These appear when time derivatives are present in the governing equations.

7.2 DIFFUSION EQUATION

Time-dependent diffusion occurs in a wide variety of physical processes. The most commonly recognizable one is the diffusion of heat. The general governing differential equation is

$$\frac{\partial \phi}{\partial t} = \frac{\partial}{\partial x}\left(D\frac{\partial \phi}{\partial x}\right) + \frac{\partial}{\partial y}\left(D\frac{\partial \phi}{\partial y}\right) + S \tag{7.1}$$

This is a special case of Eq. (2.22) in the absence of convection. Here D is the diffusion coefficient and S represents sources or sinks.

There is another class of time-dependent equations that are somewhat out of the scope of this book, but we need to consider them at least partially due to their importance. These are the equations of wave propagation, which govern a wide variety of waves, including acoustic, free surface, electromagnetic, and structural vibrations. We will write the general equation as

$$m\frac{\partial^2 \phi}{\partial t^2} + \alpha\frac{\partial \phi}{\partial t} = \frac{\partial}{\partial x}\left(k\frac{\partial \phi}{\partial x}\right) + \frac{\partial}{\partial y}\left(k\frac{\partial \phi}{\partial y}\right) + F\left(\phi, \dot{\phi}\right) \tag{7.2}$$

Here m is associated with the mass of the system, α is a viscous damping coefficient, and k (which is unity in most propagation problems) is associated with the structural stiffness in vibration problems. F is a forcing term, which is generally nonlinear.

In this chapter we will consider at least one method of solution for equations of the type of Eq. (7.2).

7.2.1 Semidiscrete Galerkin Method

Let us start with the one-dimensional heat diffusion equation

$$\rho c_v \frac{\partial T}{\partial t} = k\frac{\partial^2 T}{\partial x^2} + Q \qquad 0 < x < L \qquad t > 0 \tag{7.3}$$

The boundary conditions may, in general, be time dependent and can be any of the types considered in the previous chapters. In addition, we need an initial condition that we will write as

$$T(x,0) = T_0(x) \tag{7.4}$$

The standard weighted residuals formulation of Eq. (7.4) is

$$\int_0^L \left(\rho c_v w \frac{\partial T}{\partial t} + k \frac{\partial w}{\partial x} \frac{\partial T}{\partial x} + wQ \right) dx + w \left(-k \frac{\partial T}{\partial x} \right) \Bigg|_0^L = 0 \qquad (7.5)$$

We now assume that T can be written by separating the independent variables as

$$T(x,t) \cong \sum_j N_j(x) T_j(t) \qquad (7.6)$$

that is, the shape functions are independent of time, while the nodal values are fixed in space but can vary with time. If we freeze the time t, we are back to exactly the same situation as we had before. Therefore we can use the same shape functions as in time-independent problems to describe the solution at any fixed time t. However, the nodal values of the dependent function will be allowed to change as a function of time.

The Galerkin formulation, $w_i = N_i$, applied to Eq. (7.5) at time t is

$$\sum_j \left[\int_0^L \rho c_v N_i N_j dx \right] \dot{T}_j + \sum_j \left[\int_0^L k \frac{\partial N_i}{\partial x} \frac{\partial N_j}{\partial x} dx \right] T_j$$

$$= \int_0^L N_i Q dx - \left[N_i \left(-k \frac{\partial T}{\partial x} \right) \right]_0^L \qquad (7.7)$$

for each i, where the over dot denotes differentiation with respect to time, i.e., $\dot{T} \equiv \partial T / \partial t$.

Performing the integrations, we obtain the linear system of ordinary differential equations

$$\mathbf{C}\dot{\mathbf{T}} + \mathbf{K}\mathbf{T} = \mathbf{Q} \qquad (7.8)$$

which is referred to as the *semidiscrete Galerkin form* of Eq. (7.3).

The matrix **C**, given by $c_{ij} = \int_0^L \rho c_v N_i N_j dx$, is called the *consistent mass* matrix, or *capacitance* matrix. The term mass matrix is most commonly used, even though it is only correct to denote the matrix arising from the acceleration term in dynamic problems. Because the true consistent mass matrix and the capacitance matrix only differ by the coefficients, we choose to call the matrix **C** the mass matrix.

We must now address the solution of the system of ordinary differential equations given by Eq. (7.8). There are an abundance of methods available in the literature, particularly when the equations are stiff. Comprehensive studies have been provided by Richtmeyer and Morton (1963) and Lambert (1973). However, a majority of these methods were developed to integrate the equations obtained from finite difference approximations to the spatial derivatives. When finite element discretizations are used, the methods must be modified owing to the presence of the mass matrix. Here we will only establish the most basic methods available for the integration of Eq. (7.8) in time, which suffice in most general applications and can normally be easily adapted to more difficult situations requiring special treatment. For more comprehensive discussions and other methods not discussed here, the reader may consult the books by Zienkiewicz (1977), Bathe (1982), and Hughes (1987).

7.2.2 The θ Method

The most commonly used time integration algorithm is normally referred to as the θ method. It consists in approximating the time derivative by the backward difference

$$\dot{\mathbf{T}} \cong \frac{1}{\Delta t}\left(\mathbf{T}^{n+1} - \mathbf{T}^{n}\right) \tag{7.9}$$

where $\mathbf{T}^{n} \equiv \mathbf{T}(x, t_n)$ denotes the value of the dependent variable at time t = t_n, Δt is the time step increment, and $t_{n+1} = t_n + \Delta t$.

The temperature **T** is then defined by

$$\mathbf{T} = \theta\, \mathbf{T}^{n+1} + (1-\theta)\mathbf{T}^{n} \tag{7.10}$$

where the relaxation parameter θ is normally specified to be a value between 0 and 1 and is used to control the accuracy and stability of the algorithm. This method falls in the general category of *one-step* methods, in which the solution at each step is advanced to time t_{n+1} from known values at time step t_n.

Substituting Eqs. (7.9) and (7.10) into Eq. (7.8), we have

$$\left(\frac{1}{\Delta t}C + \theta K\right)T^{n+1} = \left(\frac{1}{\Delta t}C - (1-\theta)K\right)T^n + \theta Q^{n+1} + (1-\theta)Q^n \quad (7.11)$$

where Q has been assumed to be a function of time and approximated over the interval $t_n \le t \le t_{n+1}$ using Eq. (7.10).

From Eq. (7.11), we can calculate an approximation to the solution at time level t_{n+1} if the approximation to the solution at time t_n is known.

If linear elements are used in the space discretization and Q is independent of x, the resulting element equations are

$$\left\{\frac{\rho c_v h}{6\Delta t}\begin{bmatrix} 2 & 1 \\ 1 & 2 \end{bmatrix} + \frac{\theta k}{h}\begin{bmatrix} 1 & -1 \\ -1 & 1 \end{bmatrix}\right\}\begin{bmatrix} T_1^{n+1} \\ T_2^{n+1} \end{bmatrix}$$

$$= \left\{\frac{\rho c_v h}{6\Delta t}\begin{bmatrix} 2 & 1 \\ 1 & 2 \end{bmatrix} - \frac{(1-\theta)k}{h}\begin{bmatrix} 1 & -1 \\ -1 & 1 \end{bmatrix}\right\}\begin{bmatrix} T_1^n \\ T_2^n \end{bmatrix} \qquad (7.12)$$

$$+ \frac{\theta h Q^{n+1}}{2}\begin{bmatrix} 1 \\ 1 \end{bmatrix} + \frac{(1-\theta)h Q^n}{2}\begin{bmatrix} 1 \\ 1 \end{bmatrix}$$

where h denotes the element size.

Example 7.1

Consider the numerical solution of the equation

$$\frac{\partial T}{\partial t} = D_T \frac{\partial^2 T}{\partial x^2} \qquad 0 < x > 1$$

$$T(x,0) = 0$$

$$T(0,t) = 0 \qquad\qquad (7.13)$$

$$T(1,t) = 1$$

where $D_T = k/\rho c_v$ denotes the heat diffusivity. The analytical solution can be found (Carslaw and Jaeger, 1959) and is given by

$$T(x,t) = \frac{4}{\pi} \sum_{n=1}^{\infty} \left(\frac{1}{(2n+1)} \sin(2n+1)\pi x \ e^{-D_T (2n+1)^2 \pi^2 t} \right) \qquad (7.14)$$

Results for $D_T = .1$ using 2, 4, 8, and 16 linear elements and 1, 2, 4, and 8 quadratic elements for several values of Δt and θ are shown in Table 7.1(a,b).

With two linear elements and $\theta = 1$, the finite element equations after assembly of Eq. (7.12) are

$$\left\{ \frac{1}{12\Delta t} \begin{bmatrix} 2 & 1 & 0 \\ 1 & 4 & 1 \\ 0 & 1 & 2 \end{bmatrix} + 2D_T \begin{bmatrix} 1 & -1 & 0 \\ -1 & 2 & -1 \\ 0 & -1 & 1 \end{bmatrix} \right\} \begin{bmatrix} T_1^{n+1} \\ T_2^{n+1} \\ T_3^{n+1} \end{bmatrix}$$

$$= \frac{1}{12\Delta t} \begin{bmatrix} 2 & 1 & 0 \\ 1 & 4 & 1 \\ 0 & 1 & 2 \end{bmatrix} \begin{bmatrix} T_1^n \\ T_2^n \\ T_3^n \end{bmatrix}$$

and after applying the boundary conditions, we obtain the recurrence relation

Table 7.1a Results for $D_T = .1$ using linear elements.

$$\frac{\partial T}{\partial t} = D_T \frac{\partial^2 T}{\partial x^2} \qquad T(x, 0) = 0, \ T(0, t) = 0, \ T(1, t) = 1$$

Number of linear elements	Δt	θ	T (0.5, 0.5)	T (0.5, 1.0)
2	0.5	0	0.42000	0.46800
		0.5	0.35505	0.42194
		1	0.30469	0.37793
	0.25	0	0.32850	0.41596
		0.5	0.29810	0.38970
		1	0.27242	0.36534
	0.1	0	0.26780	0.37746
		0.5	0.25683	0.36664
		1	0.24668	0.35626
	0.01	0	0.22987	0.35229
		0.5	0.22887	0.35120
		1	0.22788	0.35012
4	0.5	0	3.31	-15.32
		0.5	0.31955	0.36289
		1	0.24046	0.32817
	0.25	0	-0.80	-4.60
		0.5	0.22307	0.33639
		1	0.10947	0.30978
	0.1	0	0.18746	0.31936
		0.5	0.17655	0.30766
		1	0.16717	0.29648
	0.01	0	0.14573	0.28975
		0.5	0.14481	0.28860
		1	0.14391	0.28745
8	0.5	0	11.07	-625.66
		0.5	0.27624	0.37223
		1	0.22563	0.31461
	0.25	0	-74.1	–
		0.5	0.21049	0.32030
		1	0.18250	0.29438
	0.1	0	–	–
		0.5	0.15616	0.29091
		1	0.14817	0.27977
	0.01	0	0.12444	0.27222
		0.5	0.12375	0.27106
		1	0.12311	0.26991
16	0.5	0	2.01	-994.73
		0.5	0.26863	0.36909
		1	0.22225	0.31127
	0.25	0	-123.59	–
		0.5	0.20241	0.31582
		1	0.17874	0.29053
	0.1	0	–	–
		0.5	0.15013	0.28660
		1	0.14401	0.27554
	0.01	0	–	–
		0.5	0.11896	0.26658
		1	0.11845	0.26544

Table 7.1b Results for $D_T = .1$ using quadratic elements.

$$\frac{\partial T}{\partial t} = D_T \frac{\partial^2 T}{\partial x^2} \quad T(x, 0) = 0, \; T(0, t) = 0, \; T(1, t) = 1$$

Number of quadratic elements	Δt	θ	$T(0.5, 0.5)$	$T(0.5, 1.0)$
1	0.5	0	0.37500	0.43750
		0.5	0.32000	0.34200
		1	0.27778	0.35185
	0.25	0	0.28906	0.38135
		0.5	0.26475	0.35769
		1	0.24400	0.33616
	0.1	0	0.23428	0.34309
		0.5	0.22573	0.33372
		1	0.21776	0.32475
	0.01	0	0.20052	0.31881
		0.5	0.19975	0.31789
		1	0.19899	0.31697
2	0.5	0	4.23	-20.73
		0.5	0.30764	0.34319
		1	0.22059	0.31211
	0.25	0	-1.20	-6.72
		0.5	0.19985	0.31883
		1	0.17725	0.29195
	0.1	0	0.16312	0.30041
		0.5	0.15235	0.28863
		1	0.14327	0.27737
	0.01	0	0.12068	0.26979
		0.5	0.11979	0.26863
		1	0.11892	0.26748
4	0.5	0	31.26	-1612.09
		0.5	0.26658	0.36905
		1	0.22100	0.31023
	0.25	0	-189.93	–
		0.5	0.20020	0.31111
		1	0.17731	0.28939
	0.1	0	–	–
		0.5	0.150335	0.28534
		1	0.14244	0.27433
	0.01	0	0.11792	0.26647
		0.5	0.11734	0.26532
		1	0.11682	0.26418
8	0.5	0	30.29	–
		0.5	0.26627	0.36798
		1	0.22114	0.31016
	0.25	0	-1455.58	–
		0.5	0.20002	0.31453
		1	0.17751	0.28926
	0.1	0	–	–
		0.5	0.14857	0.28520
		1	0.14267	0.27414
	0.01	0	–	–
		0.5	0.11743	0.26510
		1	0.11696	0.26396

Exact solution: $T(0.5, 0.5) = 0.11384$; $T(0.5, 1) = 0.26270$

$$\left(\frac{1}{3\Delta t} + 4D_T\right) T_2^{n+1} = \frac{1}{3\Delta t} T_2^n + 2D_T - \frac{1}{12\Delta t}$$

$$T_2^0 = 0$$

for T_2 as a function of time. Notice that as $\Delta t \to \infty$, we obtain

$$T_2 = \frac{1}{2}$$

which is the correct steady state result. ∎

7.2.3 Accuracy and Stability of the θ Method

In Example 7.1 we used the values $\theta = 0$, 0.5, and 1.0. These are the values most commonly assigned to θ and, except for the presence of the mass matrix, correspond to the Euler, Crank-Nicolson, and backward implicit methods, respectively. However, the appearance of the mass matrix modifies the algorithms. Therefore they are referred to as Euler-Galerkin when $\theta = 0$, Crank-Nicolson-Galerkin when $\theta = 0.5$, and backward implicit Galerkin when $\theta = 1.0$.

A truncation error analysis shows that the methods converge as first-order methods $O(\Delta t)$ when $\theta = 0.0$ and 1.0; the Crank-Nicolson-Galerkin method is second-order $O(\Delta t^2)$ in time, and for other values of θ between 0 and 1, convergence takes place at intermediate rates between first- and second-order, i.e., $O(\Delta t^r)$, $1 < r < 2$.

A stability analysis of Eq. (7.11) yields the following results (the analysis can be found in the book by Bickford, 1990):

1. If $1/2 \leq \theta < 1$, the method is *unconditionally stable*. Hence the Crank-Nicolson-Galerkin and backward implicit Galerkin methods are stable for any Δt.

2. If $0 \leq \theta < 1/2$ there is a time step limitation given by

$$\Delta t < \frac{2}{\lambda(1 - 2\theta)} \tag{7.15}$$

where λ is the largest eigenvalue of the generalized eigenvalue problem

$$(\mathbf{K} - \lambda \mathbf{C})\mathbf{X} = 0 \qquad (7.16)$$

Example 7.2

Consider once more the solution of Eq. (7.13) in Example 7.1 using two linear elements. After imposing the boundary conditions, matrices $\mathbf{C}$ and $\mathbf{K}$ reduce to 1 x 1 matrices $\mathbf{C} = [1/3]$ and $\mathbf{K} = [4D_T]$, respectively. Hence the eigenvalue problem of Eq. (7.16) has only the solution $\lambda = 12\,D_T$, and the stability limit for the Euler-Galerkin algorithm ($\theta = 0$) is given by

$$\Delta t < \frac{1}{6D_T}$$

according to Eq. (7.15). Let us rewrite the Euler-Galerkin algorithm for $T_2^n = T(0.5, t_n)$ in the form

$$T_2^{n+1} = (1 - 12\,D_T\,\Delta t)\,T_2^n + 6D_T\,\Delta t$$

and use it for solution with various values of Δt, starting from the initial condition $T^0 = 0$.

(a) Setting $\Delta t = 1/6D_T$, the recursive relation becomes

$$T_2^{n+1} = -T_2^n + 1$$

and the iterates are $T_2^0 = 0$, $T_2^1 = 1$, $T_2^2 = 0$, $T_3^2 = 1$, ..., which shows that the algorithm diverges as expected, although the iterates are bounded for the limit value of Δt.

(b) It is easily checked that if $\Delta t > 1/6D_T$, the resulting recursive relations lead to unbounded sequences (Exercise 7.6).

(c) Now we use values for which the algorithm is stable. Starting with $\Delta t > 1/8D_T$, the recursion relation is

$$T_2^{n+1} = -\frac{1}{2}T_2^n + \frac{3}{4}$$

and the time evolution is

$$T_2^0 = 0, \; T_2^1 = 0.75, \; T_2^2 = 0.375, \; T_2^3 = 0.5625, \; T_2^4 = 0.46875, \ldots, \text{ etc.}$$

Clearly as $t \rightarrow \infty$, the solution approaches the steady state limit $T_2 = 1/2$. However, because the time step is rather large, convergence is not monotonic; the approximations oscillate around the exact solution.

(d) Consider now $\Delta t = 1/24 D_T$. The algorithm is

$$T_2^{n+1} = \frac{1}{2} T_2^n + \frac{1}{4}$$

with $T_2^0 = 0, \; T_2^1 = 0.375, \; T_2^2 = 0.4375, \ldots, \; T_2^{12} = 0.49988$

We can observe a great improvement in accuracy with respect to case c). The exact solution, Eq. (7.14) at $x = 0.5$, $t = 0.5$, is $u*(0.5, 0.5) = 0.49542$. However, most important is the fact that the solution is now monotonic. ∎

Example 7.3

Let us now find the stability limit when three linear elements are used. The system of equations, after applying the boundary conditions, is

$$\frac{1}{18 \Delta t} \begin{bmatrix} 4 & 1 \\ 1 & 4 \end{bmatrix} \begin{bmatrix} T_2^{n+1} \\ T_3^{n+1} \end{bmatrix}$$

$$= \left\{ \frac{1}{18 \Delta t} \begin{bmatrix} 4 & 1 \\ 1 & 4 \end{bmatrix} - 3D_T \begin{bmatrix} 2 & -1 \\ -1 & 2 \end{bmatrix} \right\} \begin{bmatrix} T_2^n \\ T_3^n \end{bmatrix} + \begin{bmatrix} 0 \\ 3D_T \end{bmatrix}$$

To determine the stability limit, we need the eigenvalues of

$$\left| 3D_T \begin{bmatrix} 2 & -1 \\ -1 & 2 \end{bmatrix} - \frac{\lambda}{18} \begin{bmatrix} 4 & 1 \\ 1 & 4 \end{bmatrix} \right| = 0$$

These are $\lambda_1 = 54D_T$ and $\lambda_2 = 10.8D_T$. Hence for the algorithm to be stable, the time step must be such that

$$\Delta t < \frac{1}{27D_T}$$

which is considerably smaller than in the previous case. A good estimate of the stability limit is given by $\Delta t < h^2/3D_T$.

To find the maximum eigenvalue may require a significant amount of work if the matrices are large. The generalized eigenvalue problem must first be transformed into a standard eigenvalue problem, to which well-known methods such as the power method (see Isaacson and Keller, 1966) can be applied. Such transformations, as well as more advanced methods for the solution of eigenvalue problems in mechanics, are treated by Bathe (1982).

7.2.4 Mass Lumping

An important case of the θ method is the Euler-Galerkin algorithm, corresponding to $\theta = 0$. In this case, Eq. (7.11) takes the form

$$\frac{1}{\Delta t} \mathbf{C} \mathbf{T}^{n+1} = \mathbf{Q}^n + \left(\frac{1}{\Delta t} \mathbf{C} - \mathbf{K} \right) \mathbf{T}^n \qquad (7.17)$$

Notice that this is not an explicit scheme, since the mass matrix $\mathbf{C}$ must be inverted to find the solution. In fact, *fully explicit* algorithms do not arise from the standard form of the Galerkin finite element method. To obtain explicit algorithms, we must diagonalize the matrix $\mathbf{C}$. This is known as *mass lumping* and is a procedure of great practical importance in finite element modeling. It is also the subject of much controversy because of the effect that mass lumping may have on the accuracy of the algorithm.

The simplest method of mass lumping, and the most important to us, consists in defining the lumped mass matrix $\overline{\mathbf{C}}$ as

$$\overline{C} \equiv [\overline{c}_{ij}] = \begin{cases} \sum_k c_{ik} & \text{if } i = j \\ 0 & \text{if } i \neq j \end{cases}$$

The diagonal elements of $\overline{C}$ are the row sums of C, and the off-diagonal elements are zero.

Replacing the mass matrix C by the lumped mass matrix $\overline{C}$ in the Euler-Galerkin method gives

$$\frac{1}{\Delta t} \overline{C} T^{n+1} = Q^n + \left(\frac{1}{\Delta t} \overline{C} - K \right) T^n \qquad (7.18)$$

which can be solved explicitly as

$$T^{n+1} = \Delta t \, \overline{C}^{-1} \left[Q^n + \left(\frac{1}{\Delta t} \overline{C} - K \right) T^n \right] \qquad (7.19)$$

where $\overline{C}^{-1}$ is a diagonal matrix with entries $\overline{c}_{ii}^{-1} = 1 / \overline{c}_{ii}$.

Example 7.4

Let us investigate how mass lumping affects the stability of the Euler-Galerkin method used in Example 7.2. For a two-element discretization, we have $\overline{C} = [1/2]$ and $K = [4 D_T]$. The eigenvalue of $K - \lambda \overline{C}$ is $\lambda = 8 D_T$, and the stability limit is given by $\Delta t < 1/4 D_T$, as opposed to $\Delta t < 1/6 D_T$ if the consistent mass matrix C is used.

When three linear elements are used in the discretization, we have

$$\overline{C} = \begin{bmatrix} 1/3 & 0 \\ 0 & 1/3 \end{bmatrix} \qquad K = D_T \begin{bmatrix} 6 & -3 \\ -3 & 6 \end{bmatrix}$$

The roots of the determinant $|K - \lambda \overline{C}| = 0$ are $\lambda_1 = 27 D_T$ and $\lambda_2 = 9 D_T$. The stability limit is thus

$$\Delta t \le 2/27 D_T$$

which is twice as large as that obtained in Example 7.2 using the consistent mass matrix. ∎

REMARKS

1. Mass lumping can also be obtained using closed Newton-Cotes integration rules for which the nodes coincide with the quadrature points (Gray and van Genuchten, 1978). To illustrate this, consider a rectangular bilinear element as depicted in Fig. 7.1, where we show that the integration points corresponding to the trapezoidal rule agree with the location of the nodes. The trapezoidal rule applied to a function f(x,y) defined over the element gives, using Eqs (5.11) and (5.22),

$$\int_0^a \int_0^b f(x,y) \, dx \, dy \cong \frac{ab}{4}\left[f(0,0) + f(a,0) + f(0,b) + f(a,b)\right]$$

When we apply it to the mass matrix, we have

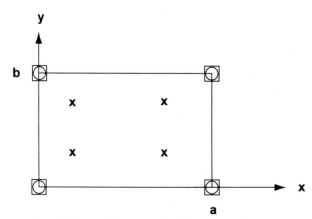

Figure 7.1 Rectangular bilinear element. Circles are nodes; crosses are Gauss integration points; squares are trapezoidal rule integration points.

$$C = \left[\int_0^a \int_0^b \rho c_v N_i N_j \, dx \, dy \right] = \frac{\rho c_v \, a \, b}{4} \begin{bmatrix} 1 & 0 & 0 & 0 \\ 0 & 1 & 0 & 0 \\ 0 & 0 & 1 & 0 \\ 0 & 0 & 0 & 1 \end{bmatrix}$$

provided that ρc_v is constant. Hence a lumped mass matrix is automatically obtained. This is a direct consequence of the fact that the shape functions satisfy the relation

$$N_i(x_j, y_j) = \delta_{ij}$$

where (x_j, y_j) are the nodal locations. The result carries over to general quadrilateral isoparametric elements, where

$$C = \int_{-1}^1 \int_{-1}^1 \rho c_v N_i N_j |J| \, d\xi \, d\eta$$

In this case the diagonal values may not be all the same – if the determinant of the Jacobian of the transformation is not constant.

A similar situation occurs if Simpson's rule is used to integrate biquadratic elements. This is left to Exercise 7.11.

2. In Example 7.4 we obtained larger stability limits using the lumped mass matrix than the consistent mass matrix. It is generally true that algorithms using the lumped mass matrix are more stable than their consistent counterparts. Furthermore, their use results in various computational advantages and simplifications. However, mass lumping can have an adverse effect on the accuracy of the algorithms, as reported by Gresho et al. (1978) and Yu and Heinrich (1986). This aspect will be discussed in more detail in Chapter 8, when we deal with convection-diffusion equations.

3. Algorithms based on the consistent mass matrix are generally more accurate than those using lumped masses. However, the consistent mass matrix may introduce undesired oscillations during the early stages of the calculation, especially if the initial data are not smooth. These oscillations disappear as time advances, but they are unacceptable in

many more complex problems involving convective transport, especially in reacting flows. This phenomenon has been called "*noise*" in the numerical solution and occurs when the time step exceeds a *critical* time step related to the solution of the equation (Wood and Lewis, 1975). Unfortunately, the critical time step is usually extremely small, and it is not realistic to attempt to eliminate the problem by making the time step smaller. Let us illustrate these facts with a simple example.

Example 7.5

We now apply the Crank-Nicolson-Galerkin method ($\theta = 1/2$) to Eq. (7.13) in Example 7.1 using three linear elements with $D_T = 1$ and the same initial and boundary conditions. The resulting system of equations can be written as (Exercise 7.12)

$$\left\{ \begin{bmatrix} 4 & 1 \\ 1 & 4 \end{bmatrix} + 27\Delta t \begin{bmatrix} 2 & -1 \\ -1 & 2 \end{bmatrix} \right\} \begin{bmatrix} T_2^{n+1} \\ T_3^{n+1} \end{bmatrix}$$

$$= \left\{ \begin{bmatrix} 4 & 1 \\ 1 & 4 \end{bmatrix} - 27\Delta t \begin{bmatrix} 2 & -1 \\ -1 & 2 \end{bmatrix} \right\} \begin{bmatrix} T_2^n \\ T_3^n \end{bmatrix} + \begin{bmatrix} 0 \\ 54\Delta t \end{bmatrix}$$

$$\text{(7.20)}$$

Recall that this method is unconditionally stable. However, if Δt is not small enough the solution will oscillate during the first time steps. If $\Delta t = 4/27$, the system becomes

$$\begin{bmatrix} 12 & -3 \\ -3 & 12 \end{bmatrix} \begin{bmatrix} T_2^{n+1} \\ T_3^{n+1} \end{bmatrix} = \begin{bmatrix} -4 & 5 \\ 5 & -4 \end{bmatrix} \begin{bmatrix} T_2^n \\ T_3^n \end{bmatrix} + \begin{bmatrix} 0 \\ 8 \end{bmatrix}$$

Inverting the matrix on the left-hand side, we can write

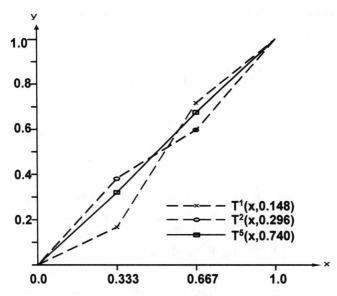

Figure 7.2 Solution to Example 7.4 with $\Delta t = 4/27$.

$$T_2^{n+1} = \frac{1}{135}\left(-33\,T_2^n + 48\,T_3^n + 24\right)$$

$$T_3^{n+1} = \frac{1}{135}\left(48\,T_2^n - 33\,T_3^n + 96\right)$$

(7.21)

from which we obtain the time history at nodes 2 and 3 given in Table 7.2. It is clear from Table 7.2 that the solution oscillates rather strongly at the beginning but the oscillations die out as time increases and the solution approaches the correct steady state. Figure 7.2 shows the first, second, and fifth time steps obtained from Eq. (7.21).

Table 7.2. Solution of Eq. (7.21) for the first five time steps.

n	T_2^n	T_3^n
1	0.17778	0.71111
2	0.38716	0.60049
3	0.29665	0.70198
4	0.35486	0.64499
5	0.32036	0.67962

For a smaller value of Δt, e.g., $\Delta t = 1/27$, the problem does not recur. In this case we have

$$T_2^{n+1} = \frac{1}{3}\left(T_2^n + T_3^n\right)$$

$$T_3^{n+1} = \frac{1}{3}\left(T_2^n + T_3^n + 1\right)$$

(7.22)

and the first five time steps are given in Table 7.3. The temperatures now increase as time increases. Notice that in Eqs. (7.22) the time step is small enough that no negative coefficients arise from the contribution of the stiffness matrix to the right-hand side.

Table 7.3. Solution of Eq. (7.22) for the first five time steps.

n	T_2^n	T_3^n
1	0.0	0.33333
2	0.11111	0.44444
3	0.18518	0.51852
4	0.23457	0.56790
5	0.26749	0.60082

We now solve the problem using a lumped mass matrix. Equation (7.20) takes the form

$$\left\{ \begin{bmatrix} 1 & 0 \\ 0 & 1 \end{bmatrix} + \frac{9\Delta t}{2} \begin{bmatrix} 2 & -1 \\ -1 & 2 \end{bmatrix} \right\} \begin{bmatrix} T_2^{n+1} \\ T_3^{n+1} \end{bmatrix}$$

$$= \left\{ \begin{bmatrix} 1 & 0 \\ 0 & 1 \end{bmatrix} + \frac{9\Delta t}{2} \begin{bmatrix} 2 & -1 \\ -1 & 2 \end{bmatrix} \right\} \begin{bmatrix} T_2^n \\ T_3^n \end{bmatrix} + \begin{bmatrix} 0 \\ 9\Delta t \end{bmatrix}$$

(7.23)

The solution for $\Delta t = 4/27$ is shown in Table 7.4 for the first five time steps.

Table 7.4. Solution of Eq. (7.23) with $\Delta t = 4/27$, for the first five time steps

n	T_2^n	T_3^n
1	0.17778	0.62222
2	0.33185	0.62815
3	0.32316	0.66884
4	0.33459	0.66652
5	0.33321	0.66701

This solution is much smoother than the one obtained using the consistent mass matrix, showing only a slight oscillation in the third significant digit. We will see in Chapter 10 that mass lumping is often used to avoid this problem in more complex transport problems.

It is important to mention that when the fully implicit Galerkin method is used, the problem of initial oscillations in time is fully suppressed by the use of mass lumping. The loss of accuracy that may arise with the use of mass lumping is discussed in Chapter 9, since it occurs in conjunction with transport by convection.

7.3 RUNGE-KUTTA METHODS

A second important family of time stepping algorithms is the explicit Runge-Kutta methods. These methods have been used extensively in compressible flow calculations and nonlinear transport (Oden and Wellford, 1972; Fletcher, 1984; Brueckner and Heinrich, 1991).

The Euler method discussed in the previous section is based on the linear Taylor polynomial approximation to the dependent variables. To improve on Euler's method we may consider more terms of the Taylor polynomial, but this has the drawback that higher order derivatives of the function would be required. The Runge-Kutta methods provide ways to improve the approximations without resorting to the higher order derivatives, which are approximated using functional values. The details of how this is done can be found in texts such as those of Isaacson and Keller (1966) or Yakowitz and Szidarovszky

(1989). We will be satisfied with presenting the formulas for the second- and fourth-order Runge-Kutta methods.

The second-order Runge-Kutta method is also known as the modified Euler method. Consider the ordinary differential equation

$$\frac{dy}{dt} = f(y,t) \tag{7.24}$$

If an approximation y_n to the solution at time $t = t_n$ is known, the approximation y_{n+1} to the solution at time $t_{n+1} = t_n + \Delta t$ is obtained in three steps using the formulae

$$k_1 = f(y_n, t_n) \tag{7.25a}$$

$$k_2 = f\left(y_n + \frac{\Delta t}{2} k_1, t_n + \frac{\Delta t}{2}\right) \tag{7.25b}$$

$$y_{n+1} = y_n + \Delta t k_2 \tag{7.25c}$$

The Taylor series expansion yields a truncation error $O(\Delta t^2)$. An algorithm with truncation error $O(\Delta t^4)$ is given by the recursion relation (classical Runge-Kutta method)

$$k_1 = f(y_n, t_n) \tag{7.26a}$$

$$k_2 = f\left(y_n + \frac{\Delta t}{2} k_1, t_n + \frac{\Delta t}{2}\right) \tag{7.26b}$$

$$k_3 = f\left(y_n + \frac{\Delta t}{2} k_2, t_n + \frac{\Delta t}{2}\right) \tag{7.26c}$$

$$k_4 = f(y_n + \Delta t k_3, t_n + \Delta t) \tag{7.26d}$$

$$y_{n+1} = y_n + \frac{\Delta t}{6}(k_1 + 2k_2 + 2k_3 + k_4) \tag{7.26e}$$

which is the fourth-order Runge-Kutta method.

Finite element discretizations typically yield systems of ordinary differential equations of the form

$$\mathbf{C}\dot{\mathbf{u}} + \mathbf{F}(\mathbf{u}) = 0$$

which, after mass lumping, can be written as

$$\dot{\mathbf{u}} = \mathbf{G}(\mathbf{u})$$

with

$$\mathbf{G} = -\overline{\mathbf{C}}^{-1}\mathbf{F}(\mathbf{u})$$

The above methods are generalized to the solution of these systems of equations simply by replacing y with $\mathbf{u}$ and f with $\mathbf{G}$ in Eqs. (7.25)–(7.26), i.e., the second-order Runge-Kutta method becomes

$$\mathbf{K}_1 = \mathbf{G}\left(\mathbf{u}^n, t_n\right) \tag{7.27a}$$

$$\mathbf{K}_2 = \mathbf{G}\left(\mathbf{u}^n + \frac{\Delta t}{2}\mathbf{K}_1, t_n + \frac{\Delta t}{2}\right) \tag{7.27b}$$

and the fourth-order method requires the additional steps

$$\mathbf{K}_3 = \mathbf{G}\left(\mathbf{u}^n + \frac{\Delta t}{2}\mathbf{K}_2, t_n + \frac{\Delta t}{2}\right) \tag{7.27c}$$

$$\mathbf{K}_4 = \mathbf{G}\left(\mathbf{u}^n + \Delta t\,\mathbf{K}_3, t_n + \Delta t\right) \tag{7.27d}$$

plus the following two equations

$$\mathbf{u}^{n+1} = \mathbf{u}^n + \Delta t\,\mathbf{K}_2 \tag{7.28a}$$

$$\mathbf{u}^{n+1} = \mathbf{u}^n + \frac{\Delta t}{6}\left(\mathbf{K}_1 + 2\mathbf{K}_2 + 2\mathbf{K}_3 + \mathbf{K}_4\right) \tag{7.28b}$$

Example 7.6

Let us apply the second-order Runge-Kutta method to the time-dependent Burgers equation

$$\frac{\partial u}{\partial t} + u\frac{\partial u}{\partial x} = \varepsilon\frac{\partial^2 u}{\partial x^2} \qquad 0 < x < 1 \qquad (7.29)$$

with boundary conditions $u(0,t) = 1$, $u(1,t) = 0$ and initial condition

$$u(x,0) = \begin{cases} 1 \text{ if } x \le \dfrac{1}{4} \\ 0 \text{ if } x > \dfrac{1}{4} \end{cases}$$

Setting $\varepsilon = 1$, the Galerkin finite element equations, using four linear elements as in Example 7.1, yield the system of equations (Exercise 7.16)

$$\frac{1}{24}\begin{bmatrix} 4 & 1 & 0 \\ 1 & 4 & 1 \\ 0 & 1 & 4 \end{bmatrix}\begin{bmatrix} \dot{u}_2 \\ \dot{u}_3 \\ \dot{u}_4 \end{bmatrix}$$

$$= -\begin{bmatrix} 8u_2 - 4u_3 - \dfrac{1}{6}u_2 + \dfrac{1}{6}u_2 u_3 + \dfrac{1}{6}u_3^2 - \dfrac{25}{6} \\ -4u_2 + 8u_3 - 4u_4 - \dfrac{1}{6}u_2^2 - \dfrac{1}{6}u_2 u_3 + \dfrac{1}{6}u_3 u_4 + \dfrac{1}{6}u_4^2 \\ -4u_3 + 8u_4 - \dfrac{1}{6}u_3^2 - \dfrac{1}{6}u_3 u_4 \end{bmatrix}$$

Next we must recast Eq. (7.28) into the form

$$\dot{\mathbf{u}} = \mathbf{G}(\mathbf{u})$$

using mass lumping. This yields

$$
\begin{bmatrix} \dot{u}_2 \\ \dot{u}_3 \\ \dot{u}_4 \end{bmatrix} = \begin{bmatrix} \dfrac{4}{5}\left(-49\,u_2 + 24\,u_3 - u_2 u_3 - u_3^2 + 25\right) \\[2mm] \dfrac{2}{3}\left(24\,u_2 - 48\,u_3 + 24 u_4 + u_2^2 + u_2 u_3 - u_3 u_4 - u_4^2\right) \\[2mm] \dfrac{4}{5}\left(24\,u_3 - 48\,u_4 + u_3^2 + u_3 u_4\right) \end{bmatrix}
$$

We can now apply the second-order Runge-Kutta method to the system of ordinary differential equations in the following way:

$$
K_1 = \begin{bmatrix} k_{12} \\ k_{13} \\ k_{14} \end{bmatrix} = \begin{bmatrix} \dfrac{4}{5}\left(-49\,u_2^n + 24\,u_3^n - u_2^n u_3^n - \left(u_3^n\right)^2 + 25\right) \\[2mm] \dfrac{2}{3}\left(24\,u_2^n - 48\,u_3^n + 24\,u_4^n + \left(u_2^n\right)^2 + u_2^n u_3^n - u_3^n u_4^n - \left(u_4^n\right)^2\right) \\[2mm] \dfrac{4}{5}\left(24\,u_3^n - 48\,u_4^n + \left(u_3^n\right)^2 + u_3^n u_4^n\right) \end{bmatrix}
$$

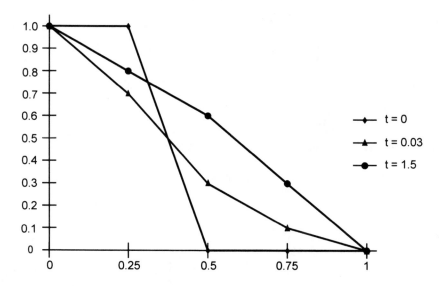

Figure 7.3 Solutions to Eq. (7.29) using the second-order Runge-Kutta method with four linear elements ($\varepsilon = 1$, $\Delta t = 0.001$).

$$
\mathbf{K}_2 = \begin{bmatrix} k_{22} \\ k_{23} \\ k_{24} \end{bmatrix} =
$$

$$
\begin{bmatrix}
\dfrac{4}{5}\left[\left(u_2^n + \dfrac{\Delta t}{2} k_{12} \right)\left(-49 - u_3^n - \dfrac{\Delta t}{2} k_{13} \right) + \left(u_3^n + \dfrac{\Delta t}{2} k_{13} \right)\left(24 - u_3^n - \dfrac{\Delta t}{2} k_{13} \right) + 25 \right] \\[4ex]
\dfrac{2}{3}\left[\begin{array}{l} \left(u_2^n + \dfrac{\Delta t}{2} k_{12} \right)\left(24 + u_2^n + \dfrac{\Delta t}{2} k_{12} + u_3^n + \dfrac{\Delta t}{2} k_{13} \right) \\[2ex] -48\left(u_3^n + \dfrac{\Delta t}{2} k_{13} \right) + \left(u_4^n + \dfrac{\Delta t}{2} k_{14} \right) \\[2ex] \bullet\left(24 - u_3^n - \dfrac{\Delta t}{2} k_{13} - u_4^n - \dfrac{\Delta t}{2} k_{14} \right) \end{array} \right] \\[8ex]
\dfrac{4}{5}\left[\begin{array}{l} \left(u_3^n + \dfrac{\Delta t}{2} k_{13} \right)\left(24 + u_3^n + \dfrac{\Delta t}{2} k_{13} + u_4^n + \dfrac{\Delta t}{2} k_{14} \right) \\[2ex] -48\left(u_4^n + \dfrac{\Delta t}{2} k_{14} \right) \end{array} \right]
\end{bmatrix}
$$

So $\mathbf{u}^{n+1}$ is given by

$$
\begin{bmatrix} u_2^{n+1} \\ u_3^{n+1} \\ u_4^{n+1} \end{bmatrix} = \begin{bmatrix} u_2^n + \Delta t k_{22} \\ u_3^n + \Delta t k_{23} \\ u_4^n + \Delta t k_{24} \end{bmatrix}
$$

Calculations performed with the above algorithm are shown in Fig. 7.3 and Table 7.5.

Table 7.5 Solution of Eq. (7.29) with $\varepsilon = 1$ and $\Delta t = 0.001$ using the second-order Runge-Kutta method and four linear elements.

t	u_2	u_3	u_4
0.00	1.0000	0.0000	0.0000
0.03	0.7095	0.3277	0.0685
1.50	0.8049	0.5670	0.2937

Notice that the greatest advantage of explicit methods is that the solution of linear systems of equations is not required, and hence large numbers of degrees of freedom can be handled without excessive memory requirements. However, the evaluations of the vector functions in Eq. (7.27) may become very computationally intensive. In large calculations, vectorization and/or parallel processing are needed to reduce the computation time, particularly in the fourth-order Runge-Kutta method.

So far, we have not discussed how we should choose the time step interval, Δt. Runge-Kutta methods are consistent and conditionally stable. However, a stability analysis can be performed only for linear equations. Nonlinear equations must be locally linearized. The linearization of a single equation of the form of Eq. (7.24) yields the model ordinary differential equation

$$\frac{dy}{dt} + \alpha y = f(t) \qquad (7.30)$$

where $\alpha = \partial f / \partial y$ and hence depends on the solution. Equation (7.30) can be used to monitor the stability of the method as the solution progresses, evaluating α at the current time. For the second-order modified Euler method of Eq. (7.25) the stability limit is given by

$$\alpha \Delta t \leq 2 \qquad (7.31)$$

while for the fourth-order Runge-Kutta scheme it is approximately

$$\alpha \Delta t \leq 2.875 \qquad (7.32)$$

A more detailed discussion is presented by Hoffman (1992).

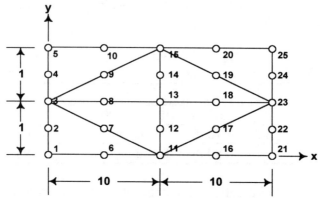

Figure 7.4 Domain and finite elements used for Couette flow.

The difficulty to calculate a time step limit is a disadvantage of the generalized Runge-Kutta methods. However, experience dictates that the stability limits for simple finite difference approximations of the linearized equations provide stable numerical approximations. We should also mention that the Runge-Kutta methods presented here are only one possibility of infinitely many that yield algorithms of the same order.

We close this section with a simple example of Couette flow calculated by Oden and Wellford (1972) using the fourth-order Runge-Kutta method and quadratic triangular elements. In this calculation the pressure was eliminated by adding "bubble" functions to the velocity shape functions. These functions are such that they vanish on the boundaries of each element and force the velocity field to satisfy the incompressibility conditions over each element. The domain used in the calculations is shown in Fig. 7.4 together with the mesh of quadratic triangular elements. The boundary conditions are

$$u(x, 0.2) = 0.1$$
$$v(x, 0.2) = 0.0$$
$$u(x, 0) = v(x, 0) = 0.0$$

and along the vertical boundaries $x = 0$ and $x = 2.0$, the natural boundary conditions are

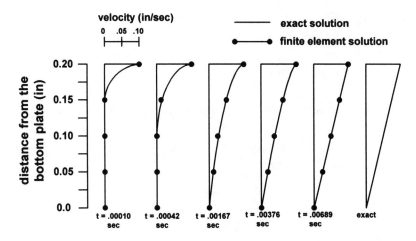

Figure 7.5 Horizontal velocity profiles at various times for transient plane Couette flow.

$$-p + \mu \frac{\partial u}{\partial x} = 0$$

in the x-momentum equation and

$$\frac{\partial v}{\partial x} = 0$$

in the y-momentum equation. The viscosity was set equal to 0.00362, and the mean density was 0.00242. The calculated horizontal velocity profiles at several times are shown in Fig. 7.5 and compared with the exact solution.

7.4 GENERALIZED NEWMARK ALGORITHMS

One of the most powerful time integration algorithms in finite element methods, and perhaps the most widely used, is the Newmark algorithm (Newmark, 1959). This algorithm was originally designed for problems of structural dynamics, and later generalized by Hughes et al. (1979) for general time-dependent problems.

Newmark methods stem from the equations of dynamic equilibrium and therefore are applicable to both parabolic equations of the form of Eq. (7.1) and second-order hyperbolic equations such as Eq. (7.2). They are based on the use of generalized physical quantities of displacements, velocities, and accelerations, which makes it easy to interpret the algorithms physically.

7.4.1 Newmark Method for Second-Order Hyperbolic Equations

Let us introduce the basic idea by means of a one-degree-of-freedom dynamic equation of the form

$$m\frac{d^2x}{dt^2} + F(x,\dot{x}) = g(t) \tag{7.33}$$

with initial conditions $x(0) = x_0$ and $dx/dt|_{x=0} = \dot{x}(0) = \dot{x}_0$. Regardless of what x represents, a temperature, concentration, velocity, or displacement, we may think of x as a generalized displacement and $\dot{x}$ and $\ddot{x}$ as generalized velocity and acceleration, respectively.

Now suppose that at time $t = t_n$ the generalized variables x_n, $\dot{x}_n$, and $\ddot{x}_n$ are known. We can then approximate their values at time $t_{n+1} = t_n + \Delta t$ using Taylor series expansions for the displacement and velocity, i.e.,

$$x^{n+1} = x^n + \Delta t\,\dot{x}^n + \frac{\Delta t^2}{2}\left[(1-2\beta)\ddot{x}^n + 2\beta\ddot{x}^{n+1}\right] \tag{7.34}$$

$$\dot{x}^{n+1} = \dot{x}^n + \Delta t\left[(1-\gamma)\ddot{x}^n + \gamma\ddot{x}^{n+1}\right] \tag{7.35}$$

The parameters β and γ were introduced by Newmark (1959) in order to include the acceleration $\ddot{x}^{n+1}$, at the end of the time interval, implicitly in the numerical scheme. This accounts for changes in the internal forces over the duration of the time step, thus resulting in a more accurate algorithm. To explain how these ideas lead to a time stepping method, first let us assume that the function $F(x,\dot{x})$ in Eq. (7.33) is linear, so that without loss of generality, we may write the equation in the form

$$m\ddot{x} + \alpha\dot{x} + ax = g \tag{7.36}$$

where α, a, and g may be functions of t. From Eqs. (7.34) and (7.35), we can write

$$x^{n+1} = p + \beta\Delta t^2 \ddot{x}^{n+1} \tag{7.37}$$

and

$$\dot{x}^{n+1} = \dot{p} + \gamma\Delta t \ddot{x}^{n+1} \tag{7.38}$$

where p and $\dot{p}$ are predictors defined as

$$p = x^n + \Delta t\dot{x}^n + \frac{1}{2}(1 - 2\beta)\Delta t^2 \ddot{x}^n \tag{7.39}$$

$$\dot{p} = \dot{x}^n + \Delta t(1 - \gamma)\ddot{x}^n \tag{7.40}$$

These can be readily calculated at $t = t_n$. Our goal is to find $x^{n+1}, \dot{x}^{n+1}$, and $\ddot{x}^{n+1}$ such that

$$m\ddot{x}^{n+1} + \alpha\dot{x}^{n+1} + ax^{n+1} = g^{n+1} \tag{7.41}$$

Substituting Eqs. (7.37) and (7.38) into Eq. (7.41), we obtain (Exercise 7.21)

$$\left(\frac{m}{\beta\Delta t^2} + \frac{\alpha\gamma}{\beta\Delta t} + a\right)\Delta p = g^{n+1} - \alpha\dot{p} - ap \tag{7.42}$$

where

$$\Delta p = x^{n+1} - p \tag{7.43}$$

Equation (7.42) is solved to obtain Δp, and from Eqs. (7.37) and (7.38), we obtain

$$x^{n+1} = p + \Delta p \tag{7.44}$$

$$\ddot{x}^{n+1} = \frac{x^{n+1} - p}{\beta \Delta t^2} \tag{7.45}$$

$$\dot{x}^{n+1} = \dot{p} + \gamma \Delta t \ddot{x}^{n+1} \tag{7.46}$$

Example 7.7

We address the solution of the simple linear equation

$$\frac{d^2 x}{dt^2} - x = 0 \quad x(0) = 1 \quad \dot{x}(0) = 0$$

The exact solution is given by $x(t) = \cosh(t)$.

We choose $\Delta t = 0.1$ and we will use Newmark's method to calculate $x(0.1)$, $\dot{x}(0.1)$ and $\ddot{x}(0.1)$. To this end, we also need to use $\ddot{x}(0) = 1$ and we will choose $\beta = 1/4$ and $\gamma = 1/2$. These choices will be discussed shortly.

From Eqs. (7.39) and (7.40) we have

$$p = x^0 + 0.1\dot{x}^0 + 0.0025\ddot{x}^0 = 1.0025$$

$$\dot{p} = \dot{x}^0 + 0.05\ddot{x}^0 = 0.05$$

Eq. (7.42) becomes

$$\left(\frac{m}{\beta \Delta t^2} + a\right) \Delta p = 399\Delta p = -ap = 1.0025$$

which yields $\Delta p = 0.002512531$ and from Eqs. (7.44)–(7.46) we get for t = 0.1 the results shown in Table 7.6.

Table 7.6 Results from Eqs. (7.44) – (7.46) for t = 0.1

	Calculated	Exact
x^1	1.005013	1.005005
$\dot{x}^1$	0.100251	0.100167
$\ddot{x}^1$	1.005013	1.005005

■

Notice that to calculate the predictors using Eqs. (7.39) and (7.40), we needed an estimate for the initial acceleration, which is not part of the problem data. This is normally obtained by replacing the initial values x_0 and $\dot{x}_0$ in the differential equation, i.e.,

$$\ddot{x}^0 = \frac{1}{m}\left[g - F\left(x^0, \dot{x}^0\right)\right] \tag{7.47}$$

If Eq. (7.33) is nonlinear, the Newmark method can be combined with the Newton-Raphson method to establish a nonlinear iteration at each time step. Following the same steps as before with the linearized operator,

$$m\ddot{x} + c\left(x^{(i)}, \dot{x}^{(i)}\right)\dot{x} + k\left(x^{(i)}, \dot{x}^{(i)}\right)x$$

where

$$c\left(x, \dot{x}\right) = \frac{\partial F}{\partial \dot{x}} \tag{7.48}$$

$$k\left(x, \dot{x}\right) = \frac{\partial F}{\partial x} \tag{7.49}$$

and $x^{(i)}, \dot{x}^{(i)}$ are the latest known approximations to x and $\dot{x}$. The following algorithm is obtained

1. Set $i = 0$ and construct the predictors $p_i, \dot{p}_i, \ddot{p}_i$ given by

$$p_0 = x^n + \Delta t \dot{x}^n + \frac{1}{2}\Delta t^2 (1 - 2\beta)\ddot{x}^n \tag{7.50}$$

$$\dot{p}_0 = \dot{x}^n + \Delta t (1 - \gamma)\ddot{x}^n \tag{7.51}$$

$$\ddot{p}_0 = 0 \tag{7.52}$$

2. Substitute into Eq. (7.33) to calculate the out-of-balance or residual vector ΔQ, given by

$$\Delta Q_i = g^{n+1} - F(p_i, \dot{p}_i) - m\ddot{p}_i \tag{7.53}$$

3. Find Δp_i from

$$\left(\frac{m}{\beta \Delta t^2} + \frac{\gamma}{\beta \Delta t} c(p_i, \dot{p}_i) + k(p_i, \dot{p}_i) \right) \Delta p_i = \Delta Q_i \tag{7.54}$$

The details on how we arrive at Eq. (7.54) are left to Exercise (7.22).

4. Correct the approximations, using Eqs. (7.37) and (7.38):

$$p_{i+1} = p_i + \Delta p_i \tag{7.55}$$

$$\ddot{p}_{i+1} = \frac{p_{i+1} - p_0}{\beta \Delta t^2} \tag{7.56}$$

$$\dot{p}_{i+1} = \dot{p}_0 + \gamma \Delta t \ddot{p}_{i+1} \tag{7.57}$$

5. Set $i = i + 1$ and go back to step 2 to calculate ΔQ_{i+1}.

6. Calculate $|\Delta Q_{i+1}|$, if $|\Delta Q_{i+1}| < \varepsilon$, where ε is a prescribed tolerance. We set

$$x^{n+1} = p_{i+1}$$

$$\dot{x}^{n+1} = \dot{p}_{i+1}$$

$$\ddot{x}^{n+1} = \ddot{p}_{i+1}$$

and the time step is completed. We can now move to the next time step.

If $|\Delta Q_{i+1}|$ is not small enough, we continue the iteration until convergence is achieved.

Example 7.8

Consider the nonlinear differential equation

$$\frac{d^2x}{dt^2} - \frac{1}{t}x\frac{dx}{dt} = 0 \quad x(1) = -1 \quad \dot{x}(1) = 2$$

which has the exact solution $x(t) = 2 \tan(ln\ t) - 1$. From the differential equation we obtain $\ddot{x}(1) = -2$. We let $\Delta t = 0.1$, $\gamma = 1/2$ and $\beta = 1/4$ as before, and $\varepsilon = 10^{-8}$. We want to calculate $x(1.1)$; for $i = 0$ we have

$$p_0 = -0.805 \quad \dot{p}_0 = 1.9 \quad \ddot{p}_0 = 0$$

The linearized functions are

$$c(x,\dot{x}) = -\frac{1}{t}x \quad k(x,\dot{x}) = -\frac{1}{t}\dot{x}$$

1. First iteration, $i = 0$

$$\Delta Q_0 = -F(p_0,\dot{p}_0) = \frac{1}{t_1}p_0\dot{p}_0 = -1.390454545$$

and

$$\left[\frac{m}{\beta\Delta t^2} + \frac{\gamma}{\beta\Delta t}\left(-\frac{1}{t_1}p_0\right) + \left(-\frac{1}{t_1}\dot{p}_0\right)\right]\Delta p_0 = \Delta Q_0$$

gives $\Delta p_0 = -0.003367459$.

Therefore we have from Eqs. (7.41)–(7.43)

$$p_1 = -0.808367459 \quad \dot{p}_1 = 1.832650815 \quad \ddot{p}_1 = -1.346983707$$

2. Second iteration, $i = 1$

$$\Delta Q_1 = \frac{1}{t_1} p_1 \dot{p}_1 - m\ddot{p}_1 = 0.000206177$$

$$\Delta p_1 = \Delta Q_1 \left/ \left(\frac{m}{\beta \Delta t^2} - \frac{\gamma}{t_1 \beta \Delta t} p_1 - \frac{1}{t_1} \dot{p}_1 \right) \right. = 0.000000499$$

and

$$p_2 = -0.80836696 \quad \dot{p}_2 = 1.832660799 \quad \ddot{p}_2 = -1.34678404$$

3. Third iteration, $i = 2$

$$\Delta Q_2 = \frac{1}{t_1} p_2 \dot{p}_2 - m\ddot{p}_2 = 5 \times 10^{-9} < \varepsilon$$

The approximate solution and the exact solution are shown in Table 7.7.

Table 7.7 Solution to Example 7.7

	Calculated	Exact
x_1	−0.808367	−0.808800
$\dot{x}_1$	1.832661	1.834799
$\ddot{x}_1$	−1.346784	−1.349078

∎

We are now ready to establish the algorithm for the systems of equations that arise from finite element discretizations of second-order hyperbolic equations of the form of Eq. (7.2). We will rewrite Eq. (7.2) in the more general form

$$m \frac{\partial^2 \phi}{\partial t^2} + N(\phi, \dot{\phi}) = f \tag{7.58}$$

where N may be a nonlinear function of ϕ and its first derivative. A discretization of Eq. (7.58) using the Galerkin method produces a system of ordinary differential equations of the form

$$\mathbf{M}\ddot{\phi} + \mathbf{N}(\phi, \dot{\phi}) = \mathbf{F} \tag{7.59}$$

where $\mathbf{M}$ is the mass matrix. The generalization of the algorithm for the one-degree-of-freedom Eq. (7.33) can be stated as follows:

1.) Predictors
    $i = 0$

$$p^0 = f_n + \Delta t \dot{f}_n + \frac{\Delta t^2}{2}(1 - 2\beta)\ddot{f}_n \tag{7.60}$$

$$\dot{p}^0 = \dot{f}_n + \Delta t(1 - \gamma)\ddot{f}_n \tag{7.61}$$

$$\ddot{p}^0 = 0 \tag{7.62}$$

2. Calculate the residual vector

$$\Delta \mathbf{Q}^i = \mathbf{F}(t_{n+1}) - \mathbf{N}(\mathbf{p}^i, \dot{\mathbf{p}}^i) - \mathbf{M}\ddot{\mathbf{p}}^i \tag{7.63}$$

3. Calculate the tangent matrices

$$\mathbf{C} = \frac{\partial \mathbf{N}}{\partial \dot{\mathbf{p}}} \qquad \left[c_{ij} \right] = \left[\frac{\partial N_i}{\partial \dot{p}_j} \right] \tag{7.64}$$

$$\mathbf{K} = \frac{\partial \mathbf{N}}{\partial \mathbf{p}} \qquad \left[k_{ij} \right] = \left[\frac{\partial N_i}{\partial p_j} \right] \qquad (7.65)$$

4. Solve for Δp_i from

$$\left[\frac{1}{\beta \Delta t^2} \mathbf{M} + \frac{\gamma}{\beta \Delta t} \mathbf{C}\left(\mathbf{p}^i, \dot{\mathbf{p}}^i \right) + \mathbf{K}\left(\mathbf{p}^i, \dot{\mathbf{p}}^i \right) \right] \Delta \mathbf{p}^i = \Delta \mathbf{Q}^i \qquad (7.66)$$

5.) Correctors

$$\mathbf{p}^{i+1} = \mathbf{p}^i + \Delta \mathbf{p}^i \qquad (7.67)$$

$$\ddot{\mathbf{p}}^{i+1} = \frac{1}{\beta \Delta t^2} \left(\mathbf{p}^{i+1} - \mathbf{p}^0 \right) \qquad (7.68)$$

$$\dot{\mathbf{p}}^{i+1} = \dot{\mathbf{p}}^0 + \gamma \Delta t \ddot{\mathbf{p}}^{i+1} \qquad (7.69)$$

6. Go back to step 2 and evaluate

$$\Delta \mathbf{Q}^{i+1} = \mathbf{F}\left(t_{n+1} \right) - \mathbf{N}\left(\mathbf{p}^{i+1}, \dot{\mathbf{p}}^{i+1} \right) - \mathbf{M} \ddot{\mathbf{p}}^{i+1} \qquad (7.70)$$

7. Check convergence

(a.) If $\left\| \Delta \mathbf{Q}^{i+1} \right\| < \varepsilon$, then
    set

$$\phi_{n+1} = \mathbf{p}^{i+1}$$

$$\dot{\phi}_{n+1} = \dot{\mathbf{p}}^{i+1}$$

and

$$\ddot{\phi}_{n+1} = \ddot{\mathbf{p}}^{i+1}$$

and start the next time step.

(b.) If $\left\|\Delta\mathbf{Q}^{i+1}\right\| > \varepsilon$ set $i = i + 1$ and continue with steps 3 through 7.

REMARKS

1. The convergence check, may be performed using a criterion different from that in step 7. However, the one in step 7 appears to be the most appropriate, since once $\Delta\mathbf{Q}$ becomes very small, $\Delta\mathbf{p}$ must also approach zero. We may also use different norms to measure $\Delta\mathbf{Q}$. The most commonly used are the Euclidean norm

$$\left\|\Delta\mathbf{Q}\right\|_E = \left(\sum_{i=1}^{n}(\Delta\mathbf{Q})_i^2\right)^{\frac{1}{2}} \tag{7.71}$$

and the maximum norm

$$\left\|\Delta\mathbf{Q}\right\|_M = \max_{i=1,n}\left|(\Delta\mathbf{Q})_i\right| \tag{7.72}$$

where n denotes the length of the vector, and $(\Delta\mathbf{Q})_i$ its components.

2. The matrices $\mathbf{C}$ and $\mathbf{K}$ in Eqs. (7.64) and (7.65) are evaluated as *infrequently* as possible because most of the computational cost is associated with forming these matrices and solving the system Eq. (7.66). In practice, the coefficient matrix in Eq. (7.66) is formed once during the first iteration ($i = 0$) and kept constant during subsequent iterations. These strategies are, however, problem dependent and must be determined by the user.

Example 7.9

Consider the one-dimensional Sine-Gordon equation

$$\frac{\partial^2 u}{\partial t^2} - \frac{\partial^2 u}{\partial x^2} - \sin u = 0 \qquad a < x < b \tag{7.73}$$

with boundary conditions

$$\frac{\partial u}{\partial x} = 0 \qquad x = a \qquad x = b \tag{7.74}$$

and initial conditions

$$u(x,0) = f(x) \tag{7.75}$$

$$\dot{u}(x,0) = g(x) \tag{7.76}$$

We assume that we have discretized the interval $a \le x \le b$ using linear elements of size h. It is not possible to find the element equations for a direct application of Galerkin's method to Eq. (7.73), without previous linearization. Hence we will introduce a *product approximation* to the nonlinear term. The idea is to approximate the sine function in the form

$$\sin u \cong \sum_i N_i \sin u_i \tag{7.77}$$

as opposed to the impractical form

$$\sin u \cong \sin\left(\sum_i N_i u_i\right) \tag{7.78}$$

These types of approximations have been analyzed by Christie et al. (1981). It has been shown that its use causes no loss of accuracy (see Exercise 7.23), while simplifying the formulation of nonlinear problems.
The element equations become

$$\frac{h}{6}\begin{bmatrix} 2 & 1 \\ 1 & 2 \end{bmatrix}\begin{bmatrix} \ddot{u}_1 \\ \ddot{u}_2 \end{bmatrix} + \frac{1}{h}\begin{bmatrix} 1 & -1 \\ -1 & 1 \end{bmatrix}\begin{bmatrix} u_1 \\ u_2 \end{bmatrix} - \frac{h}{6}\begin{bmatrix} 2\sin u_1 + \sin u_2 \\ \sin u_1 + 2\sin u_2 \end{bmatrix} = \begin{bmatrix} 0 \\ 0 \end{bmatrix}$$

The equations are independent of $\dot{u}$. Therefore the predictors take the form

$$\mathbf{p}^0 = \begin{bmatrix} p_1^0 \\ p_2^0 \end{bmatrix} = \begin{bmatrix} u_1^n + \dfrac{1}{2}(1-2\beta)\Delta t^2 \ddot{u}_1 \\ u_2^n + \dfrac{1}{2}(1-2\beta)\Delta t^2 \ddot{u}_2 \end{bmatrix}$$

$$\ddot{\mathbf{p}}^0 = \begin{bmatrix} 0 \\ 0 \end{bmatrix}$$

The element tangent matrix $\mathbf{C} = \mathbf{0}$. The matrix $\mathbf{K}$ is given by

$$\mathbf{K} = \begin{bmatrix} \dfrac{1}{h} - \dfrac{h}{3}\cos u_1 & -\dfrac{1}{h} - \dfrac{h}{6}\cos u_2 \\ -\dfrac{1}{h} - \dfrac{h}{6}\cos u_1 & \dfrac{1}{h} - \dfrac{h}{3}\cos u_2 \end{bmatrix}$$

The element out-of-balance vector is

$$\Delta \mathbf{Q}^i = \frac{h}{6}\begin{bmatrix} 2\sin p_1^i + \sin p_2^i \\ \sin p_1^i + 2\sin p_2^i \end{bmatrix} - \frac{1}{h}\begin{bmatrix} p_1^i - p_2^i \\ p_2^i - p_1^i \end{bmatrix} - \frac{h}{6}\begin{bmatrix} 2\ddot{p}_1^i + \ddot{p}_2^i \\ \ddot{p}_1^i + 2\ddot{p}_2^i \end{bmatrix}$$

and the element equations for $\Delta \mathbf{p}^i$ are

$$\left\{ \frac{h}{6\beta\Delta t^2}\begin{bmatrix} 2 & 1 \\ 1 & 2 \end{bmatrix} + \begin{bmatrix} \dfrac{1}{h} - \dfrac{h}{3}\cos u_1 & -\dfrac{1}{h} - \dfrac{h}{6}\cos u_2 \\ -\dfrac{1}{h} - \dfrac{h}{6}\cos u_1 & \dfrac{1}{h} - \dfrac{h}{3}\cos u_2 \end{bmatrix} \right\}\begin{bmatrix} \Delta p_1^i \\ \Delta p_2^i \end{bmatrix} = \Delta \mathbf{Q}^i$$

Notice that because the first derivative does not appear in the equation, the parameter γ is also unnecessary in this case. The solution will proceed with the assembled global equations.

Solutions to Eqs. (7.73)–(7.76) under a variety of different initial conditions are given by Argyris and Haase (1987). Results using 400

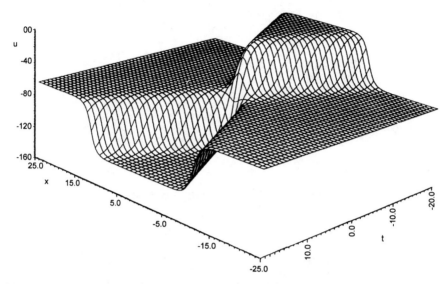

Figure 7.6 Solution of the Sine-Gordon equation using linear elements (Reprinted from *Computer Methods in Applied Mechanics and Engineering*, Vol. 61, J. Argyris and M. Haase, An Engineer's Guide to Solution Phenomena, pp. 71-122, 1987, with permission from Elsevier Science).

linear elements in the interval $-25 \leq x \leq 25$, and for $t_0 = -20 \leq t \leq 20$, for $f(x)$ and $g(x)$ chosen such that the exact solution is

$$u(x,t) = 4 \arctan\left[\frac{1}{2}\cosh(1.1547x)\,\mathrm{cosech}(0.57735t)\right]$$

are shown in Fig. 7.6.

Argyris et al. (1991) applied the Newmark algorithm to the solution of the two-dimensional Sine-Gordon equation. ∎

REMARKS

1. It should now be clear that as the equations become more complicated, nonlinear and time dependent, we are also faced with choices and a variety of possibilities on how to formulate the finite element approximations. Here we chose to use the product approximation to simplify the calculations. In Chapter 10 we will see how other modifications can be made to eliminate the nonlinear iteration when solving the time dependent Navier-Stokes equations.

2. In general, the standard Galerkin approximation of a nonlinear function f(u) of the dependent variable u is given by

$$f(u) \cong f\left(\sum_n N_i u_i\right) \tag{7.79}$$

That is, the function u is approximated using shape functions and the function evaluated at the approximate value. The product approximation, on the other hand, uses the form

$$f(u) \cong \sum_i N_i f(u_i) \tag{7.80}$$

Further application and analysis are found in the work by Christie et al. (1981).

3. Unconditional stability requires that $\gamma \geq 1/2$, the value $\gamma = 1/2$ results in a scheme second order accurate in time. For $\gamma > 1/2$, we only have first-order accuracy in general. For the numerical scheme to be unconditionally stable (in the linearized sense), β must be chosen such that $\beta \geq \gamma/2$. If $\beta < \gamma/2$, stability is conditional. Furthermore, the value

$$\beta = \frac{1}{4}\left(\gamma + \frac{1}{2}\right)^2 \tag{7.81}$$

maximizes the high frequency numerical dissipation, which can be desirable in many problems. These are the reasons why the values $\gamma = 1/2$ and $\beta = 1/4$ were chosen in the previous examples. More detailed discussions and results related to the accuracy and stability of Newmark algorithms are found in the works by Newmark (1959) and Hughes et al. (1979).

7.4.2 Generalized Newmark Method for Parabolic Equations

Many of the equations of interest, such as convective transport equations and the Navier-Stokes equations, are parabolic, and the Newmark method must be modified when applied to them. Let us write the general equation in the form

$$\frac{\partial \phi}{\partial t} + N(\phi) = f \tag{7.82}$$

where ϕ and $\dot{\phi}$ will be looked at as generalized velocity and acceleration, respectively; generalized displacements do not appear in the formulation. For a one-degree-of-freedom system, the Newmark method takes the following form (see Exercises 7.25 and 7.26).

1. Predictors, $i = 0$

$$p_0 = \phi_n + (1 - \gamma) \Delta t \, \dot{\phi}_n \tag{7.83}$$

$$\dot{p}_0 = 0 \tag{7.84}$$

2. Out-of-balance vector

$$\Delta Q^i = f_{n+1} - N(p_i) - \dot{p}_i \tag{7.85}$$

3. Linearized operator

$$A(p_i) = \frac{1}{\gamma \Delta t} + k(p_i) \tag{7.86}$$

where

$$k(\phi) = \frac{d\,f}{d\,\phi}$$

4. $\qquad\qquad \Delta p_i = \Delta Q_i / A(p_i) \tag{7.87}$

5. Correctors

$$p_{i+1} = p_i + \Delta p_i \tag{7.88}$$

$$\dot{p}_{i+1} = (p_{i+1} - p_0) / \gamma \Delta t \tag{7.89}$$

It is a simple matter (Exercise 7.28) to show that if Eq. (7.82) is linear, the algorithm is identical to the θ method. Therefore it provides a generalization of the θ method to nonlinear equations. We also note that in this case, the parameter β does not enter the calculations because the generalized displacements do not appear in the equation.

For the general nonlinear systems of equations arising in finite element discretizations of the form

$$\mathbf{C}\dot{\boldsymbol{\phi}} + \mathbf{N}(\boldsymbol{\phi}) = \mathbf{F} \tag{7.90}$$

the algorithm can be stated as follows;

1. Set $i = 0$, predictors

$$\mathbf{p}^0 = \boldsymbol{\phi}_n + (1-\gamma)\Delta t\,\dot{\boldsymbol{\phi}}_n \tag{7.91}$$

$$\dot{\mathbf{p}}^0 = 0 \tag{7.92}$$

2. Out-of-balance-vector

$$\Delta\mathbf{Q}^i = \mathbf{F}_{n+1} - \mathbf{N}(\mathbf{p}^i) - \mathbf{C}\dot{\mathbf{p}}^i \tag{7.93}$$

3. Tangent operator

$$\mathbf{K} = \frac{\partial\mathbf{N}}{\partial\mathbf{p}} \qquad [k_{ij}] = \left[\frac{\partial N_i}{\partial p_j}\right] \tag{7.94}$$

4. Find the increment $\Delta\mathbf{p}^i$ from

$$\left[\frac{1}{\gamma\,\Delta t}\mathbf{C} + \mathbf{K}(\mathbf{p}^i)\right]\Delta\mathbf{p}^i = \Delta\mathbf{Q}^i \tag{7.95}$$

5. Corrector

$$\mathbf{p}^{i+1} = \mathbf{p}^i + \Delta\mathbf{p}^i$$

$$\dot{\mathbf{p}}^{i+1} = \frac{1}{\gamma\,\Delta t}\left(\mathbf{p}^{i+1} - \mathbf{p}^0\right)$$

(7.96)

Steps 6 and 7 remain the same as in the case of second order hyperbolic equations described in Section 7.4.1. To illustrate the use of the algorithm let us apply it to the solution of the nonlinear Burgers equation as we did before in Example 7.6 using the second-order Runge-Kutta method.

Example 7.10

We look at the solution of Burgers equation:

$$\frac{\partial u}{\partial t} + u\frac{\partial u}{\partial x} = \varepsilon\frac{\partial^2 u}{\partial x^2}$$

(7.97)

$$u(x,0) = \begin{cases} 1 & \text{if } x \leq \dfrac{1}{4} \\[2mm] 0 & \text{if } x > \dfrac{1}{4} \end{cases} \qquad u(0,t) = u(1,t) = 0 \qquad \varepsilon = 1$$

Using four linear elements, we obtain the system of nonlinear equations

$$\mathbf{C}\dot{\mathbf{u}} + \mathbf{N}(\mathbf{u}) = 0$$

where

$$\mathbf{C} = \frac{1}{24}\begin{bmatrix} 4 & 1 & 0 \\ 1 & 4 & 1 \\ 0 & 1 & 4 \end{bmatrix}$$

$$\mathbf{N(u)} = \begin{bmatrix} 8u_2 - 4u_3 - \dfrac{1}{6}u_2 + \dfrac{1}{6}u_2u_3 + \dfrac{1}{6}u_3^2 - \dfrac{1}{6} \\[2mm] -4u_2 + 8u_3 - 4u_4 - \dfrac{1}{6}u_2^2 - \dfrac{1}{6}u_2u_3 + \dfrac{1}{6}u_3u_4 + \dfrac{1}{6}u_4^2 \\[2mm] -4u_3 + 8u_4 - \dfrac{1}{6}u_3^2 - \dfrac{1}{6}u_3u_4 \end{bmatrix}$$

and

$$\mathbf{u} = \begin{bmatrix} u_2 \\ u_3 \\ u_4 \end{bmatrix}$$

Therefore

$$\Delta Q^i = \begin{bmatrix} 8p_2^i - 4p_3^i - \dfrac{1}{6}p_2^i + \dfrac{1}{6}p_2^i p_3^i + \dfrac{1}{6}\left(p_3^i\right)^2 - \dfrac{1}{6} \\[2mm] -4p_2^i + 8p_3^i - 4p_4^i - \dfrac{1}{6}\left(p_2^i\right)^2 - \dfrac{1}{6}p_2^i p_3^i + \dfrac{1}{6}p_3^i p_4^i + \dfrac{1}{6}\left(p_4^i\right)^2 \\[2mm] -4p_3^i + 8p_4^i - \dfrac{1}{6}\left(p_3^i\right)^2 - \dfrac{1}{6}p_3^i p_4^i \end{bmatrix}$$

$$-\frac{1}{24}\begin{bmatrix} 4\dot{p}_2^i + \dot{p}_3^i \\ \dot{p}_2^i + 4\dot{p}_3^i + \dot{p}_4^i \\ \dot{p}_3^i + 4\dot{p}_4^i \end{bmatrix}$$

The tangent matrix $\mathbf{K}$ is

$$\mathbf{K} = \begin{bmatrix} 8 - \dfrac{1}{6}u_2 + \dfrac{1}{6}u_3 & -4 + \dfrac{1}{6}u_2 + \dfrac{1}{3}u_3 & 0 \\[2ex] -4 - \dfrac{1}{3}u_2 - \dfrac{1}{6}u_3 & 8 - \dfrac{1}{6}u_2 + \dfrac{1}{6}u_4 & -4 + \dfrac{1}{6}u_3 + \dfrac{1}{3}u_4 \\[2ex] 0 & -4 - \dfrac{1}{3}u_3 - \dfrac{1}{6}u_4 & 8 - \dfrac{1}{6}u_3 \end{bmatrix}$$

and the increments are obtained from

$$\left\{ \frac{1}{\gamma \Delta t} \begin{bmatrix} 4 & 1 & 0 \\ 1 & 4 & 1 \\ 0 & 1 & 4 \end{bmatrix} \right.$$

$$\left. + \begin{bmatrix} 8 - \dfrac{1}{3}p_2^i + \dfrac{1}{6}p_3^i & -4 + \dfrac{1}{6}p_2^i + \dfrac{1}{3}p_3^i & 0 \\[2ex] -4 - \dfrac{1}{3}p_2^i - \dfrac{1}{6}p_3^i & 8 - \dfrac{1}{6}p_2^i + \dfrac{1}{6}p_4^i & -4 + \dfrac{1}{6}p_3^i + \dfrac{1}{3}p_4^i \\[2ex] 0 & -4 - \dfrac{1}{3}p_3^i - \dfrac{1}{6}p_4^i & 8 - \dfrac{1}{6}p_3^i \end{bmatrix} \right\} \Delta \mathbf{p}^i = \Delta \mathbf{F}^i$$

Solutions obtained with $\gamma = 1/2$ and $\Delta t = 0.005$ are shown in Fig. 7.7. ■

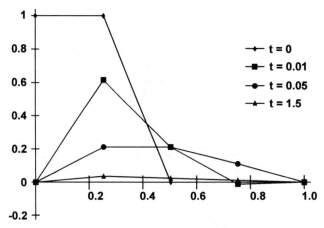

Figure 7.7 Solution of Eq. (7.97) using the generalized Newmark method and four linear elements.

We look at one more example that will serve as an introduction to the next chapter. This is again solution of Burgers equation, Eq. (7.97), but this time with boundary conditions u(0,t) = 1 and u(1,t) = 0. The initial condition is u(x,0) = 0. This problem has an analytical solution at steady state that is easy to compute. A numerical steady-state solution calculated using linear elements (Exercise 7.29) with $\Delta x = \Delta t = \varepsilon = 0.1$ is shown in Table 7.8 (at every other node for conciseness). The results show a solution essentially accurate to two significant digits. However, if we change ε to 0.01, the situation changes dramatically, as shown in Table 7.9, where we see that a highly oscillatory numerical approximation has been obtained. This is not an error – this is the solution one gets with any time-dependent or direct steady state solver (Exercise 7.30) that does not include a special treatment for convection–dominated problems, which is the subject of Chapter 8. We illustrate this further in Fig. 7.8, where we show a solution using $\Delta x = 0.5$, $\Delta t = 0.05$, and $\varepsilon = 0.0025$ at time t = 1.0 when the wave front is halfway into the domain and at time t = 3, which is essentially steady state. The oscillation problem occurs early in the solution and, as opposed to what we observed before in the diffusion equation, does not disappear with time but becomes amplified. The origin of these oscillations is very different from that of the oscillations discussed above in the context of the diffusion equation – it is related to the relative importance between the convective and diffusion terms in the equation. Because one of our main interests in this book is the solution of convective flows, we will devote the next two chapters to dealing with this situation.

Table 7.8 Solution of Burgers' equation with $\varepsilon = \Delta t = h = 0.1$;
$u(x,0 = 0)$; $u(0,t) = 1$; $u(1,t) = 0$

x	Finite element solution	Analytical solution
0.0	1.000	1.000
0.2	1.000	0.999
0.4	0.997	0.995
0.6	0.970	0.964
0.8	0.770	0.762
0.9	0.464	0.462
1.0	0.000	0.000

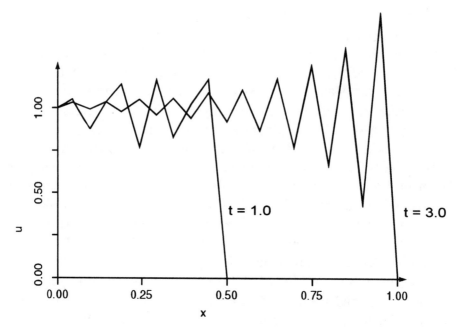

Figure 7.8 Time dependent solution to Burgers equation at t = 1.0 and t = 3.0 for $\Delta x = \Delta t = 0.05$ and $\varepsilon = 0.0025$.

Table 7.9 Solution of Burgers equation at time t = 2.9, with $\varepsilon = 0.01$ and $h = \Delta t = 0.10$; $u(x,0) = 1$; $u(0,t) = 1$

x	Finite element solution	Analytical solution
0.0	1.000	1.000
0.1	1.032	1.000
0.2	0.985	1.000
0.3	1.057	1.000
0.4	0.945	1.000
0.5	1.101	1.000
0.6	0.854	1.000
0.7	1.213	1.000
0.8	0.648	1.000
0.9	1.457	1.000
1.0	0.000	0.000

7.5 CLOSURE

We have established the basic algorithms most commonly used in finite element simulations of second-order hyperbolic and parabolic equations that arise in fluid flow, heat, and mass transport.

The reader should be aware that this subject has been the object of intensive work by many researchers over the years; there is an enormous amount of literature and methods that has been developed. We have looked in some detail at three of these methods: the θ method, the Runge-Kutta methods, and the generalized Newmark algorithms. The first two apply only to parabolic equations, while the third is also applicable to (in fact, it was developed for) second-order hyperbolic equations.

With this chapter we also conclude the presentation of the basic material necessary for any finite element approximation of time-dependent, nonlinear, second-order partial differential equation. In the following chapters we will develop more specialized tools to treat the Navier-Stokes equations including convective heat and mass transfer.

REFERENCES

Argyris, J. H. and Haase, M. (1987) "An Engineer's Guide to Solution Phenomena: Application of the Finite Element Method," *Comput. Methods Appl. Mech. Eng.*, Vol. 61, pp. 71–122.

Argyris, J. H., Haase, M. and Heinrich, J. C. (1991) "Finite Element Approximations to Two-Dimensional Sine-Gordon Solution," *Comput. Methods Appl. Mech. Eng.*, Vol. 86, pp. 1–26.

Bathe, K. J. (1982) *Finite Element Procedures in Engineering Analysis*. Englewood Cliffs, N. J.: Prentice-Hall.

Bickford, W. B. (1990) *A First Course in the Finite Element Method*. Homewood, Ill.: Irwin.

Brueckner, F. P. and Heinrich, J. C. (1991) "Petrov-Galerkin Finite Element Model for Compressible Flow," *Int. J. Numer. Methods Eng.*, Vol. 32, pp. 255–274.

Carslaw, H. S. and Jaeger, J. C. (1959) *Conduction of Heat in Solids*. Oxford: Clarendon Press.

Christie, I., Griffiths, D. F., Mitchell, A. R. and Sanz-Cerna, J. M. (1981) "Product Aproximation for Nonlinear Problems in the Finite Element Method," *IMA J. Numer. Anal.*, Vol. 1, pp. 253–266.

Fletcher, C. A. J. (1984) *Computational Galerkin Methods.* N. Y.: Springer-Verlag.

Gray, W. G. and van Genuchten, M. Th. (1978) "Economical Alternative to Gaussian Quadrature over Isoparametric Quadrilaterals"" *Int. J. Numer. Methods Enr.*, Vol. , pp. 1478–1484.

Gresho, P. M., Lee, R. L. and Sani, R. L. (1978) "Advection Dominated Flows with Emphasis on the Consequences of Mass Lumping," pp. 335–350 in *Finite Elements in Fluids*, Vol. 3 (Gallagher et al. Editors).

Hoffman, J. D. (1992) *Numerical Methods for Engineers and Scientists,* New York.: McGraw-Hill.

Hughes, T. J. R. (1987) *The Finite Element Method Linear Static and Dynamic Finite Element Analysis.* Englewood Cliffs, N. J.: Prentice-Hall.

Hughes, T. J. R., Pister, K. S. and Taylor, R. L. (1979) "Implicit Explicit Finite Elements in Nonlinear Transient Analysis," *Comput. Methods Appl, Mech. Eng.*, Vols. 17 and 18, pp. 159–182.

Isaacson, E. and Keller, H. B. (1966) *Analysis of Numerical Methods.* New York: John Wiley and Sons.

Lambert, J. D. (1973) *Computational Methods in Ordinary Differential Equations.* London: John Wiley and Sons.

Newmark, N. M. (1959) "A Method of Computation for Structural Dynamics," *Trans. ASCE*, EM 3, Vol. 85, pp. 67–94.

Oden, J. T. and Wellford, Jr., L. C. (1972) "Analysis of Viscous Flow by the Finite Element Method," *AIAA J.*, Vol. 10, pp. 1590–1599.

Richtmeyer, R. D. and Morton, K. W. (1963) *Difference Methods for Initial Value Problems,* 2nd ed. New York: Wiley Interscience.

Wood, W. L. and Lewis, R. W. (1975) "A Comparison of Time Marching Schemes for the Transient Heat Conduction Equation," *Int. J. Numer. Metods Eng.*, Vol. 9, pp. 679–689.

Yakowitz, S. and Szidarovszky, F. (1989) *An Introduction to Numerical Computations,* 2nd ed. New York: Macmillan.

Yu, C. C. and Heinrich, J. C. (1986) "Petrov-Galerkin Methods for the Time-Dependent Convective Transport Equation," *Int. J. Numer. Methods Eng.*, Vol. 23, pp. 883–901.

Zienkiewicz, O. C. (1977) *The Finite Element Method*, 3rd ed. London: McGraw-Hill.

EXERCISES

7.1   Use the θ method to find the solution to the equation

$$\frac{\partial u}{\partial t} = \frac{1}{2}\frac{\partial^2 u}{\partial x^2} \qquad 0 < x < 1 \quad t > 0$$

$$u(x,0) = 100$$
$$u(0,t) = u(1,t) = 0$$

at time t = 0.5. The exact solution is given by

$$u*(x,t) = \sum_{n=1,3,5,\ldots}^{\infty}(400/n\pi)(\sin n\pi x)\exp\left(-n^2\pi^2 t/4\right)$$

Experiment with different number of elements and time steps to obtain a solution such that $\left|u_i(0.5) - u*(x_i,0.5)\right| < 0.01$ at all nodes x_i.

7.2   Find the stability limit for the problem of Example 7.2 using four linear elements.

7.3   Repeat Exercise 7.2 using two quadratic elements.

7.4   The stability limit for the Euler algorithm applied to Eq. (7.13) in conjunction with a central difference approximation to the second derivative with respect to x is given by (Richtmeyer and Morton, 1963)

$$\Delta t \le \frac{(\Delta x)^2}{2\alpha}$$

Compare with the stability limit obtained for the Euler-Galerkin method using a lumped mass matrix.

7.5   Define the θ method for the first-order hyperbolic equation

$$\frac{\partial u}{\partial t} = a \frac{\partial u}{\partial x} \qquad u(0,t) = u_0 \qquad u(x,0) = f(x)$$

where a is a constant. Determine the stability limit of the Euler-Galerkin method with lumped mass matrix and linear elements.

7.6   Check that if $\Delta t = 1/5\alpha$ or $\Delta t = 1/4\alpha$ in Example 7.2, the solution becomes unbounded as u increases.

7.7   Determine the stability limits in Example 7.2 if the θ method is used with $\theta = 1/4$, using a consistent mass matrix and lumped mass matrices.

7.8   Repeat Exercise 7.7 using two quadratic elements.

7.9   Find the consistent mass matrix for a bilinear rectangular element.

7.10   Find the consistent mass matrix for the biquadratic Lagrangian element.

7.11   Show that the lumped mass matrix for the biquadratic element is obtained when the integrals are evaluated using Simpson's rule.

7.12   Starting from Eq. (7.13), use the data in Example 7.5, assemble three linear elements and apply the boundary conditions to deduce Eq. (7.20).

7.13   Show that Eq. (7.21) converges to the correct steady state solution $T_2 = 1/3$ and $T_3 = 2/3$ as t→∞.

7.14   Consider a time-space finite element as defined in the figure shown below. Construct the bilinear time-space shape functions and find the element equations for the diffusion equation, Eq. (7.13).

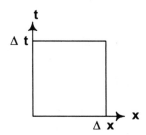

7.15 Show that in Problem 7.14 the resulting algorithm corresponds to the θ method with $\theta = 2/3$. For further variations on space-time elements, consult Zienkiewicz (1977).

7.16 Derive the nonlinear finite element equation system in Example 7.6. Remember that $\dot{u}_1 = \dot{u}_2 = 0$, $u_1 = 1$, and $u_5 = 0$.

7.17 Use the second-order Runge-Kutta method to solve Eq. (7.29) for $-1 < x < 1$ with boundary conditions $u(-1,t) = 1.0$, $u(1,t) = 0$, and initial condition

$$u(x,0) = \begin{cases} 1 & -1 \le x \le 0 \\ 0 & 0 < x \le 1 \end{cases}$$

with $\varepsilon = 0.1$. Use 10 and 20 linear elements, and calculate up to $t = 0.5$.

7.18 Repeat Exercise 7.17 using 5 and 10 quadratic elements.

7.19 Use the second- and fourth-order Runge-Kutta methods to solve the one-dimensional linearized compressible flow equations setting $c = \rho_0 = 1$

$$\frac{\partial \rho}{\partial t} + \rho_0 \frac{\partial u}{\partial x} = 0$$

$$\frac{\partial u}{\partial t} + \frac{c^2}{\rho_0} \frac{\partial \rho}{\partial x} = 0$$

over $0 < x < 1$, with initial condition

$$\left.\begin{array}{l} \rho(x,0) = 0 \\ u(x,0) = 1 \end{array}\right\} 0 < x < 1$$

and boundary conditions

$$\left.\begin{array}{l} \rho(0,t) = 0 \\ u(1,t) = 0 \end{array}\right\} t > 0$$

7.20 Use the second- and fourth-order Runge-Kutta method to solve the pendulum equation (von Karman and Biot, 1940)

$$I\frac{d^2\theta}{dt^2} = -mgL\sin\theta$$

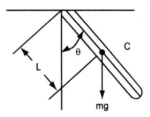

where m = 1 kg is the mass, g = 10m/s^2 is gravity, C is the center of gravity, and I the moment of inertia. Use I = 5 kg m² and L = 0.5 m.

7.21 Derive Eq. (7.42) from Eqs. (7.37), (7.38), and (7.41).

7.22 Derive Eq. (7.54). First define

$$Q(x,\dot{x},\ddot{x}) = m\ddot{x} + F(x,\dot{x}) - G = 0$$

Linearize and use Eqs. (7.37) and (7.38). Remember that Eq. (7.42) requires $\ddot{p}_0 = 0$. Hint: Expand F in Taylor series up to linear terms.

7.23 Consider the differential equation

$$\frac{d^2u}{dx^2} = e^u \qquad u(0) = u(1) = 0$$

with solution u(x) = −*ln* 2 + 2 *ln*{1.33sec[0.665(x−0.5)]}. Solve using the standard Galerkin approximation,

$$e^u \cong \exp\left(\sum_i N_i u_i\right)$$

and the product approximation,

$$e^u = \sum_i N_i e^{u_i}$$

Using 4 and 10 linear elements, compare the errors in the approximations at x = 0.5 (see Christie et al., 1981).

7.24   Repeat Exercise 7.23 usng 2,3,4, and 5 quadratic elements. The behavior of the error will differ from that for linear elements.

7.25   For the one-degree-of-freedom linear equation (see Eq. (7.36)),

$$m\frac{\partial^2 \phi}{\partial t^2} + \alpha\frac{\partial \phi}{\partial t} + k\phi = f(t)$$

show that the Newmark method can be formulated in terms of the generalized velocity $\dot{\phi}$ in the form

$$A\,\Delta\dot{p} = \Delta Q$$

where the predictors and ΔQ remain the same as before A is given by

$$A = \frac{m}{\gamma\,\Delta t} + \alpha\frac{\Delta t \beta}{\gamma}k$$

and the correctors are now

$$\dot{\phi}_{n+1} = \dot{p} + \Delta \dot{p}$$

$$\ddot{\phi}_{n+1} = \left(\dot{\phi}_{n+1} - \dot{p} \right) / \gamma \Delta t$$

$$\phi_{n+1} = p + \beta \Delta t^2 \ddot{\phi}_{n+1}$$

7.26   Show that if ϕ does not appear in the equation, the algorithm in Exercise 7.25 is identical to that defined in Section 7.4.2 for Eq. (7.82) with $\dot{\phi}$ and ϕ replaced by $\ddot{\phi}$ and $\dot{\phi}$, respectively.

7.27   Generalize the above algorithm to a nonlinear equation of the form of Eq. (7.33) and show that if ϕ does not enter the formulation, it reduces to the algorithm of Section 7.4.2.

7.28   Show that the Newmark method as defined by Eqs. (7.83)–(7.89) is identical to the θ method if Eq. (7.82) is linear, i.e., of the form

$$\dot{u} + \alpha u = f$$

7.29   Obtain the results of Table 7.8 using the Newmark method.

7.30   Reproduce the results of Table 7.9. Use any finite element scheme that has been discussed so far.

EIGHT

STEADY STATE CONVECTIVE TRANSPORT

8.1 OVERVIEW

Most of the discussions in the previous chapters have been restricted to diffusion equations in which the differential operator is self-adjoint and for which a rich mathematical theory is available. We now turn our attention to transport by convection. This is an important transport mechanism in situations where the physical system under consideration has a fluid component, and in many cases the dominant one.

The presence of convection introduces specific difficulties in the numerical solution of the governing differential equations, which will be addressed in this chapter. The difficulties stem from the directional, nonsymmetric nature of the convection terms. Mathematically, the differential operators are non-self-adjoint when described using an Eulerian frame of reference, which corresponds to a stationary observer watching a fluid flow by in some direction.

First, we will explain the numerical problems that arise from the presence of the convective terms using a simple, one-dimensional convection-diffusion equation. A solution will be proposed and then extended and developed to treat equations in two and three space dimensions as well as the Navier-Stokes and other nonlinear equations. Here we will discuss only the use of linear, bilinear, and trilinear elements. The interested reader should consult the references given here for techniques involving other types of elements.

8.2 ONE-DIMENSIONAL CONVECTION-DIFFUSION

Consider a fluid moving with a constant velocity u in the x-direction, which carries a scalar species ϕ that obeys Fick's diffusion law. The flux of ϕ due to convection is given by $u\phi$, and the diffusive flux is $-Dd\phi/dx$, where D is the diffusion coefficient. If these are the only transport mechanisms, in the absence of sources or sinks, the equation of conservation of the species ϕ becomes

$$-\frac{d}{dx}\left(D\frac{d\phi}{dx}\right) + u\frac{d\phi}{dx} = 0 \qquad (8.1)$$

Equation (8.1) provides the most basic form to express the relation between transport by convection and diffusion and serves as a model equation for a great variety of situations. We will assume that it is defined over an interval $0 < x < L$ with appropriate boundary conditions.

REMARKS

1. If $u = 0$ in Eq. (8.1), the equation is a purely elliptic boundary value problem; if $D = 0$, Eq. (8.1) represents a first-order initial value problem. As long as $D \neq 0$, the character of the equation will remain that of a boundary value problem. However, if $u \gg D$, it will behave mostly as a hyperbolic equation. This competition between the elliptic and hyperbolic character of the equation is a source of problems in the numerical solution.

2. When u dominates D, usually there are still localized regions where the second-order term is important, such as sharp fronts, boundary layers, or internal boundary layers. It becomes necessary to watch for changes in the dominant scales in different parts of the domain, and we need to keep track of the behavior of the numerical algorithm locally.

The weighted residuals method applied to Eq. (8.1) yields the weak formulation

$$\int_0^L \left(D\frac{dw}{dx}\frac{d\phi}{dx} + wu\frac{d\phi}{dx}\right)dx + w\left(-D\frac{d\phi}{dx}\right)\Bigg|_0^L = 0 \qquad (8.2)$$

where w denotes the weighting function. We now discretize the interval $0 < x \leq L$ using linear elements of size $\Delta x = h$ and apply the Galerkin method to Eq. (8.2). The resulting element stiffness matrix is (Exercise 8.1)

$$
\mathbf{k}_e = \begin{bmatrix} \dfrac{D}{h} - \dfrac{u}{2} & -\dfrac{D}{h} + \dfrac{u}{2} \\[2mm] -\dfrac{D}{h} - \dfrac{u}{2} & \dfrac{D}{h} + \dfrac{u}{2} \end{bmatrix} \tag{8.3}
$$

Hence the assembled stiffness matrix has the form

$$
\mathbf{K} = \begin{bmatrix} \dfrac{D}{h} - \dfrac{u}{2} & -\dfrac{D}{h} + \dfrac{u}{2} & 0 & 0 & & \\[2mm] -\dfrac{D}{h} - \dfrac{u}{2} & \dfrac{2D}{h} & -\dfrac{D}{h} + \dfrac{u}{2} & 0 & & \\[2mm] 0 & -\dfrac{D}{h} - \dfrac{u}{2} & \dfrac{2D}{h} & -\dfrac{D}{h} + \dfrac{u}{2} & & \\[2mm] & & & & & \\[2mm] & & & -\dfrac{D}{h} - \dfrac{u}{2} & \dfrac{2D}{h} & -\dfrac{D}{h} + \dfrac{u}{2} \\[2mm] & & & 0 & -\dfrac{D}{h} - \dfrac{u}{2} & \dfrac{D}{h} + \dfrac{u}{2} \end{bmatrix}
$$

$$\tag{8.4}$$

The matrix in Eq. (8.4) differs from that obtained in Section 3.6.2, Eq. (3.72) for pure conduction, in two very important ways.

(a) It is unsymmetric; the lack of symmetry stems from the contributions of the convective term.

(b) As convection dominates and $u/2$ becomes larger than D/h, the matrix is no longer diagonally dominant. As a consequence, many iterative solvers cannot be used in these systems of linear equations.

From Eq. (8.4), it follows that the difference equation for an interior node x has the form

$$\left(-\frac{D}{h} - \frac{u}{2}\right)\phi_{i-1} + \frac{2D}{h}\phi_i + \left(-\frac{D}{h} + \frac{u}{2}\right)\phi_{i+1} = 0 \qquad (8.5)$$

We will introduce the cell Péclet number γ, which is defined as

$$\gamma = \frac{u\,h}{D} \qquad (8.6)$$

and rewrite Eq. (8.5) in the form

$$\left(-1 - \gamma/2\right)\phi_{i-1} + 2\phi_i + \left(-1 + \gamma/2\right)\phi_{i+1} = 0 \qquad (8.7)$$

The solution to this difference equation can be easily obtained (Exercise 8.2) and is given as

$$\phi_i = A + B\left(\frac{2+\gamma}{2-\gamma}\right)^i \qquad (8.8)$$

where A and B are constants that depend on the boundary conditions. Comparing Eq. (8.8) to the exact solution for Eq. (8.1), given by

$$\phi(x) = c_1 + c_2\,e^{\frac{u}{D}x} \qquad (8.9)$$

we observe that Eq. (8.8) correctly represents the exponential term in Eq. (8.9) by means of the (1,1) Padé rational approximation (Varga, 1962), i.e.,

$$e^{\frac{u}{D}x_i} = e^{\gamma i} \cong \left(\frac{2+\gamma}{2-\gamma}\right)^i + O(h^3) \qquad (8.10)$$

with $x_i = ih$. However, if $\gamma > 2$ in Eqs. (8.8) and (8.10), the denominator becomes negative and the numerical solution will be oscillatory. This is clearly illustrated in Fig. 8.1, where the solution to Eq.(8.1), using 10

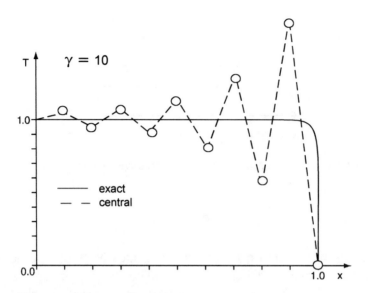

Figure 8.1 Solution to Eq. (8.1) with boundary conditions $\phi(0) = 1$, $\phi(1) = 0$, and $\gamma = 10$. Solid line is the analytical solution; dashed line the Galerkin finite element solution.

linear elements and the boundary conditions $\phi(0) = 1$, $\phi(1) = 0$ is shown for $\gamma = 10$.

REMARKS

1. An error analysis of the above numerical solution yields

$$\left\| \phi - \phi^h \right\|_\infty = O\left(h^2\right) \tag{8.11}$$

However, it is easily verified that as $|\gamma|$ becomes large, Eq. (8.10) exhibits an incorrect asymptotic behavior. To illustrate this, rewrite Eq. (8.7) as

$$\frac{1}{\gamma}\left(-\phi_{i-1} + 2\phi_i - \phi_{i+1}\right) + \frac{1}{2}\left(\phi_{i+1} - \phi_{i-1}\right) = 0$$

and take the limit as $\gamma \to \infty$. The difference equation reduces to

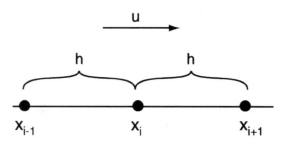

Figure 8.2 Convective transport over two consecutive elements.

$$\phi_{i+1} = \phi_{i-1}$$

and results in a solution in which odd nodes have one value and even nodes may have a different value. The particular values must be determined by the boundary conditions. Therefore the boundary condition at $x = L$ affects the solution to the left of $x = L$, but there is no physical means of propagation to the left when u is positive.

In practice, the values of γ are usually large, e.g., $O(10^4)$ is not unusual. This imposes a very restrictive condition on the size of h to satisfy the inequality $\gamma < 2$.

2. When γ is large, physically we are dealing with a mesh length scale that is too large compared to the diffusion length scale. Figure 8.2 shows two consecutive elements of size h and a positive convective velocity u. The Galerkin approximation to the convective term in Eq. (8.5) is

$$u\frac{d\phi}{dx}\bigg|_{x=x_i} \cong \frac{u}{2}(\phi_{i+1} - \phi_{i-1}) \tag{8.12}$$

Equation (8.12) states that the value of ϕ at node x_i depends on the value at node x_{i+1}, which is not plausible if the convective velocity is transporting fluid from node x_i toward node x_{i+1}.

This nonphysical numerical representation can only be remedied if h is made smaller than the diffusion length scale, i.e., $\gamma < 2$, to make it possible for the effect of perturbations at node x_{i+1} to be felt at node x_i.

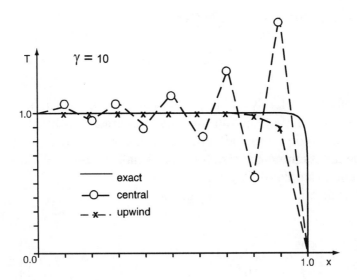

Figure 8.3 Solution to Eq. (8.14) with boundary conditions $\phi(0) = 1$, $\phi(1) = 0$, and $\gamma = 10$. Solid line is the analytical solution; dashed line the central difference solution; dash-dot line the upwind difference solution.

3. Those readers familiar with finite difference methods will recognize that Eq. (8.5), resulting from the linear finite element discretization of Eq. (8.1), is identical to the central finite difference approximation to that equation. In fact, finite difference users ran into this problem long before the first finite element applications to convective equations were attempted. This led to the development of *upwind* finite difference algorithms (Roache, 1972), in which the first derivative term in Eq. (8.1) is approximated by a backward derivative with respect to the velocity direction, that is,

$$
u\frac{d\phi}{dx}\bigg|_{x=x_i} \cong
\begin{cases}
\dfrac{u}{h}(\phi_i - \phi_{i-1}) & u > 0 \\[2mm]
\dfrac{u}{h}(\phi_i - \phi_{i+1}) & u < 0
\end{cases}
\tag{8.13}
$$

Assuming again that $u > 0$, this approximation leads to a difference equation for node x_i of the form (Exercise 8.3)

$$
(1+\gamma)\phi_{i-1} - (2+\gamma)\phi_i + \phi_{i+1} = 0
\tag{8.14}
$$

which has the solution (Exercise 8.4)

$$\phi_i = A + B(1+\gamma)^i \tag{8.15}$$

Equation (8.15) is clearly stable for all values of γ. Upwinding appeared to provide a simple solution to this numerical problem. However, numerical approximations, although stable, turn out to be unacceptably inaccurate unless the mesh size h is very small. Figure 8.3 illustrates this for the same example of Fig. 8.2.

A truncation error analysis gives

$$\left[\left(-D\frac{d^2\phi}{dx^2}+u\frac{d\phi}{dx}\right)_{x=x_i} - \left\{\frac{D}{h^2}(-\phi_{i-1}+2\phi_i-\phi_{i+1})+\frac{u}{h}(\phi_i-\phi_{i-1})\right\}\right] = \frac{uh}{2}\phi_i^{(2)}$$

$$-\frac{h^2}{6}\left[u\phi_i^{(3)}+\frac{D}{2}\phi_i^{(4)}\right] + \quad \text{HOT}$$

$$\tag{8.16}$$

where $\phi_i^{(n)} \equiv d^n\phi / dx^n|_{x=x_i}$ and HOT stands for higher order terms, both in powers of h and the order of derivatives of ϕ.

The leading term on the right-hand side depends on h, rendering the method a first-order method. The situation is further complicated by the presence of the second derivative of ϕ evaluated at $x = x_i$ in the leading term of the truncation error. To interpret the approximation correctly (Roach, 1972), we must move the leading term of the truncation error in Eq. (8.16) to the left-hand side and regard the difference equation, Eq. (8.14), as a second-order approximation to the modified equation

$$-\left(D+\frac{uh}{2}\right)\frac{d^2\phi}{dx^2} + u\frac{d\phi}{dx} = 0 \tag{8.17}$$

where the term uh/2 is called an *artificial numerical diffusion* and will, in general, be much larger than D.

4. From the above discussion, it is clear that the upwind differencing method achieved stability by means of adding an artificial numerical diffusion to the system. On the other hand, numerical solutions obtained using the Galerkin method with $\gamma < 2$ result in approximations that are always *underdiffused*. This fact can be easily verified through simple numerical examples (Exercises 8.6 – 8.8) and points to the fact that numerical methods of the finite element type introduce artificial numerical diffusion when both convection and diffusion transport appear in the equations. This will be the basis for the development of Petrov-Galerkin methods in the next section.

8.3 PETROV-GALERKIN METHOD

To obtain improved algorithms that will render accurate solutions for larger values of the parameter γ, we consider the Galerkin solution of the modified convection-diffusion equation

$$-\left(D + \frac{\alpha\, uh}{2}\right)\frac{d^2\phi}{dx^2} + u\frac{d\phi}{dx} = 0 \tag{8.18}$$

where α is a parameter such that $0 \le \alpha \le 1$. Notice that if $\alpha = 0$, we recover the Galerkin solution, which is underdiffused, and if $\alpha = 1$, we will have the upwind solution, which is overdiffused. Therefore it is reasonable to expect that for some value of α between 0 and 1 we will obtain the correct solution. We will refer to the term $\alpha uh/2$ as the *balancing diffusion*.

The Galerkin discretization of Eq. (8.18) using linear elements of size $\Delta x = h$ yields the difference equation (Exercise 8.11)

$$\left[1 + \frac{\gamma}{2}(\alpha + 1)\right]\phi_{i-1} - 2\left(1 + \frac{\alpha\gamma}{2}\right)\phi_i + \left[1 + \frac{\gamma}{2}(\alpha - 1)\right]\phi_{i+1} = 0 \tag{8.19}$$

with solution

$$\phi_i = A + B\left[\frac{2 + \gamma(\alpha + 1)}{2 + \gamma(\alpha - 1)}\right]^i \tag{8.20}$$

from which the condition on α for the solution to be non-oscillatory follows, i.e.,

$$\alpha \geq 1 - \frac{2}{\gamma} \tag{8.21}$$

The value of α for which equality holds is called the *critical* value of α and is denoted by α_{cr}

$$\alpha_{cr} \equiv 1 - \frac{2}{\gamma} \tag{8.22}$$

To investigate this algorithm further, we use Taylor series expressions to obtain the truncation error TE defined by

$$TE \equiv \left[-\frac{d^2\phi}{dx^2} + \frac{\gamma}{h}\frac{d\phi}{dx} \right]_{x=x_i} \tag{8.23}$$

$$-\frac{1}{h^2}\left\{ -\left[1 + \frac{\gamma}{2}(\alpha+1)\right]\phi_{i-1} + 2\left(1 + \frac{\alpha\gamma}{2}\right)\phi_i - \left[1 + \frac{\gamma}{2}(\alpha-1)\right]\phi_{i+1} \right\}$$

We obtain (Exercise 8.14)

$$TE \equiv \frac{\alpha\gamma}{2}\phi_i^{(2)} + \frac{2}{h^2}\left(1 + \frac{\alpha\gamma}{2}\right)\left[\frac{h^4}{4!}\phi_i^{(4)} + \frac{h^6}{6!}\phi_i^{(6)} + \frac{h^8}{8!}\phi_i^{(8)} + \cdots \right] \tag{8.24}$$

$$-\frac{\gamma}{h^2}\left[\frac{h^3}{3!}\phi_i^{(3)} + \frac{h^5}{5!}\phi_i^{(5)} + \frac{h^7}{7!}\phi_i^{(7)} + \cdots \right]$$

We now use Eq. (8.1) written in the form

$$\phi_i^{(2)} = \frac{\gamma}{h}\phi_i^{(1)} \tag{8.25}$$

to express the higher order derivatives of ϕ by means of the recursive relation

$$\phi_i^{(n)} = \left(\frac{\gamma}{h}\right)^{n-2} \phi_i^{(2)} \qquad n = 3,4,5,\ldots \qquad (8.26)$$

Substituting into Eq. (8.24) yields

$$TE = \left\{\frac{\alpha\gamma}{2} + 2\left(1+\frac{\alpha\gamma}{2}\right)\left[\frac{\gamma^2}{4!}+\frac{\gamma^4}{6!}+\frac{\gamma^6}{8!}+\cdots\right] - \gamma\left[\frac{\gamma}{3!}+\frac{\gamma^3}{5!}+\frac{\gamma^5}{7!}+\cdots\right]\right\}\phi_i^{(2)}$$
$$(8.27)$$

Hence the total truncation error can be expressed as a numerical diffusion; manipulating Eq. (8.27) further (Exercise 8.15), we obtain

$$TE = \left[\frac{1}{\gamma^2}\left\{2\left(1+\frac{\alpha\gamma}{2}\right)\tanh\frac{\gamma}{2} - \gamma\right\}\sinh\gamma\right]\phi_i^{(2)} \qquad (8.28)$$

which is the total discretization error in the approximation.

REMARKS

1. Let us set $\alpha = 0$ in Eq. (8.28); that is, we look at the error in the Galerkin method. The function

$$f(\gamma) = \left(2\tanh\frac{\gamma}{2} - \gamma\right)\frac{\sinh\gamma}{\gamma^2} \qquad (8.29)$$

can be rewritten in the form

$$f(\gamma) = \left(\frac{\tanh\gamma/2}{\gamma/2} - 1\right)\frac{\sinh\gamma}{\gamma}$$

where $\sin(h\gamma/\gamma) > 0$ for all γ and $\tanh[(\gamma/2)/(\gamma/2)] - 1 < 0$ for all γ. Thus $f(\gamma) < 0$ for all γ, and the Galerkin method is underdiffused.

2. If we set TE = 0 and solve for α, we obtain

$$\alpha = \coth\frac{\gamma}{2} - \frac{2}{\gamma} \qquad (8.30)$$

This value of α is often referred to as the optimal value, as it provides the exact amount of balancing diffusion to counteract the numerical diffusion. We will simply denote it by α. It can be easily verified through numerical examples that its use yields exact nodal values when the coefficients are constant, and yields superconvergent solutions in the general case.

Example 8.1

In Fig. 8.4 we show the solution to the equation

$$-\frac{d^2\phi}{dx^2} + \frac{60}{x}\frac{d\phi}{dx} = 0 \quad 1 < x < 2 \quad \phi(1) = 1 \quad \phi(2) = 0 \qquad (8.31)$$

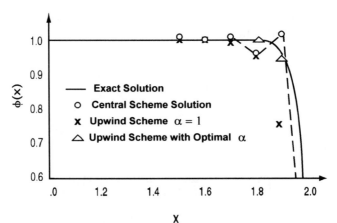

Figure 8.4 Solution to Eq. (8.31) using 10 linear elements.

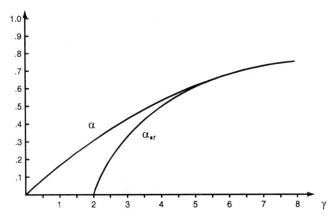

Figure 8.5 Plot of α and α_{cr} as a funtion of γ.

obtained using a uniform mesh with h = 0.1. The parameter α is calculated using the average velocity over each element. This yields a solution exact to four significant digits. It is left to Exercise 8.8 to reproduce the results of Fig. 8.4. ∎

3. To calculate α from Eq. (8.30) introduces an additional computational expense. Figure 8.5 shows α and α_{cr} as a function of γ. We observe that as γ increases, α approaches α_{cr} very rapidly. In fact, for $\gamma > 8.0$ the extra numerical diffusion introduced using γ_{cr} rather than γ is less than 0.1%. Moreover, as γ approaches 0.0, when $\gamma < 0.1$, the amount of balancing diffusion required in Eq. (8.18) is again less than 0.1%. Therefore a practical formula for α is given by

$$\alpha = \begin{cases} 0.0 & \gamma < 0.1 \\ \coth\dfrac{\gamma}{2} - \dfrac{2}{\gamma} & 0.1 \le 8.0 \\ 1 - \dfrac{2}{\gamma} & \gamma > 8.0 \end{cases} \tag{8.32}$$

4. The name Petrov-Galerkin is given to algorithms that utilize weighting functions that are different from the shape functions in the weighted residuals formulation. To derive a Petrov-Galerkin formulation

from the balancing diffusion concepts discussed above, we write the Galerkin weak form of Eq. (8.18) as

$$\int_0^L \left\{ D \frac{dN_i}{dx} \frac{d\phi}{dx} + u \left(N_i + \frac{\alpha h}{2} \frac{dN_i}{dx} \right) \frac{d\phi}{dx} \right\} dx = 0 \qquad (8.33)$$

and set

$$w_i = N_i + \frac{\alpha h}{2} \frac{dN_i}{dx} \qquad (8.34)$$

The weak form (8.33) can now be written as

$$\int_0^L \left\{ D \frac{dw_i}{dx} \frac{d\phi}{dx} + u w_i \frac{d\phi}{dx} \right\} dx = 0 \qquad (8.35)$$

with the perturbed weighting function w_i given by Eq. (8.34).

Equations (8.34) and (8.35) define our basic Petrov-Galerkin method for the one-dimensional convection-diffusion equation.

5. The weighting functions in Eq. (8.34) were first proposed by Hughes and Brooks (1979) and Kelly et al. (1980). Many other schemes based on perturbed shape functions have been proposed in the literature; a review of these is found in the works by Heinrich and Zienkiewicz (1979a,b). Here we will concentrate on this particular form because of its direct physical interpretation in terms of adding a balancing diffusion to the numerical scheme, and the simplicity of the formulation in terms of shape functions and their derivatives. The references given here should be enough to guide interested readers into any and all aspects of Petrov-Galerkin methods for convective transport.

6. Notice that Eq. (8.35) can be written directly from Eq. (8.33) only for linear elements, for which the second derivative of the shape functions vanishes. Also, the fact that the weighting functions are not continuous raises questions about the convergence of the method. These have been addressed by Hughes and Brooks (1982). Theoretical studies on the convergence properties of Petrov-Galerkin methods using the continuous quadratic perturbation functions first introduced by Christie et al. (1976) were performed by Griffiths and Lorenz (1978), who showed second-order asymptotic convergences as long as $\alpha \to 0$ as $h \to 0$.

Further mathematical studies have been done by Barrett and Morton (1980) and by Babuska and Szymczak (1982).

7. For quadratic elements, Petrov-Galerkin methods have been devised by Heinrich (1980); Heinrich and Zienkiewicz, (1977). These elements require the introduction of two parameters to obtain optimal solutions and are, in general, expensive and not as practical for convection–diffusion problems. This is in contrast with the pure diffusion case, where the quadratic elements are near optimal, as was discussed in Section 4.2. Petrov-Galerkin methods for cubic elements were developed by Christie and Mitchell (1978). Because of the excellent accuracy that can be achieved with linear elements, the use of higher order elements in convection-diffusion equations becomes hard to justify. Unless very special requirements must be satisfied that linear elements cannot fulfill, the added complexity and cost are generally deciding factors against the use of even quadratic elements. We will therefore concentrate on the use of linear elements and their bilinear and trilinear extensions to two and three dimensions.

8. Let us now consider Eq. (8.1) with a nonzero source term, i.e.,

$$-\frac{d}{dx}\left(D\frac{d\phi}{dx}\right) + u\frac{d\phi}{dx} = S$$

The weighted residual form is

$$\int_0^L \left\{ D\frac{dw}{dx}\frac{d\phi}{dx} + uw\frac{d\phi}{dx} \right\} dx = \int_0^L wS\,dx \qquad (8.36)$$

In a Galerkin formulation, the weighting functions will be chosen equal to the shape functions, i.e., $w = N_i$ while a Petrov-Galerkin formulation uses the modified weighting functions of Eq. (8.34). The following example proposed by Leonard (1979) shows the importance of the Petrov-Galerkin formulation, as opposed to the simpler added balancing dissipation approach, in which the right-hand side term of Eq. (8.36) would be weighted by $w = N_i$ as in a Galerkin formulation.

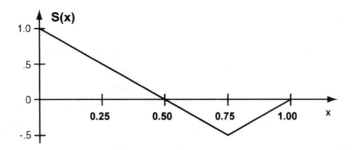

Figure 8.6 Source term for problem of Example 8.2.

Example 8.2

Solve the equation

$$\frac{d\phi}{dx} = S(x) \quad 0 < x \le 1.5 \tag{8.37}$$

$$\phi(0) = 0$$

where $S(x)$ is shown in Fig. 8.6 and is given by

$$S(x) = \begin{cases} 1 - 2x & 0.0 \le x \le 0.75 \\ 2(x - 1) & 0.75 \le x \le 1.0 \\ 0 & x > 1.0 \end{cases} \tag{8.38}$$

Two numerical solutions on a mesh size $h = 0.1$ are shown in Fig. 8.7. One is based on the added balancing diffusion only, in which the source term is weighted using the shape functions, and shows a very significant approximation error. The other uses the Petrov-Galerkin formulation, with

$$w_i = N_i + \frac{h}{2}\frac{dN_i}{dx} \tag{8.39}$$

and produces the exact solution (Exercise 8.16). ∎

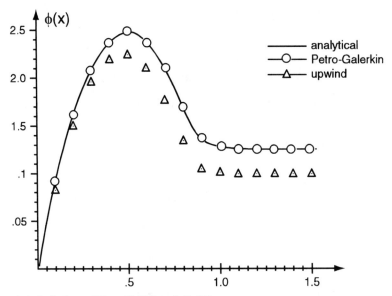

Figure 8.7 Solution of Eqs. (8.37) and (8.38).

To further stress the importance of the consistent weighting of the right-hand side using the Petrov-Galerkin functions, consider a linear function $S(x) = ax$ and a finite element discretization of the interval $0 \leq x \leq L$ using a regular mesh of size h. The analytical solution to Eq. (8.37) can be written at the nodal points x_i in the form

$$\phi(x_i) = \phi(x_{i-1}) + \frac{a}{2}\left(x_i^2 - x_{i-1}^2\right) \tag{8.40}$$

The Petrov-Galerkin discretization of Eq. (8.37) yields (Exercise 8.17)

$$\phi_i = \phi_{i-1} + \frac{a h}{2}\left(\frac{5}{6}x_{i-1} + \frac{4}{3}x_i - \frac{1}{6}x_{i+1}\right) \tag{8.41}$$

A little further manipulation (Exercise 8.18) transforms Eq. (8.41) into the form

$$\phi_i = \phi_{i-1} + \frac{a}{2}\left(x_i^2 - x_{i-1}^2\right) \tag{8.42}$$

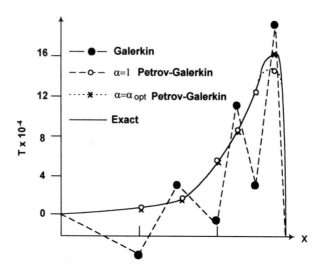

Figure 8.8 One-dimensional problem with unequal mesh spacing and source term (Eq. 8.43).

Therefore the Petrov-Galerkin solution is exact whenever the source term is linear or piecewise linear as long as there are nodes located at the points of slope discontinuity.

Example 8.3

Consider the solution of the equation

$$-\frac{d^2T}{dx^2} + 200\frac{dT}{dx} = x^2 \qquad 0 < x < 1$$

$$(8.43)$$

$$T(0) = T(1) = 0$$

Table 8.1 shows the solutions obtained using the Galerkin method ($\alpha = 0$), full up-winding ($\alpha = 1$), and Petrov-Galerkin ($\alpha = \coth \gamma/2 - 2/\gamma$). Two meshes were used, a regular mesh of size h = 0.05 and a six element irregular mesh with the nodes at the positions indicated in Table 8.1. The solution obtained with the irregular mesh is shown in Fig. 8.8. The much better accuracy of the Petrov-Galerkin method is clearly shown even in the presence of a quadratic source term and a variable mesh.

Table 8.1 Solutions to Eq. (8.43) using linear elements

x	Analytical Solution	Regular mesh h = 0.05			Irregular mesh		
		$\alpha = 0$	$\alpha = 1$	P-G	$\alpha = 0$	$\alpha = 1$	P-G
0.1	0.00000	0.00000	0.00000	0.00000	-	-	-
0.3	0.00005	0.00004	0.00005	0.00005	-0.00062	0.00004	0.00004
0.4	0.00011	0.00010	0.00011	0.00011	-	-	-
0.5	0.00021	0.00018	0.00021	0.00021	-	-	-
0.55	0.00028	0.00033	0.00028	0.00028	0.00031	0.00029	0.00029
0.6	0.00037	0.00030	0.00037	0.00037	-	-	-
0.7	0.00058	0.00043	0.00058	0.00058	-0.00014	0.00059	0.00059
0.8	0.00087	0.00053	0.00087	0.00087	0.00105	0.00087	0.00087
0.9	0.00124	0.00048	0.00122	0.00124	0.00031	0.00123	0.00124
0.95	0.00145	0.000258	0.00130	0.00145	0.00206	0.00131	0.00146

P-G, Petrov-Galerkin method, $\alpha = \coth(\gamma / 2) - (2 / \gamma)$ ∎

9. Problems involving reaction as well as diffusion of the form

$$-D\frac{d^2\phi}{dx^2} + u\frac{d\phi}{dx} + a\phi = S \qquad (8.44)$$

can also lead to numerical difficulties, particularly when the reaction term $a\phi$ is important. Idelsohn et al. (1996) have shown that stable solutions can be obtained by means of two stabilizing parameters. A different approach proposed by Harari and Hughes (1994) uses only one parameter. The interested reader should refer to these papers for details or numerical solutions of equations of the form of Eq. (8.44). We will now move on to multidimensional convection-diffusion equations.

8.4 PETROV-GALERKIN METHOD IN TWO DIMENSIONS

We will still restrict ourselves to the time independent equations. We write the convection-diffusion equation in the form

$$-\nabla^T \mathbf{D} \nabla \phi + \mathbf{V} \cdot \nabla \phi = 0 \qquad (8.45)$$

where ∇ is the gradient operator in Cartesian coordinates,

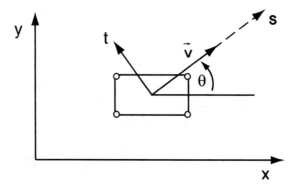

Figure 8.9 Rectangular bilinear element.

$$\nabla_{xy} = \begin{bmatrix} \dfrac{\partial}{\partial x} \\[2ex] \dfrac{\partial}{\partial y} \end{bmatrix}$$

the diffusion matrix for an isotropic material is given as

$$\mathbf{D} = D\begin{bmatrix} 1 & 0 \\ 0 & 1 \end{bmatrix} \tag{8.46}$$

and

$$\mathbf{V} = \begin{bmatrix} u \\ v \end{bmatrix}$$

is the convective velocity vector that we will assume to be constant, without loss of generality.

Consider a rectangular bilinear element as shown in Fig. 8.9, where **V** represents the average fluid velocity within the element. We introduce a local rotation of the operator in Eq. (8.45) to a new coordinate system s-t such that s is aligned with the direction of the velocity **V** and t is orthogonal to s. In the new coordinate system, Eq. (8.45) becomes

$$-\nabla_{st}^T \mathbf{D} \nabla_{st} \phi + |\mathbf{V}| \frac{\partial \phi}{\partial s} = 0 \qquad (8.47)$$

where $|\mathbf{V}|$ is the magnitude of the velocity vector. The gradient operators in the two coordinate systems are related by

$$\nabla_{st} = \mathbf{T} \nabla_{xy} \qquad (8.48)$$

where

$$\mathbf{T} = \begin{bmatrix} \cos\theta & \sin\theta \\ -\sin\theta & \cos\theta \end{bmatrix} \qquad (8.49)$$

Rewriting Eq. (8.47) in the form

$$-\frac{\partial}{\partial s}\left(D\frac{\partial\phi}{\partial s}\right) + |\mathbf{V}|\frac{\partial\phi}{\partial s} = \frac{\partial}{\partial t}\left(D\frac{\partial\phi}{\partial t}\right) \qquad (8.50)$$

we see that in this coordinate system there is a convection-diffusion problem in the s-direction only. Therefore a balancing diffusion should be added only in the direction of flow. Thus we introduce an *anisotropic* balancing diffusion of the form

$$\mathbf{D}^1 = D^1 \begin{bmatrix} 1 & 0 \\ 0 & 0 \end{bmatrix} = \frac{\alpha|\mathbf{V}|\bar{h}}{2}\begin{bmatrix} 1 & 0 \\ 0 & 0 \end{bmatrix} \qquad (8.51)$$

where the length $\bar{h}$ will be given later. Clearly, the parameter α will be determined as before from Eq. (8.30) or Eq. (8.32), where the parameter γ is now given by

$$\gamma = \frac{|\mathbf{V}|h}{D} \qquad (8.52)$$

In the s-t system we obtain

$$-\nabla^T\left(\mathbf{D}+\mathbf{D}^1\right)\nabla\phi +|\mathbf{V}|\frac{\partial\phi}{\partial s} = 0 \qquad (8.53)$$

and after rotating back to the x-y coordinate system, we have (Exercise 8.20)

$$-\left\{\frac{\partial}{\partial x}\left(D\frac{\partial}{\partial x}\right)+\frac{\partial}{\partial x}\left[\frac{\alpha\,u\,\overline{h}}{2}\left(\frac{u}{|\mathbf{V}|}\frac{\partial\phi}{\partial x}+\frac{v}{|\mathbf{V}|}\frac{\partial\phi}{\partial y}\right)\right]+\frac{\partial}{\partial y}\left(D\frac{\partial\phi}{\partial y}\right)\right.$$

$$(8.54)$$

$$\left.+\frac{\partial}{\partial y}\left[\frac{\alpha\,v\,\overline{h}}{2}\left(\frac{u}{|\mathbf{V}|}\frac{\partial\phi}{\partial x}+\frac{v}{|\mathbf{V}|}\frac{\partial\phi}{\partial y}\right)\right]\right\}+u\frac{\partial\phi}{\partial x}+v\frac{\partial\phi}{\partial y}=0$$

The weak Galerkin formulation can be written as

$$\int_{\Omega}\left\{\frac{\partial N_i}{\partial x}\left[D\frac{\partial\phi}{\partial x}+\frac{\alpha\,u\,\overline{h}}{2|\mathbf{V}|}\left(u\frac{\partial\phi}{\partial x}+v\frac{\partial\phi}{\partial y}\right)\right]+\frac{\partial N_i}{\partial y}\left[D\frac{\partial\phi}{\partial y}+\frac{\alpha\,v\,\overline{h}}{2|\mathbf{V}|}\left(u\frac{\partial\phi}{\partial x}+v\frac{\partial\phi}{\partial y}\right)\right]\right.$$

$$\left.+N_i\left(u\frac{\partial\phi}{\partial x}+v\frac{\partial\phi}{\partial y}\right)\right\}d\Omega -\int_{\Gamma}N_i D\frac{\partial\phi}{\partial n}d\Gamma = 0$$

$$(8.55)$$

and rearranging, we obtain

$$\int_{\Omega}\left\{D\left(\frac{\partial N_i}{\partial x}\frac{\partial\phi}{\partial x}+\frac{\partial N_i}{\partial y}\frac{\partial\phi}{\partial y}\right)+u\left[N_i+\frac{\alpha\,\overline{h}}{2|\mathbf{V}|}\left(u\frac{\partial N_i}{\partial x}+v\frac{\partial N_i}{\partial y}\right)\right]\frac{\partial\phi}{\partial x}\right.$$

$$(8.56)$$

$$\left.+v\left[N_i+\frac{\alpha\,\overline{h}}{2|\mathbf{V}|}\left(u\frac{\partial N_i}{\partial x}+v\frac{\partial N_i}{\partial y}\right)\right]\frac{\partial\phi}{\partial y}\right\}d\Omega -\int_{\Gamma}N_i D\frac{\partial\phi}{\partial n}d\Gamma = 0$$

This last expression suggests that the Petrov-Galerkin weights take the form

$$w_i = N_i + \frac{\alpha \overline{h}}{2|\mathbf{V}|}\left(u\frac{\partial N_i}{\partial x} + v\frac{\partial N_i}{\partial y}\right) \qquad (8.57)$$

REMARKS

1. It is clear that the weighting functions in Eq. (8.57) provide a natural extension of the one-dimensional weighting functions given by Eq. (8.34). If $v = 0$, Eq. (8.57) reduces to Eq. (8.36), and the same is true if $u = 0$ and convection is parallel to the y-axis.

2. When Green's theorem in the plane is applied to Eq. (8.54), the full perturbed weighting function should appear in the line integral of Eq. (8.55). However, the perturbation to the shape function was omitted, leaving the full Petrov-Galerkin weights to act only in the interior of the domain Ω. The reasons for this have been explained by Hughes and Brooks (1982) and are consistent with the fact that the perturbations are C^{-1} functions. Therefore they do not act along boundaries.

3. The most important feature of the present Petrov-Galerkin method is that it does not introduce any *cross-flow diffusion*, that is, diffusion in the direction orthogonal to the flow, which is a serious problem with most upwind finite difference schemes as well as some other types of finite element schemes. This has been documented by de Vahl Davis and Mallison (1976) in the context of the driven cavity flow, where very different solutions are obtained when different upwind schemes or different meshes are used. In the present method, balancing diffusion is introduced only in the direction of flow. For this reason, it is also called streamline-upwind-Petrov-Galerkin, or the SUPG method.

Example 8.4

Consider the advection of a cosine hill in a rotating field (Hughes and Brooks, 1982) as shown in Fig. 8.10. The flow field is a circular vortex around the origin defined by

$$u = -y$$
$$v = x$$

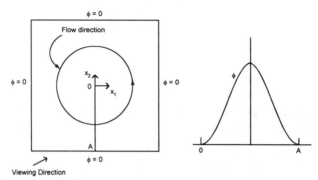

Figure 8.10 Problem statement for advection in a rotating flow field. (From *A Theoretical Framework for Petrov-Galerkin Methods with Discontinuous Weighting Functions: Applications to the Streamline Upwind Procedure* by T. J. R. Hughes and A. Brooks. Copyright J. Wiley and Sons Ltd. Reproduced with permission).

along the outside boundary, defined by $|x| = 2$ and $|y| = 2$, we set $\phi = 0$, and on the "internal boundary" $\overline{OA}$ we prescribe ϕ as a cosine as shown in Fig. 8.10. Set D = 0 so the solution is a pure convection of the hill around the origin. The solutions obtained with the Galerkin, Petrov-Galerkin, and full upwind methods are shown in Fig. 8.11. The Petrov-Galerkin solution is excellent, the Galerkin solution oscillates, and the upwind solution is grossly overdiffused. ∎

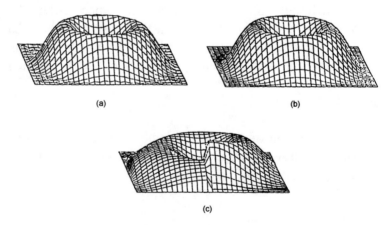

Figure 8.11 Advection of a rotating flow field (elevations of ϕ) for (a) Galerkin weighting, (b) Petrov-Galerkin weighting, and (c) upwind weighting. (From *A Theoretical Framework for Petrov-Galerkin Methods with Discontinuous Weighting Functions: Applications to the Streamline Upwind Procedure* by T. J. R. Hughes and A. Brooks. Copyright J. Wiley and Sons Ltd. Reproduced with permission).

4. In two- (and three-) dimensional calculations, it is often more practical to use the Petrov-Galerkin weights in the convective and source terms only, while using the shape functions as weights in the diffusion term. Although this makes no difference in one dimension, as noted in Remark 6 of Section 8.3, in two dimensions the cross derivatives of the shape functions are not zero, but a constant over the element. Therefore the resulting form of the discretized equation will not be the same. The differences, however, are not important and do not affect the accuracy of the approximations.

5. The element length $\bar{h}$ in the direction of flow is given by (Heinrich and Yu, 1988)

$$\bar{h} = \frac{1}{|\mathbf{V}|}\left(|h_1| + |h_2|\right) \tag{8.58}$$

where

$$\begin{aligned} h_1 &= \mathbf{a} \cdot \mathbf{V} \\ h_2 &= \mathbf{b} \cdot \mathbf{V} \end{aligned} \tag{8.59}$$

and the vectors $\mathbf{a}$ and $\mathbf{b}$ are given by

$$\begin{aligned} a_1 &= \frac{1}{2}\left(x_2 + x_3 - x_1 - x_4\right) \\ a_2 &= \frac{1}{2}\left(y_2 + y_3 - y_1 - y_4\right) \\ b_1 &= \frac{1}{2}\left(x_3 + x_4 - x_1 - x_2\right) \\ b_2 &= \frac{1}{2}\left(y_3 + y_4 - y_1 - y_2\right) \end{aligned} \tag{8.60}$$

Vectors $\mathbf{a}$ and $\mathbf{b}$ are the vectors that join the midpoints of opposite sides in a quadrilateral element, as shown in Fig. 8.12; h_1 and h_2 are the projections of these vectors in the direction of flow.

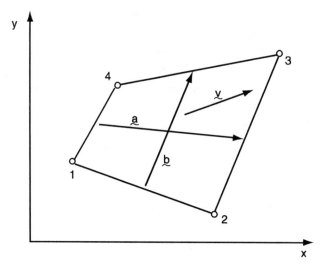

Figure 8.12 General bilinear quadrilateral element.

6. Algorithms for convective flows have also been developed for triangular elements. Among others, the work of Tabata (1977), Kikuchi and Ushijima (1982) and Mizukami and Hughes (1985) are representative of the techniques employed in conjunction with triangular elements. We will not develop methods based on triangular elements here, as the bilinear isoparametric element suffices and is more accurate in most applications involving convective flows.

7. A variety of methods have been proposed for convection-dominated problems in the past years. Many of these methods are similar to the Petrov-Galerkin method presented here. The so-called discontinuity capturing algorithms (Mizukami and Hughes, 1985; Hughes et al. 1986; Galeao et al., 1988) introduce a second diffusion in a direction orthogonal to the gradient of the unknown variable $\nabla\phi$. As a result, the problem becomes nonlinear, and a second parameter, besides α, must be calculated, for which a uniquely defined criterion is not available. However, these methods have been shown to be very effective in sample calculations.

Another family of algorithms that generalize the Petrov-Galerkin methods is called the Galerkin-least-squares (GLS) method (Hughes et al., 1989) and will be discussed next. Notice that all the methods presented so far have been devised for the case of time-independent equations. Application of the Petrov-Galerkin algorithms to time-dependent problems is addressed in Chapter 9.

8. Our Petrov-Galerkin procedure can be considered a particular case of the more general GLS formulations that were first introduced by Hughes et al. (1989). To illustrate the GLS method, let us go back to the one-dimensional convection-diffusion equation, (Eq. (8.36). From Eq. (8.20), the residual $R(\phi,x)$ is given by

$$R = -\frac{d}{dx}\left(D\frac{d\phi}{dx}\right) + u\frac{d\phi}{dx} - S \tag{8.61}$$

and the least squares formulation requires the minimization of the functional

$$I = \int_\Omega R^2 d\Omega \tag{8.62}$$

over the space of trial functions. If ϕ is approximated using shape functions, $\phi = N_j\phi_j$, the conditions for a minimum are given by

$$\frac{\partial}{\partial\phi_i}\int_\Omega\left[\left(-\frac{d}{dx}\left(D\frac{dN_j}{dx}\right) + u\frac{dN_j}{dx}\right)\phi_j - S\right]^2 dx = 0$$

which leads to the finite element equations

$$\int_\Omega\left[u\frac{dN_i}{dx} - \frac{d}{dx}\left(D\frac{dN_i}{dx}\right)\right]\left[\left(-\frac{d}{dx}D\frac{dN_j}{dx} + u\frac{dN_j}{dx}\right)\phi_j - S\right]dx = 0 \tag{8.63}$$

On the other hand, the Galerkin formulation of Eq. (8.36) is

$$\int_\Omega N_i\left[\left(-\frac{d}{dx}D\frac{dN_j}{dx} + u\frac{dN_j}{dx}\right)\phi_j - S\right]dx = 0 \tag{8.64}$$

The Galerkin-least-squares formulation consists of satisfying a linear combination of Eqs. (8.63) and (8.64), that is,

$$\int_{\Omega} \left[N_i + \tau \left(u \frac{dN_i}{dx} - D \frac{d^2 N_i}{dx^2} \right) \right] \left[\left(-\frac{d}{dx} \left(D \frac{dN_j}{dx} \right) + u \frac{dN_j}{dx} \right) \phi_j - S \right] dx = 0$$

(8.65)

This last equation clearly constitutes a Petrov-Galerkin formulation with weighting function

$$w_i = N_i + \tau \left(u \frac{dN_i}{dx} - D \frac{d^2 N_i}{dx^2} \right)$$

(8.66)

If linear elements are used, the second derivative of the shape function in Eq. (8.66) vanishes, so we recover the Petrov-Galerkin formulation by setting

$$\tau = \frac{\alpha h}{2 u}$$

(8.67)

This method offers a variety of possibilities, can be applied to many different elements, and has been analyzed by Hughes et al. (1989). However, as discussed above, we must have a criterion to choose τ, which is not always immediately available. An even more general formulation, the Generalized Galerkin-least-squares (GGLS) method has been proposed by Idelsohn et al. (1995) to deal with time-dependent convection-diffusion problems in which two parameters are determined.

8.5 PETROV-GALERKIN METHOD IN THREE DIMENSIONS

The extension of the Petrov-Galerkin method to three dimensions is carried out in the same way as for the two-dimensional case and is left to Exercise 8.23. The resulting weighting functions are

$$w_i = \frac{\alpha \bar{h}}{2 |V|} \left(u \frac{\partial N_i}{\partial x} + v \frac{\partial N_i}{\partial y} + w \frac{\partial N_i}{\partial z} \right)$$

(8.68)

where $\mathbf{V} = [u \ v \ w]^T$ and N_i are the trilinear shape functions for the eight-noded isoparametric brick element. The parameters α and γ are as defined in Eqs. (8.30) or (8.32) and (8.52), and $\bar{h}$ is calculated from (Pepper and Heinrich, 1993)

$$\bar{h} = \frac{1}{|\mathbf{V}|} \left(|h_1| + |h_2| + |h_3| \right) \tag{8.69}$$

where

$$h_1 = \mathbf{a} \cdot \mathbf{V}$$

$$h_2 = \mathbf{b} \cdot \mathbf{V} \tag{8.70}$$

$$h_3 = \mathbf{c} \cdot \mathbf{V}$$

and the vectors **a, b,** and **c** go through the midpoints of opposite sides, as shown in Figure 8.13 and are given by

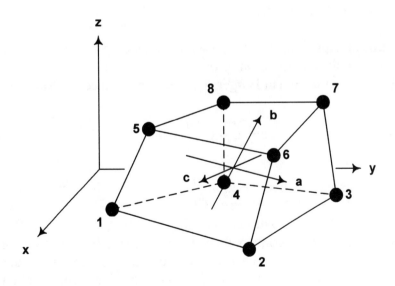

Figure 8.13 General trilinear hexahedral element.

$$a_1 = \frac{1}{4}\left(x_2 + x_3 + x_6 + x_7 - x_1 - x_4 - x_5 - x_8\right)$$

$$a_2 = \frac{1}{4}\left(y_2 + y_3 + y_6 + y_7 - y_1 - y_4 - y_5 - y_8\right)$$

$$a_3 = \frac{1}{4}\left(z_2 + z_3 + z_6 + z_7 - z_1 - z_4 - z_5 - z_8\right)$$

$$b_1 = \frac{1}{4}\left(x_5 + x_6 + x_7 + x_8 - x_1 - x_2 - x_3 - x_4\right)$$

$$b_2 = \frac{1}{4}\left(y_5 + y_6 + y_7 + y_8 - y_1 - y_2 - y_3 - y_4\right) \tag{8.71}$$

$$b_3 = \frac{1}{4}\left(z_5 + z_6 + z_7 + z_8 - z_1 - z_2 - z_3 - z_4\right)$$

$$c_1 = \frac{1}{4}\left(x_1 + x_2 + x_5 + x_6 - x_3 - x_4 - x_7 - x_8\right)$$

$$c_2 = \frac{1}{4}\left(y_1 + y_2 + y_5 + y_6 - y_3 - y_4 - y_7 - y_8\right)$$

$$c_3 = \frac{1}{4}\left(z_1 + z_2 + z_5 + z_6 - z_3 - z_4 - z_7 - z_8\right)$$

8.6 NONLINEAR EQUATIONS

In nonlinear equations the Petrov-Galerkin method must be combined with the nonlinear iteration procedure. This can lead to numerical instabilities. Let us revisit Burgers equation introduced in Example 6.1,

$$\varepsilon \frac{d^2 u}{dx^2} + u \frac{du}{dx} = 0 \tag{8.72}$$

It was established early on (Heinrich and Zienkiewicz, 1979a,b; Hughes et al., 1979) that the use of Petrov-Galerkin in this equation and in the Navier-Stokes equations, in conjunction with the Newton-Raphson method, leads to slow convergence and even instability when convection becomes dominant. Heinrich and Zienkiewcz (1979a,b) concluded that the use of direct iteration was effective and converged rapidly. Hughes et al. (1979) used a modified quadrature upwind technique to resolve the problem.

Example 8.5

Consider the solution of Eq. (8.72) with boundary conditions $u(0) = 1$, $u(1) = 0$. The analytical solution is given by

$$u(x) = A\left(\frac{1 + Be^{Ax}}{1 - Be^{Ax}}\right) \tag{8.73}$$

where A and B are the integration constants. Setting $\varepsilon = 0.1$ and using 10 linear elements, we solve the equation using the Galerkin and Petrov-Galerkin methods together with a direct iteration and a Newton-Raphson iteration. The results are summarized in Table 8.2. The convergence criterion is that the maximum change at any node between two consecutive iterations be less than 10^{-4}.

Table 8.2 Results for one-dimensional Burgers equation

x	Direct Iteration		Newton-Raphson		Analytical Solution
	Galerkin	Petrov-Galerkin	Galerkin	Petrov-Galerkin	
.2	1.000	0.999	1.000	0.999	0.999
.4	0.997	0.995	0.997	0.995	0.995
.6	0.969	0.963	0.970	0.964	0.964
.8	0.766	0.759	0.770	0.762	0.762
.9	0.460	0.459	0.464	0.462	0.462

The solutions using the Galerkin method are not oscillatory, even though the cell Péclet number is 10 in some elements. Near the boundary layer at $x = 1$, the average cell Peclet number is only 0.231. An important difference between linear and non linear transport equations is that the cell Péclet number is always variable for the nonlinear problems, and it also depends on the solution. Therefore, it is not possible to predict the onset of oscillation a priori for the nonlinear Burgers equation or the Navier-Stokes equations.

Notice also that the Galerkin solution using Newton-Raphson is underdiffused. This method yields a more accurate solution than the direct iteration method. However, convergence for the Newton-Raphson method becomes more difficult as diffusion becomes smaller.

In Table 8.3 we show the number of iterations required by each method to obtain the solutions in Table 8.2, and how this number varies as ε becomes smaller.

Table 8.3: Number of iterations needed for convergence.

ε^{-1}	Direct iteration		Newton-Raphson	
	Galerkin	Petrov-Galerkin	Galerkin	Petrov-Galerkin
5	5	5	4	4
10	6	6	5	5
10^2	6*	4	6*	6
10^3	**	4	16*	7
10^6	**	4	**	7

* Solution is oscillatory
** Solution does not converge after 20 iterations ∎

It is interesting to notice how the Newton-Raphson Petrov-Galerkin method becomes less efficient as convection dominates. In fact, in simulations of the Navier-Stokes equations, it fails to converge for strongly convection dominated problems (Heinrich and Zienkiewicz, 1979a, b; Hughes et al., 1979).

When direct steady state solutions to the Navier-Stokes equations are sought, the Newton-Raphson method is our method of preference. However, the Petrov-Galerkin scheme must be modified or the calculation must be done starting from low values of Re and increasing slowly. Each time Re is increased, the converged solution at the previous Re is used as the initial condition in the calculation. We will discuss this further in subsequent chapters, but we must mention that there is a danger involved if this technique is used. In many flow problems, as the Re (or other governing parameters) is increased, the solution reaches bifurcation points where it changes to another solution that is physically more stable. However, the numerical solutions may continue to follow the branch that is physically unstable, leading to unrealistic results.

For this reason, for many simulations we prefer to use a time-dependent algorithm to reach steady state. As we discuss in the next chapter, in a time-dependent simulation of the Navier-Stokes equations the convective terms may be calculated explicitly, and a Petrov-Galerkin method as presented in Section 8.4 may be applied directly. There are two advantages to this: first, if there are instabilities and bifurcations in the flow, they will be captured by the numerical method; second, we need not modify the Petrov-Galerkin formulations obtained for linear equations.

Example 8.6

Let us use the direct substitution method in Eq. (8.72) with a uniform mesh of size h. At each iteration, we know the function $u^n(x)$, and these are used to find the new iterate $u^{n+1}(x)$ from the weak form

$$\int_0^1 \left(\varepsilon \frac{dw^n}{dx} \frac{du^{n+1}}{dx} + w^n u^n \frac{du^{n+1}}{dx} \right) dx = 0 \qquad (8.74)$$

This equation differs from the weak form corresponding to the Galerkin formulation, in that the weighting functions change at each iteration with the cell Péclet number γ^n, given by

$$\gamma^n = \frac{\bar{u}^n h}{\varepsilon} \qquad (8.75)$$

where $\bar{u}^n$ is the average of u^n over each element. The weighting functions in the Petrov-Galerkin method are

$$w_i^n = N_i + \frac{\alpha^n h}{2} \frac{dN_i}{dx} \qquad (8.76)$$

with

$$\alpha^n = \coth \frac{\gamma^n}{2} - \frac{2}{\gamma^n} \qquad (8.77)$$

∎

Here we have assumed that α is given by the same expression as for the linear problem, although this is not obvious. Heinrich and Envia

(1982) showed that using Taylor series expansions of the same form as those leading to Eq. (8.28), at least the first two leading terms in the truncation error of the nonlinear equation are identical to those of the linear case. This result, plus extensive numerical experimentation, is the basis for the use of Eq. (8.77) to obtain the Petrov-Galerkin parameter α in the numerical solution of the Burgers and Navier-Stokes equations.

Example 8.7

We now apply the Newton-Raphson iteration to Eq. (8.7). The Petrov-Galerkin formulation can be written in the form (see Eq. (6.20)

$$\int_0^1 \left(\varepsilon \frac{dw}{dx} \frac{d\Delta u}{dx} + wu^n \frac{d\Delta u}{dx} + w \frac{du^n}{dx} \Delta u \right) dx \qquad (8.78)$$

This is an equation of the convection-diffusion-reaction type, and the use of the Petrov-Galerkin method alone may not be sufficient to solve it accurately enough for convergence to occur. One possible generalization of the algorithm is suggested by Idelsohn et al. (1996) in the context of time-dependent equations and is applicable to equations of the form of Eq. (8.78). Another simpler modification to the Petrov-Galerkin method consists in using only the shape functions as weights for the third term in Eq. (8.78). The effect of this modification is left to Exercise 8.25. ∎

8.7 CLOSURE

We have introduced a finite element formulation capable of producing stable and accurate solutions to convection-diffusion equations. The formulation derived herein is only one of many that have been successfully used in convective transport problems. It is in our opinion the most practical and does not involve user-defined parameters. In the next chapter we will extend the Petrov-Galerkin algorithms to time-dependent problems and later on we will apply them to flow calculations. It should now be clear to the reader that introduction of convection in the transport equations creates a number of difficulties, not just in their numerical solution but in the analytical solution of the equations as well. This will become even more complex when we add time dependency in the next chapter. However, we will show that the present method can be extended to time-dependent problems and that adequate algorithms can be developed using the same framework as we did in this chapter.

REFERENCES

Babuska, I. and Szymczak, W. G. (1982) "An Error Analysis for the Finite Element Method Applied to Convection Diffusion Problems," *Comput. Methods Appl. Mech. Eng.*, Vol. 31, pp. 19–42.

Barrett, J. W. and Morton, K. W. (1980) "Optimal Finite Element Solutions to Diffusion-Convection Problems in One Dimension," *Int. J. Numer. Methods Eng.*, Vol. 14, pp. 1457–1474.

Barrett, K. E. (1977) "Finite Element Analysis of Flow Between Rotating Discs Using Exponentially Weighted Basis Functions," *Int. J. Numer. Methods Eng.*, Vol. 14, pp. 1457–1474.

Christie, I., Griffiths, D. R., Mitchell, A. R. and Zienkiewicz, O. C. (1976) "Finite Element Methods for Second Order Differential Equation with Significant First Derivatives," *Int. J. Numer. Methods Eng*, Vol. 10, pp. 1389–1396.

Christie, I. and Mitchell, A. R. (1978) "Upwinding of High Order Galerkin Methods in Conduction-Convection Problems," *Int. J. Numer. Methods Eng.*, Vol. 12, pp. 1764–1771.

de Vahl Davis, G. and Mallinson, C. D. (1976) "An Evaluation of Upwind and Central Difference Approximations by a Study of Recirculating Flow," *Comput. Fluids*, Vol. 4, pp. 29–43.

Griffiths, D. F. and Lorenz, J. (1978) "An Analysis of the Petrov-Galerkin Finite Element Method," *Comput. Methods Appl. Mech. Eng.* Vol. 14, pp. 39–64.

Harari, I. and Hughes, T. J. R. (1994) "Stabilized Finite Element Method for Steady Advection Diffusion with Production," *Comput. Methods Appl. Mech. Eng.*, Vol. 115, pp. 165–191.

Heinrich, J. C. (1980) "On Quadratic Elements in Finite Element Solutions of Steady-State Convection Diffusion Equation," *Int. J. Numer. Methods Eng.*, Vol. 15, pp. 1041–1052.

Heinrich, J. C. and Envia, E. (1982) "Finite Element Techniques in Transport Phenomena," pp. 14–27 to14–40.

Heinrich, J. C. and Yu, C. C. (1988) "Finite Element Simulations of Buoyancy Driven Flows with Emphasis on Natural Convection in a Horizontal Circular Cylinder," *Comput. Methods Appl. Mech. Eng.*, Vol. 69, pp. 1–27.

Heinrich, J. C. and Zienkiewicz, O. C. (1977) "Quadratic Finite Element Schemes for Two-Dimensional Convective Transport Problems," *Int. J. Numer. Methods Eng.*, Vol. 11, pp. 1831–1844.

Heinrich, J. C. and Zienkiewicz, O. C. (1979a) "The Finite Element Method and 'Upwinding' Techniques in the Numerical Solution of Convection Dominated Flow Problems," pp. 105–126 in *Finite Element Methods for Convecion Dominated Flows*, (T. J. R. Hughes, editor). AMD-Vol. 34, New York: ASME.

Heinrich, J. C. and Zienkiewicz, O. C. (1979b) "Solution of Non-Linear Second Order Equations with Significant First Derivatives by a Petrov-Galerkin Method," *Numerical Analysis of Singular Perturbation Problems* (P. Hemker and J. J. Miller, editors). London: Academic Press.

Hemker, P. W. (1977) "A Numerical Study of Stiff Two-Point Boundary Value Problelms." Materialisch Centrum, Amsterdam.

Hughes, T. J. R. and Brooks, A. (1979) "A Multi-Dimensional Upwind Scheme with No Crosswind Diffusion," pp. 19–35 in *Finite Elements for Convection Dominated Flow*, AMD-Vol. 34. (T. J. R. Hughes editor). New York: ASME.

Hughes, T. J. R. and Brooks, A. (1982) "A Theoretical Frame Work for Petrov-Galerkin Methods with Discontinuous Weighting Functions: Application to the Streamline Upwind Procedure," pp. 47–65 in *Finite Elements in Fluids*, Vol. 4 (R. Gallagher, et al. editors). Wiley and Sons.

Hughes, T. J. R., Franca, L. P. and Hulbert, G. (1989) "A New Finite Element Formulation for Computational Fluid Dynamics VIII: The Galerkin-Least-Squares Method for Advective-Diffusive Equations," *Comput. Methods Appl. Mech. Eng.*, Vol. 73, pp. 173–189.

Hughes, T. J. R., Liu, W. K. and Brooks, A. (1979) "Finite Element Analysis of Incompressible Viscous Flows by the Penalty Function Formulation," *J. Comput. Phys.*, Vol. 30, pp. 1–60.

Hughes, T. J. R., Mallet, M. and Mizukami, A. (1986) "A New Finite Element Formulation for Computational Fluid Dynamics II:

Beyond SVPG," *Comput. Methods Appl. Mech. Eng.*, Vol. 54, pp. 341–355.

Idelsohn, S. R., Heinrich, J. C. and Oñate, E. (1995) "Petrov-Galerkin Methods for the Transient Advective-Diffusinve Equation with Sharp Gradients," *Int. J. Numer. Methods Eng.*, Vol. 39, pp. 1455–1473.

Idelsohn, S. R., Nigro, N., Storti, M. and Buscaglia, G. (1996) "A Petrov-Galerkin Formulation for Advection-Reaction-Diffusion Problems," *Comput. Methods Appl. Mech. Eng.*, Vol. 136, pp. 27–46.

Isaacson, E. and Keller, H. B. (1966) *Analysis of Numerical Methods.* New York: John Wiley and Sons.

Kelly, D. W., Nakazawa, S., Zienkiewicz, O. C. and Heinrich, J. C. (1980) "A Note on Upwinding and Aeristropic Balancing Dissipation in Finite Element Approximations to Convective Diffusion Problems," *Int. J. Numer. Methods Eng.*, Vol. 15, pp. 1705–1711.

Kikuchi, F. and Ushijima, T. (1982) "Theoretical Analysis of Some Finite Element Methods for Convective Diffusion Equations," in *Finite Elements in Fluids*, Vol. 4 (R. Gallagher et al. editors). New York: Wiley and Sons.

Leonard, B. P. (1979) "A Survey of Finite Differences of Opinions on Numerical Muddling of the Incomprehensible Defective Confusion Equation," pp. 12–17 in *Finite Elements for Convection Dominated Flow*, AMD-Vol. 34, T. J. R. Hughes editor). New York: ASME.

Mizukami, A. and Hughes, T. J. R. (1985) "A Petrov-Galerkin Finite Element Method for Convection-Dominated Flows: An Accurate Upwinding Technique for Satisfying the Maximum Principle," *Comput. Methods Appl. Mech. Eng.*, Vol. 50, pp. 181–193.

Pepper, D. W. and Heinrich, J. C. (1993) "Transient Natural Convection Within a Sphere Using a 3-D Finite Element Method," pp. 369–378 in *Finite Elements in Fluids* (K. Morgan et al., editors). Swansea: Pineridge Press.

Roache, P. J. (1972) *Computational Fluid Dynamics.* Albuquerque, N.M.: Hermosa Publishers.

Tabata, M. (1977) "A Finite Element Approximation Corresponding to the Upwind Finite Differencing," *Mem. Numer. Math.*, Vol. 4, pp. 47–63.

Varga, R. S. (1962) *Matrix Iterative Analysis*. Englewood Cliffs, N.J.: Prentice-Hall.

Zienkiewicz, O. C. and Heinrich, J. C. (1978) "The Finite Element Method and Convection Problems in Fluid Mechanics," pp. 1–22 in *Finite Elements in Fluids*, Vol. 3, (R. Gallagher et al., editors). London: Wiley and Sons.

EXERCISES

8.1   Derive the element stiffness matrix given in Eq. (8.3).

8.2   In Eq. (8.7), assume solutions of the form $\phi_i = Cr^i$, where C is an arbitrary constant. Substitute ϕ_i into Eq. (8.7) to find the characteristic equation for r and obtain Eq. (8.8). For a general methodology, consult Isaacson and Keller (1966).

8.3   Use Eq. (8.13) and a central difference approximation to the diffusion term in Eq. (8.1) to derive Eq. (8.14).

8.4   Using the method described in Exercise 8.2 above, derive Eq. (8.15).

8.5   Use Taylor series to derive Eq. (8.16).

8.6   Consider the following one-dimensional convection-diffusion equation

$$-\frac{d^2\phi}{dx^2} + 200\frac{d\phi}{dx} = x^2$$

$$\phi(0) = 1, \; \phi(1) = 0$$

(a) Find the analytical solution.
(b) Find a stable Galerkin finite element solution.
(c) Find an upwind finite differences solution.

Your Galerkin finite element solution will be slightly larger than the analytical solution indicating that it is underdiffused.

8.7   Repeat Exercise 8.6 for the equation

$$-\frac{d^2\phi}{dx^2} + \frac{60}{x}\frac{d\phi}{dx} = x^2$$

$$\phi(1) = 1, \ \phi(2) = 0$$

Use an irregular mesh with nodes at x = 1.0, x = 1.3, x = 1.55, x = 1.7, x = 1.8, x = 1.9, x = 1.95 and x = 2.0.

8.8   Repeat Exercise 8.6 for the equation

$$-\frac{d^2\phi}{dx^2} + \frac{60}{x}\frac{d\phi}{dx} = 0$$

$$\phi(1) = 1 \quad \phi(2) = 0$$

Use h = 0.1.

8.9   Find the Galerkin finite element solution to the equation

$$-\frac{d^2\phi}{dx^2} + 50\frac{d\phi}{dx} = 0$$

$$\phi(0) = 0 \quad \frac{d\phi}{dx}\bigg|_{x=2} = 1$$

using 10 linear elements, and compare to the exact solution.

8.10   Repeat Exercise 8.9 for the equation

$$-\frac{d^2\phi}{dx^2} + 50\frac{d\phi}{dx} = 100$$

$$\phi(0) = 0 \qquad \frac{d\phi}{dx}\bigg|_{x=2} = 0$$

Exercises 8.9 and 8.10 are discussed by Zienkewicz and Heinrich (1978).

8.11   Derive Eq. (8.19) from Eq. (8.18) using linear elements.

8.12   Use the method described in Exercise 8.2 to derive Eq. (8.20).

8.13   Derive an equivalent expression for Eq. (8.20) when u is negative.

8.14   Using Taylor series expansions for ϕ_{i+1} and ϕ_{i-1}, derive Eq. (8.24) from Eq. (8.23).

8.15   In Eq. (8.24) add the extra term $-\phi^{(2)} + \gamma / h\phi^{(1)} = 0$ to the right-hand side and modify Eq. (8.27) to be able to obtain Eq. (8.28). (You will also need the half-angle formulae $\cosh\gamma = 2\sinh^2\gamma/2 + 1$ and $\sinh\gamma = 2\sinh\gamma/2\cosh\gamma/2$).

8.16   Show that the exact solution to Eqs. (8.37) and (8.38) is given by

$$\phi(x) = \begin{cases} x - x^2 & 0 \le x \le 0.75 \\ 1.125 + x^2 - 2x & 0.75 \le x \le 1.0 \\ 0.125 & x \ge 1.0 \end{cases}$$

and show that the Petrov-Galerkin finite element solution is exact, provided that a node is placed at x = 0.75.

8.17   In the limit when D goes to zero in Eq. (8.36), we have $\gamma \to \infty$ and therefore $\alpha = 1$ in the absence of diffusion. Write

$$w_i = N_i + \frac{\alpha h}{2} \frac{d N_i}{d x}$$

and show that the discrete difference equations corresponding to Eq. (8.37) take the form

$$-\frac{1}{2}(1+\alpha)\phi_{i-1} + \alpha\,\phi_i + \frac{1}{2}(1-\alpha)\phi_{i+1}$$

$$= \frac{h}{2}\left\{ \left(\frac{1}{3}+\frac{\alpha}{2}\right)S_{i-1} + \frac{4}{3}S_i + \left(\frac{1}{3}-\frac{\alpha}{2}\right)S_{i+1}\right\}$$

Hence, for $\alpha = 1$ and $S(x) = ax$, we get Eq. (8.41). If in the above equation we set $\alpha = 1$ on the left-hand side and $\alpha = 0$ on the right-hand side, we have the added diffusion method shown in Fig. 8.7, which compares to the classical upwind differences scheme.

8.18 Replace $h = x_i - x_{i-1}$ and $x_{i+1} = x_i + h = 2x_i - x_{i-1}$ in Eq. (8.41) and derive Eq. (8.42).

8.19 Show that the coordinate transformation in Fig. 8 .9 is defined by

$$\begin{bmatrix} x \\ y \end{bmatrix} = \begin{bmatrix} \cos\theta & -\sin\theta \\ \sin\theta & \cos\theta \end{bmatrix} \begin{bmatrix} s \\ t \end{bmatrix}$$

and derive Eq. (8.47).

8.20 Use the results of Exercise 8.19 and the fact that

$$\cos\theta = \frac{u}{|V|} \qquad \sin\theta = \frac{v}{|V|}$$

to derive Eq. (8.54) from Eq. (8.53).

8.21   Use the Galerkin-least-squares method to solve the equation in Exercise 8.6 using quadratic elements. How do you determine τ?

8.22   Formulate the Galerkin-least-squares scheme for the diffusion-reaction equation

$$-\frac{d^2\phi}{dx^2} + \phi = 100$$

$$\phi(0) = \phi(1) = 0$$

and solve using 10 linear elements. To find appropriate values for τ, consult Idelsohn et al. (1996).

8.23   For the three-dimensional convection-diffusion equation

$$-\nabla D\nabla\phi + \mathbf{V} \cdot \nabla\phi = 0$$

perform a local rotation to coordinates r-s-t, where r is in the direction $\mathbf{v}$. Add an anisotropic diffusion in the direction of r only and rotate back to obtain the weighting functions given by Eq. (8.68).

8.24   Find the analytical solution in Example 8.5 for $\varepsilon = 0.2$ and $\varepsilon = 0.1$.

8.25   In the weak form of the Burgers equation obtained using the Newton-Raphson method given in Eq. (8.78), solve for the case $\varepsilon = 0.05$ and boundary conditions $u(0) = 1$, $u(1) = 0$, using 10 and 20 linear elements, the Galerkin, Petrov-Galerkin, and inconsistent Petrov-Galerkin, where in the term $wdu^n/dx\,\Delta u$ the shape functions are used as weights, and in the second term $wu^n d\Delta u/dx$ the weights are the perturbed Petrov-Galerkin weights.

8.26   For piecewise quadratic functions, a Petrov-Galerkin method is defined using (Heinrich and Zienkiewicz, 1977).

$$w_i(x) = N_i(x) - \alpha F(x) \qquad \text{i is a corner node}$$

$$w_i(x) = N_i(x) + 4\beta F(x) \qquad \text{i is an interior midnode}$$

where the perturbation function F(x) is given over the interval $0 \le x \le h$ by

$$F(x) = \frac{5}{2}\left(\frac{x}{h}\right)\left[2\left(\frac{x}{h}\right)^2 - 3\left(\frac{x}{h}\right) + 1\right]$$

and the parameters α and β are obtained from

$$\beta = \coth\frac{\gamma}{4} - \frac{4}{\gamma}$$

where

$$\alpha = 2\left(\tanh\frac{\gamma}{2}\right)\left(1 + \frac{3\beta}{\gamma} + \frac{12}{\gamma^2}\right) - \frac{12}{\gamma} - \beta$$

Apply the algorithm defined above to the equation in Exercises 8.6–8.10. Compare to solutions using linear elements.

8.27   Can you derive a Petrov-Galerkin method for quadratic functions using weights of the form

$$w_i(x) = N_i(x) + a\frac{dN_i}{dx} \qquad \text{corner nodes}$$

$$w_i(x) = N_i(x) + b\frac{dN_i}{dx} \qquad \text{mid nodes?}$$

8.28   Show that for linear elements there are infinitely many perturbation functions that when added to the shape function produce weights yielding the exact solution to Eq. (8.1) at the nodes. The only conditions these perturbations must satisfy are

$$i) \quad \int_0^h F(x)\,dx = \frac{h}{2}$$

ii)   F must be symmetric about $x = \dfrac{h}{2}$.

8.29   Show that the shape functions given by

$$N_i(x) = \begin{cases} \dfrac{(x - x_{i-1})}{\theta h} & x_{i-1} \le x \le x_{i-1} + \theta h \\[2mm] 1 & x_{i-1} + \theta h \le x \le x_i \\[2mm] \dfrac{(x_i + \theta h - x)}{\theta h} & x_i \le x \le x_i + \theta h \\[2mm] 0 & \text{otherwise} \end{cases}$$

as shown in the following figure

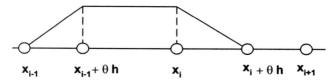

x_{i-1} \qquad $x_{i-1} + \theta h$ \qquad x_i \qquad $x_i + \theta h$ \quad x_{i+1}

yield Eq. (8.19) when applied to discretize Eq. (8.1) using a Galerkin formulation. Here θ takes the place of $(1 - \alpha)$.

8.30   Another set of shape functions that can be used to find solutions of Eq. (8.1) is given in the interval $-1 \le \xi \le 1$ by

$$N_i(\xi) = \begin{cases} \dfrac{e^{2\gamma\xi} - e^{-2\gamma}}{1 - e^{-2\gamma}} & -1 \le \xi \le 0 \\[2mm] \dfrac{e^{2\gamma} - e^{2\gamma\xi}}{e^{2\gamma} - 1} & 0 \le \xi \le 1 \\[2mm] 0 & \text{otherwise} \end{cases}$$

(see Hemker (1977) and Barrett (1977) in regard to these shape functions). Apply these functions to solve Eq. (8.1).

TIME-DEPENDENT CONVECTION-DIFFUSION

9.1 OVERVIEW

In the previous chapter we looked at the difficulties involved in the numerical solution to stationary transport equations dominated by convection. In the time-dependent case, we not only need to avoid oscillations in the numerical approximations, but we must also make sure that we can track the solution accurately as perturbations move over the computational mesh. We must make sure that the amplitude of the waves is not modified by the numerical scheme and that no phase lag is introduced. First, we will explain the concepts of numerical damping and phase lag in the numerical solutions and the techniques available to analyze them. Then we will extend the Petrov-Galerkin method to the time-dependent case, in such a way that those errors are minimized. We will discuss the main properties of the algorithms from both a theoretical and a practical point of view.

9.2 TIME-DEPENDENT CONVECTION

We now turn our attention to the one-dimensional equation

$$\frac{\partial \phi}{\partial t} + u\frac{\partial \phi}{\partial x} - \frac{\partial}{\partial x}\left(D\frac{\partial \phi}{\partial x}\right) = S \qquad (9.1)$$

with initial boundary condition

$$\phi(x,0) = f(x) \tag{9.2}$$

We do not need to specify boundary conditions yet; this simplifies the discussion. The accuracy of the numerical solution to Eq. (9.1) depends both on the time integration method and the Petrov-Galerkin treatment of convection. They cannot be separated, and two numerical difficulties arise that were nonexistent in the diffusion equations of Chapter 7, namely, numerical damping and numerical phase lag. Before we define specific algorithms, we will discuss these properties separately, but first, let us illustrate the problems.

Example 9.1

Consider Eq. (9.1) with $D = 0$, $u = 0.25$, and $S = 0$ and the initial and boundary conditions

$$\phi(x,0) = e^{-800(x-u)^2}$$
$$\phi(0,t) = \phi(2,t) = 0$$

The Crank-Nicolson-Galerkin method applied to Eq. (9.1) using linear elements, $\theta = 1/2$ in Eq. (7.11), with a uniform mesh of size $\Delta x = h = 0.025$ and $\Delta t = 0.09$ is shown in Fig. 9.1, where we set $\phi = 0$ if $|\phi| < 10^{-10}$ in the initial condition. In this figure, the initial condition is shown as the distribution with a peak of 1; the numerical results at times $t = 2.07$ and 4.05 are shown to the right. If we consider that the analytical solution is a simple translation of the initial spike, given by

$$\phi(x,t) = e^{-800[x-(t+1)u]^2}$$

the results shown in Fig. 9.1 are very discouraging. This example illustrates the main problems in time-dependent solutions of convective-transport problems; the amplitude of the initial perturbation has decreased by about 40%. The numerical solution is out of phase, the

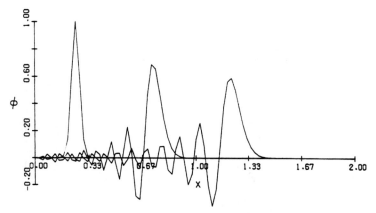

Figure 9.1 Pure advection of a Gaussian wave form using the Crank-Nicolson-Galerkin method.

tip of the spike should be at x = 0.768 at t = 2.07 and at x = 1.26 at t = 4.05. Instead, the numerical solution is maximum at x = 0.73 and at x = 1.22, respectively. Finally, there is an enormous amount of numerical dispersion or "noise" trailing behind the numerical solution. ■

Notice that the above is not an unstable behavior. In fact, the solution is not becoming unbounded but is breaking up into a number of small waves that travel downstream. The problems that we observe will be explained below. Let us look at one more example that includes diffusion.

Example 9.2

We find the solution of Eq. (9.1) with S = 0, u = 0.25, and D = 0.0003125 using the space-time elements introduced in Exercises 7.14 and 7.15 and shown in Fig 9.2. The shape functions are given by

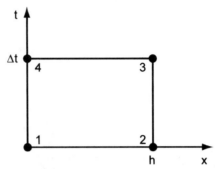

Figure 9.2 Time-space bilinear element.

$$N_1(x,t) = \left(1 - \frac{x}{h}\right)\left(1 - \frac{t}{\Delta t}\right)$$

$$N_2(x,t) = \frac{x}{h}\left(1 - \frac{t}{\Delta t}\right)$$

$$N_3(x,t) = \frac{x}{h}\frac{t}{\Delta t} \tag{9.3}$$

$$N_4(x,t) = \left(1 - \frac{x}{h}\right)\frac{t}{\Delta t}$$

The formulation of the algorithm is left to Exercise 9.2. The analytical solution is given by

$$\phi(x,t) = \frac{1}{\sqrt{1+t}} e^{-[x-(t+1)u]^2 / [4D(t+1)]} \tag{9.4}$$

and the initial and boundary conditions are the same as in Example 9.1 The solution using $h = 0.025$ and $\Delta t = 0.08$ is shown in Fig. 9.3; the initial condition and both the analytical and numerical solutions are shown at time $t = 2.0$ and $t = 4.0$. Notice that in this problem, $\gamma = 20$, but the numerical solutions are not oscillatory. However, the amplitude of the Gaussian wave calculated numerically at $t = 2.0$ is only 60% of the correct value; at $t = 4.0$, it is only 55% of what it should be. The solution is clearly overdamped; moreover, some phase lag can be observed in Fig 9.3. Notice that in this case there is no numerical dispersion at the

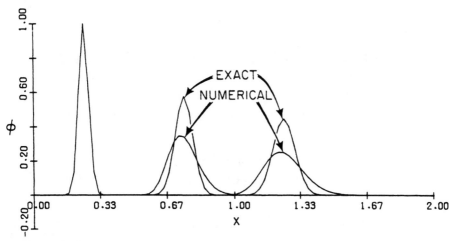

Figure 9.3 Convection-diffusion of a Gaussian wave form with bilinear time-space elements.

trailing edge of the wave form. Different numerical algorithms may exhibit quite different behaviors even though they are essentially the same. In the first example, we used the θ method with θ = 1/2, while the present algorithm is also obtained from the θ method when θ = 2/3 (Exercise 9.3).   ∎

We will look at ways to analyze the behavior of the methods and how to improve them.

9.2.1 Numerical Damping

It is best to introduce the concept of numerical damping through the solution of the equation

$$\frac{\partial \phi}{\partial t} + u \frac{\partial \phi}{\partial x} = 0 \tag{9.5}$$

that is, the purely convective problem with D = 0 in Eq. (9.1). Let us write the solution in the form of a complex Fourier series,

$$\phi = \sum_{n=-\infty}^{\infty} \Phi_n = \sum_{n=-\infty}^{\infty} c_n e^{i\alpha_n(x-at)} \tag{9.6}$$

where the coefficients c_n and the wave numbers α_n are determined from the initial condition Eq. (9.2). We define the *analytical amplification factor* (AAF) as the ratio of the amplitude of the n-th Fourier component at time $t + \Delta t$, to its amplitude at time t (Lapidus and Pinder, 1982):

$$A A F = \left| \frac{\Phi_n(x, t + \Delta t)}{\Phi_n(x, t)} \right| \tag{9.7}$$

Therefore the amplitude of the n-th component increases, remains the same, or decreases over an interval Δt, according to whether the AAF is greater than, equal to, or less than 1. In this case, invoking linearity, so that we may consider only one component at a time, we have

$$\Phi_n(x, t + \Delta t) = c_n e^{i\alpha_n[x-a(t+\Delta t)]} \tag{9.8}$$

and after some simple manipulation, we get

$$A A F = \left| \frac{\Phi_n(x, t + \Delta t)}{\Phi_n(x, t)} \right| = \left| e^{i\alpha_n \Delta t} \right| = 1 \tag{9.9}$$

Therefore the numerical solution of Eq. (9.1), in the absence of numerical error, must not change the amplitude of the Fourier components of the solution with time.

To analyze the behavior of numerical algorithms, we write the numerical solution as a discrete Fourier expansion of the form (Richtmeyer and Morton, 1957)

$$\phi(x_j, t_n) \equiv \phi_j^n = \sum_{k=-K}^{K} a_k e^{ikj\Delta x \xi^n} \tag{9.10}$$

where the a_k are to be determined from the initial conditions, ξ is constant and k assumes integer values; i is the purely imaginary unit, $i = \sqrt{-1}$. We can show that Eq. (9.10) provides the exact solution to the difference equation generated by the numerical method, where the a_k are the Fourier coefficients of the initial condition; see Exercises 9.4 – 9.6. Defining the *numerical amplification factor* (NAF) as the ratio of the amplitude of the n-th Fourier component at time $t + \Delta t$, to its amplitude at time t, we obtain

$$NAF = \left| \frac{\phi_j^{n+1}}{\phi_j^n} \right| = |\xi| \tag{9.11}$$

We can now define the *damping* in the numerical algorithm as the error γ_d in the amplification factor for a component n. Taking the ratio of the NAF to the AAF, one obtains

$$\gamma_d = \frac{NAF}{AAF} = \frac{|\xi|}{1} = |\xi| \tag{9.12}$$

In order to propagate the solution without amplification error, we must have $|\xi| = 1$.

Example 9.3

The Crank-Nicolson-Galerkin method applied to Eq. (9.5) using a uniform mesh of linear elements produces element equations of the form

$$\left(\frac{h}{6\Delta t} \begin{bmatrix} 2 & 1 \\ 1 & 2 \end{bmatrix} + \frac{u}{4} \begin{bmatrix} -1 & 1 \\ -1 & 1 \end{bmatrix} \right) \begin{bmatrix} \phi_1^{n+1} \\ \phi_2^{n+1} \end{bmatrix} = \left(\frac{h}{6\Delta t} \begin{bmatrix} 2 & 1 \\ 1 & 1 \end{bmatrix} - \frac{u}{4} \begin{bmatrix} -1 & 1 \\ -1 & 1 \end{bmatrix} \right) \begin{bmatrix} \phi_1^n \\ \phi_2^n \end{bmatrix} \tag{9.13}$$

and from here, assembling two elements, we find the difference equation for node j as

$$\frac{2}{3}\left[\left(\phi_{j-1}^{n+1} + 4\phi_j^{n+1} + \phi_{j+1}^{n+1}\right) - \left(\phi_{j-1}^n + 4\phi_j^n + \phi_{j+1}^n\right)\right]$$

(9.14)

$$+ \frac{u\Delta t}{h}\left[\left(\phi_{j+1}^{n+1} - \phi_{j-1}^{n+1}\right) + \left(\phi_{j+1}^n - \phi_{j-1}^n\right)\right] = 0$$

replacing $\phi_j^n = \xi^n e^{ik jh}$ and making use of Euler's formula, $e^{i\alpha} = \cos\alpha +$ $i\sin\alpha$, we finally obtain

$$\xi = \frac{4 + 2\cos kh - 3i\dfrac{u\Delta t}{h}\sin kh}{4 + 2\cos kh + 3i\dfrac{u\Delta t}{h}\sin kh}$$

(9.15)

■

REMARKS

1. The argument kh can essentially attain any value. It is customary to replace it by $\theta = kh$, noting that k is a wave number, i.e., k $= 2\pi/L$ where L is the wavelength. We can express θ as a function of $\Delta x/L$; in this case

$$\theta = 2\pi\left(\frac{\Delta x}{L}\right)$$

(9.16)

This allows us to conveniently plot the magnitude of ξ versus the number of elements per wavelength.

2. In our example above, the expression obtained for ξ is of the form $\xi = Z/\bar{Z}$, where $Z = (4 + 2\cos kh) - i(3u\Delta t/h\sin kh)$, and therefore $|\xi| = 1$ always. Thus the Crank-Nicolson-Galerkin algorithm for purely convective flow has no damping. This is both good and bad; it is good because the amplitude of each Fourier component should be preserved, but it is bad because if any perturbations are introduced in the calculation, there is no mechanism to eliminate them.

3. Before we go any further, we need to introduce the Courant number, c, defined as

$$c = \frac{u \, \Delta t}{h}$$

which is the nondimensional parameter that governs Eq. (9.15).

Let us now reexamine Example 9.1, where the Crank-Nicolson-Galerkin method was used with rather disastrous results. If the mesh and time step are designed so that $c = 1$, a particle in the fluid will travel exactly the distance h in one time step. Therefore the peak of the Gaussian wave will move from one node to the next, and the numerical results would be excellent (Exercise 9.10). However, the Courant number used was $c = 0.9$, and therefore the peak could not be captured by the mesh except every 10 time steps. Because mass must be conserved, the numerical scheme can now do one of two things.

(i) Assume that the peak is attained at the node closest to it, thus changing the speed of propagation of the wave.

(ii) Redistribute the excess mass over the domain, the difficulty being that the algorithm does not have long-term memory (it only remembers from t to $t + \Delta t$), and so once the amplitude decreases, it cannot come back up again.

What actually happens is a combination of the two. This introduces the dispersion and damping observed in Fig. 9.1 and eventually destroys the solution. Because there is no mechanism for the algorithm to damp the oscillations, they just stay there as in Fig. 9.1, and the amplitude continues to decay. Some damping in the numerical method is sometimes very convenient to eliminate undesired perturbations.

Example 9.4

We apply the Petrov-Galerkin method defined by Eqs. (8.3) and (8.34) to Eq. (9.5). Time is discretized using the backward implicit θ method, where $\theta = 1$, and the mass matrix is lumped. The difference equations produced by this algorithm are (Exercise 9.12)

$$\phi_j^{n+1} + c\left(\phi_j^{n+1} - \phi_{j-1}^{n+1}\right) = \phi_j^n \tag{9.17}$$

and replacing ϕ_j^n by $\xi^n e^{ik jh}$ as before, we obtain

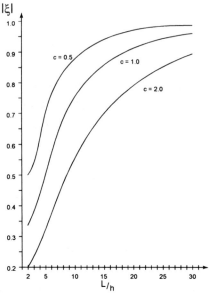

Figure 9.4 Damping characteristics of the Petrov-Galerkin backward-implicit method with lumped mass.

$$\xi = \frac{1}{1 + c\left(1 - \cos\theta + i\sin\theta\right)} \tag{9.18}$$

and therefore

$$|\xi| = \left[\left(1 + c\right)^2 + c^2 - 2c\left(1 + c\right)\cos\theta\right]^{-\frac{1}{2}} \tag{9.19}$$

Figure 9.4 shows $|\xi|$ versus wavelength/h for three Courant numbers $c = 0.5$, 1.0 and 2.0. ∎

REMARKS

1. This algorithm is subject to considerable damping. For example, for $c = 1$ the numerical amplification factor $|\xi|$ is 0.577 if six elements per wavelength are used, and 0.862 if we describe the wave with 15 elements. The latter is a very fine resolution, and the amplitude

would still be reduced by 14% in one time step. The use of mass lumping usually introduces significant damping.

2. The damping is reduced as the Courant number decreases. This requires a reduction in the time step and the corresponding additional expense. Notice that for $c = 2$ the damping is very large. In general, we should not calculate with Courant numbers greater than 1 if the time evolution needs to be described accurately. However, if only steady state solutions are sought, a time-dependent algorithm can provide a convenient way to get to steady state in nonlinear problems. In those cases the extra damping provides added stability to the algorithm.

3. In Chapter 7 we stated that the θ method is unconditionally stable if $\theta \geq 1/2$; therefore the algorithm used in Example 9.4 is unconditionally stable. Because the numerical amplification factor is $|\xi|$ if $|\xi| < 1$, a perturbation in the numerical solution will decrease in amplitude and disappear with time. Therefore the algorithm will be stable. On the other hand, if $|\xi| > 1$, any perturbation will be amplified in time, and the algorithm will be unstable. This is called the von Neumann stability criterion. The stability of an algorithm is governed by $|\xi|$ in such a way that

$$|\xi| \begin{cases} < 1 & \text{stable} \\ = 1 & \text{neutrally stable} \\ > 1 & \text{unstable} \end{cases}$$

Therefore, for a stable algorithm $|\xi| \leq 1$ must be satisfied. This criterion for stability is only useful for linear equations but can be extended to systems of equations as well (Richtmeyer and Morton, 1957).

Example 9.5

Let us now use the θ method with $\theta = 0$ to discretize Eq. (9.5) in the algorithm of Example 9.4. The difference equation becomes

$$\phi_j^{n+1} = c\left(\phi_{j-1}^n - \phi_j^n\right) + \phi_j^n \tag{9.20}$$

and $|\xi|$ can be found to be given by (Exercise 9.13)

$$|\xi| = \left(1 - 4c(1-c)\sin^2 \frac{\theta}{2}\right)^{\frac{1}{2}} \qquad (9.21)$$

The stability condition $|\xi| \le 1$ becomes

$$-1 \le 1 - 4c(1-c)\sin^2 \frac{\theta}{2} \le 1 \qquad (9.22)$$

and after some algebra, shows that the first inequality is always satisfied, while the second is satisfied only if $c \le 1$. Therefore the scheme is stable if and only if

$$\Delta t \le \frac{h}{u}$$

∎

von Neumann's method is readily extended to two dimensions, expressing the Fourier components of the numerical solution in the form

$$\phi\left(x_j, y_\ell, t_n\right) \equiv \phi_{i\ell}^n = \xi^n e^{ikj\Delta x} e^{im\ell\Delta y} \qquad (9.23)$$

The damping characteristics of the algorithm can also be obtained once ξ has been found.

9.2.2 Phase Error

Phase error is the error associated with the translational velocity of each Fourier component of the numerical solution. After one time step, the phase angle of a Fourier component of the numerical solution is given by

$$\tau = \tan^{-1}\left(\frac{I_m \xi}{R_e \xi}\right) \qquad (9.24)$$

where $I_m \xi$ and $R_e \xi$ denote the imaginary and real parts of the complex number ξ, respectively.

On the other hand, in one time step the analytical wave moves a distance τ_o given by $\tau_o = 2\pi/N$, where N is the number of time steps required to advance one full wavelength. Therefore we must have

$$N = \frac{L}{u \, \Delta t} \qquad (9.25)$$

and, replacing the expression for τ_o above, we obtain

$$\tau_0 = c\theta \qquad (9.26)$$

The phase error Θ can then be defined as

$$\Theta = \tau_0 - \tau = c\theta - \tan^{-1}\left(\frac{I_m \, \xi}{R_e \, \xi}\right) \qquad (9.27)$$

Example 9.6

The numerical phase angle for the algorithm defined by Eq. (9.20) is given by

$$\tau = \tan^{-1}\left(\frac{-c \sin \theta}{1 - c(1 - \cos \theta)}\right) \qquad (9.28)$$

The phase error Θ as a function of the wavelength to mesh size ratio L/h is shown in Fig. 9.5 for c = 0.25, 0.5, 0.75, and 1.0. The results show that for c = 0.25 and c = 0.75, the phase error is large, requiring about 20 elements per wavelength to eliminate it. However, for c = 0.5 and c =1.0, the algorithm is exactly in phase.

REMARKS

1. In a uniform mesh of size h, the shortest wavelength that can be represented is L = 2h. However, this provides a very inaccurate approximation to such short waves. In practice, the numerical algorithms are constructed in such a way that the shortest wavelengths of interest can be reasonably well approximated using about six elements. From

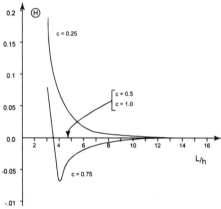

Figure 9.5 Phase error versus wavelength to mesh size ratio (L/h) for the Euler-Petrov-Galerkin method with lumped mass.

this point of view, the algorithms examined in Examples 9.4 through 9.6 leave much to be desired. In the next section we will look into constructing better algorithms.

2. The use of irregular meshes in hyperbolic problems must be treated carefully. As we move from a finer to a coarser mesh, the latter will not be able to carry the shortest wavelength perturbations that are possible in the finer mesh. As a result, some reflection of the shorter waves will occur as the mesh coarsens. Methods to avoid these internal reflections are discussed by Lick (1989) and in general, they consist in modifying the algorithm at the positions where the mesh changes, in order to make the mesh interface appear as an open boundary to the shorter waves.

We will now extend the Petrov-Galerkin methods to time-dependent equations and analyze them according to their stability, damping and phase-error properties.

9.3 PETROV-GALERKIN METHOD FOR TIME-DEPENDENT CONVECTION-DIFFUSION

We return to the numerical solution of Eq. (9.1). There are many ways in which we may attempt to obtain improved algorithms to approximate their solutions. If we stick to the philosophy of generating Petrov-Galerkin schemes in which the weighting functions are obtained

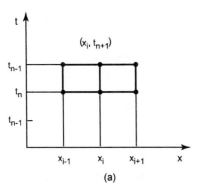
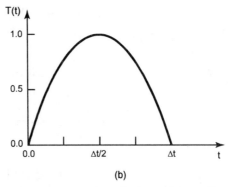

(a)   (b)

Figure 9.6 Space-time elements. (a) Element arrangement required to advance from time t_n to t_{n+1} at node x_i. (b) Quadratic time variation of the weighting functions.

by perturbing the shape functions using the derivatives of the same, we are led to consider methods based on space-time elements. This has the inconvenience that only implicit algorithms can be obtained. However, it has the advantage that the structure of the numerical schemes can be easily interpreted physically and we can construct highly accurate, easy to implement single-step methods.

9.3.1 Quadratic in Time, Linear in Space Weights

In the previous section we saw that the finite element method using bilinear time-space shape functions was overly diffusive, and this remains true when combined with a Petrov-Galerkin weighting. In order to improve the accuracy in time, we can construct weighting functions that are parabolic in time, as shown in Fig. 9.6b, where the time variation $T(t)$ is given by

$$T(t) = \frac{4t}{\Delta t}\left(1 - \frac{t}{\Delta t}\right) \tag{9.29}$$

The weighting functions are numbered as shown in Fig. 9.2 and become $N_i(x)T(t)$. Notice that only the weighting functions at nodes 3 and 4 are required to define the algorithm.

In the one-dimensional case, we interpreted the truncation error in the difference equations as an added diffusion, and we obtained improved algorithms by introducing a balancing diffusion dependent on a parameter α. In this case we will proceed in the same fashion and add a balancing dispersion term of the form $\beta d(\partial^3 \phi / \partial x^2 \partial t)$ as well, where the coefficient d must be proportional to $uh\Delta t$ for dimensional consistency. Therefore we now look at the solution of the modified equation

$$\frac{\partial \phi}{\partial t} + u \frac{\partial \phi}{\partial x} - \left(D + \frac{\alpha u h}{2}\right) \frac{\partial^2 \phi}{\partial x^2} + \beta d \frac{\partial^3 \phi}{\partial x^2 \partial t} = 0 \qquad (9.30)$$

We apply the Galerkin method and operate on the weak form as was done in Section 8.3 for the steady state equation (Exercise 9.23). We obtain the Petrov-Galerkin weights

$$w_i(x,t) = M_i(x,t) + \frac{\alpha h}{2} \frac{\partial M_i(x,t)}{\partial x} + \frac{\beta h \Delta t}{4} \frac{\partial^2 M_i(x,t)}{\partial x \partial t} \qquad (9.31)$$

The functions $M_i(x,t)$ in $0 \le x \le h$, $0 \le t \le \Delta t$ are

$$M_1(x,t) = 4\left(1 - \frac{x}{h}\right) \frac{t}{\Delta t} \left(1 - \frac{t}{\Delta t}\right)$$

$$M_2(x,t) = 4 \frac{x}{h} \frac{t}{\Delta t} \left(1 - \frac{t}{\Delta t}\right)$$

$$M_3(x,t) = M_2(x,t) \qquad (9.32)$$

$$M_4(x,t) = M_1(x,t)$$

Function $M_1(x,t)$ is shown in Fig. 9.7. The weighting functions become

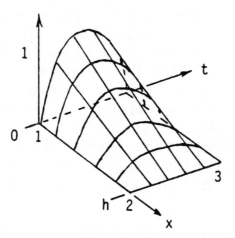

Figure 9.7 Quadratic in time, linear in space function $M_1(x,t)$.

$$
w_1(x,t) = w_4(x,t) = 4\left(1-\frac{x}{h}\right)\frac{t}{\Delta t}\left(1-\frac{t}{\Delta t}\right) - 2\alpha\frac{t}{\Delta t}\left(1-\frac{t}{\Delta t}\right) - \beta\left(1-\frac{2t}{\Delta t}\right)
$$

$$
w_2(x,t) = w_3(x,t) = 4\frac{x}{h}\frac{t}{\Delta t}\left(1-\frac{t}{\Delta t}\right) + 2\alpha\frac{t}{\Delta t}\left(1-\frac{t}{\Delta t}\right) + \beta\left(1-\frac{2t}{\Delta t}\right)
$$

$$(9.33)$$

To determine the values of α and β, we proceed in the same way as we did for the steady state case. First, we write the difference equation for node i. This turns out to be

$$
\frac{1}{9\Delta t}\left[\left(\phi_{i-1}^{n+1} - \phi_{i-1}^{n}\right) + 4\left(\phi_{i}^{n+1} - \phi_{i}^{n}\right) + \left(\phi_{i+1}^{n+1} - \phi_{i+1}^{n}\right)\right]
$$

$$
-\frac{\alpha}{6\Delta t}\left[\left(\phi_{i+1}^{n+1} - \phi_{i+1}^{n}\right) - \left(\phi_{i-1}^{n+1} - \phi_{i-1}^{n}\right)\right] + \frac{u}{6h}\left[\left(\phi_{i+1}^{n+1} + \phi_{i+1}^{n}\right) - \left(\phi_{i-1}^{n+1} + \phi_{i-1}^{n}\right)\right]
$$

$$
-\left[\frac{u}{6h}(\alpha-\beta) + \frac{D}{3h^2}\right]\left(\phi_{i+1}^{n+1} - 2\phi_{i}^{n+1} + \phi_{i-1}^{n+1}\right)
$$

$$
-\left[\frac{u}{6h}(\alpha+\beta) + \frac{D}{3h^2}\right]\left(\phi_{i+1}^{n} - 2\phi_{i}^{n} + \phi_{i-1}^{n}\right) = 0
$$

$$(9.34)$$

The derivation of the above equation is left to Exercise 9.24. We now find the truncation error and manipulate it in a fashion similar to that used in the steady state case utilizing the original differential equation to replace higher order derivatives. We can then write the truncation error as a series, where the leading term is the diffusion term containing ϕ_{xx} and the rest are dispersion terms of the form $\phi_{tx}, \phi_{txx}, \phi_{txxx}, \ldots$, where the subscripts indicate differentiation, e.g., $\phi_{txx} = \partial^3\phi/\partial t\partial x^2$. The process becomes quite involved, and we will not repeat it here. Details can be found in the works by Yu (1986) and Yu and Heinrich (1986).

The truncation error TE can be calculated to be

$$
\begin{aligned}
\text{TE} = &\left(\frac{2D}{3\gamma}\right)\sinh\gamma\left[1-\left(\alpha+\frac{2}{\gamma}\right)\tanh\frac{\gamma}{2}\right]\left(\phi_i^{n+1}\right)_{xx} \\
&+\left(\frac{2h}{3\gamma^2}\right)\left[\sinh\gamma-\left(\alpha+\frac{2}{\gamma}\right)(\cosh\gamma-1)\right]\left(\phi_i^{n+1}\right)_{tx} \\
&+\left(\frac{h^2}{3}\right)\left[\frac{\alpha}{\gamma}+\frac{\beta c}{2}-\frac{c^2}{6}+\frac{2}{\gamma^3}\sinh\gamma\left(1-\tanh\frac{\gamma}{2}\right)-\frac{4}{\gamma^4}(\cosh\gamma-1)\right]\left(\phi_i^{n+1}\right)_{txx} \\
&+\left(\frac{h^3}{3}\right)\left[\frac{c^2}{3}\left(\frac{1}{\gamma}-\frac{c-\alpha}{4}\right)-\frac{\alpha}{2}\left(\frac{1}{6}-\frac{c}{\gamma}-\frac{2}{\gamma^2}\right)-\frac{1}{6\gamma}+\frac{\beta c^2}{4}+\frac{2}{\gamma^4}\sinh\gamma\left(1-\tanh\frac{\gamma}{2}\right)\right. \\
&\left.-\frac{4}{\gamma^5}(\cosh\gamma-1)\right]\left(\phi_i^{n+1}\right)_{txxx} + \text{HOT}
\end{aligned}
$$

(9.35)

Notice that β does not appear in the first two terms. Moreover, if we choose α so as to eliminate the diffusion error, the second term also vanishes, and α turns out to be

$$
\alpha = \coth\frac{\gamma}{2} - \frac{2}{\gamma}
$$

as before. Replacing α in Eq. (9.35) gives

$$TE = \left(\frac{h^2}{3}\right)\left[\frac{\alpha}{\gamma} + \frac{\beta c}{2} - \frac{c^2}{6}\right]\left(\phi_i^{n+1}\right)_{txx}$$

$$+ \left(\frac{h^3}{3}\right)\left[\frac{c^3}{3\gamma} - \frac{c^2}{12} - \frac{1}{6\gamma} - \frac{\alpha}{12} + \frac{\alpha c}{2\gamma} + \frac{\alpha c^3}{12} + \frac{\alpha}{\gamma^2} + \frac{\beta c^2}{4}\right]\left(\phi_i^{n+1}\right)_{txxx} + HOT$$

$$(9.36)$$

and we can now choose β so as to eliminate the dispersive term ϕ_{txx}. We find

$$\beta = \frac{c}{3} - \frac{2\alpha}{\gamma c} \tag{9.37}$$

Notice that α remains a function of γ only while β is a function of α, γ, and c. The parameter β as a function of γ is shown in Fig. 9.8 for several values of the Courant number c.

The resulting algorithm is, in general, third-order accurate in space and second order in time. As we will see later, it has excellent amplitude and phase angle conservation properties.

REMARKS

1. If $\beta = 0$, the algorithm reduces to applying the Petrov-Galerkin weights developed for the steady state equation with a second-order time-stepping scheme, and in this case it is only second-order accurate in space.

2. We can choose the Courant number so that the next term in the truncation error also vanishes, obtaining a scheme that is fourth order in time. However, this is not a practical thing to do, since in general, we must work with a variable Courant number. Details are discussed by Yu and Heinrich (1986).

3. The present algorithm reduces to the one of Chapter 8 when the solution approaches steady state. This follows directly from Eq. (9.34) when the terms representing the time derivatives are set to zero.

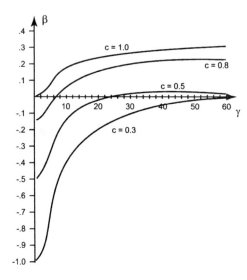

Figure 9.8 Parameter β as a function of γ for various values of the Courant number, c.

4. In the limiting case when $\gamma \to 0$, the expression for β is undefined. We must then realize that physically u must go to zero and Eq. (9.34) becomes independent of β. In this case the algorithm reduces to the Crank-Nicolson-Galerkin scheme because $\gamma \to 0$ if $u \to 0$.

5. If $\gamma \to \infty$, thus $\alpha \to 1$ and $\beta \to c/3$, as shown in Fig. 9.8. This is the purely convective case, and Eq. (9.36) becomes

$$\frac{1}{18\Delta t}\left[5\left(\phi_{i-1}^{n+1} - \phi_{i-1}^{n}\right) + 8\left(\phi_{i}^{n+1} - \phi_{i}^{n}\right) - \left(\phi_{i+1}^{n+1} - \phi_{i+1}^{n}\right)\right]$$
$$+ \frac{u}{3h}\left[\phi_{i}^{n+1} - \phi_{i-1}^{n+1} + \frac{\Delta t}{6h}\left(\phi_{i+1}^{n+1} - 2\phi_{i}^{n+1} + \phi_{i-1}^{n+1}\right)\right. \qquad (9.38)$$
$$\left. + \phi_{i}^{n} - \phi_{i-1}^{n} - \frac{\Delta t}{6h}\left(\phi_{i+1}^{n} - 2\phi_{i}^{n} + \phi_{i-1}^{n}\right)\right] = 0$$

which has a truncation error

$$TE = \frac{h^3}{36}(c^2 - 1)(\phi_{i}^{n+1})_{txxx} + HOT \qquad (9.39)$$

and is therefore third-order accurate. In the case c = 1, the leading term in Eq. (9.39) vanishes, and the algorithm is superconvergent. If $\beta = 0$, we revert back to a second-order scheme.

6. There are several Petrov-Galerkin algorithms that have been proposed for the transient problems, e.g. Westerink and Shea (1989), Cardle (1995), Perrochet (1995) and Idelsohn et al. (1996). Most of these formulations follow arguments similar to the ones presented here and have their own distinct advantages. A "best" scheme is not achievable using simple elements and one-step methods. However, the one presented here leads to robust algorithms that will perform satisfactorily under most practical circumstances.

7. Various combinations of algorithms using bilinear space-time weights, lumped mass matrix, and inconsistent weighting of the convective term were analyzed by Yu (1986). The main conclusions are as follows.

(i) Consistent mass and weighting should be used for optimal accuracy. Surprisingly, mass lumping seems to improve accuracy in the purely convective case when solved using bilinear time-space weights. However, for quadratic in time weighting functions the consistent mass matrix is superior. This was also confirmed by numerical experiments.

(ii) The way to obtain the Petrov-Galerkin weighting functions is not unique. Any function of time that is symmetric about t = Δt/2 and not constant can be used to replace Eq. (9.29).

9.3.2 Stability Analysis

The amplification factor corresponding to the difference equation generated by the Petrov-Galerkin method, Eq. (9.34), is given by (Exercise 9.30)

$$\xi = \frac{\left[\frac{2}{9}\left(1+2\cos^2\frac{\theta}{2}\right) - \frac{2c}{3}\left(\alpha+\beta+\frac{2}{\gamma}\right)\sin^2\frac{\theta}{2}\right] - i\left[\left(\frac{c+\alpha}{3}\right)\sin\theta\right]}{\left[\frac{2}{9}\left(1+2\cos^2\frac{\theta}{2}\right) + \frac{2c}{3}\left(\alpha-\beta+\frac{2}{\gamma}\right)\sin^2\frac{\theta}{2}\right] + i\left[\left(\frac{c-\alpha}{3}\right)\sin\theta\right]} \quad (9.40)$$

and it follows that if $\beta = 0$, the algorithm is unconditionally stable for all α.

If $\beta \neq 0$, $|\xi| \leq 1$ only if $c \leq 1$, so the method is conditionally stable. Improved accuracy in the approximation of the time evolution can only be obtained if $c < 1$. The damping and phase error are shown in Figs. 9.9 and 9.10, respectively, for the case $\gamma = 20$, $c = 0.8$, when $\alpha = \beta = 0$, $\alpha \neq 0$, $\beta = 0$, and for $\alpha \neq 0$, $\beta \neq 0$. The Petrov-Galerkin method shows some damping for small values of L/h; however, this is only about 3% when L/h = 6 and less than 1% when L/h = 10. On the other hand, Fig. 9.10 shows practically no phase error when L/h = 5.

REMARKS

1. The dissipation and phase errors shown in Figs. 9.9 and 9.10 are not after one time step as in sections 9.2.1 and 9.2.2, but after the Fourier component has traveled a full wavelength. Therefore these errors are very small when compared to those of the other algorithms considered earlier in the chapter.

2. Notice that when $\alpha \neq 0$ and $\beta = 0$, the amplitude and phase errors are not much larger than for the full Petrov-Galerkin with $\alpha \neq 0$ and $\beta \neq 0$. For many practical problems we can use the Petrov-Galerkin weights in conjunction with a second order in time algorithm of the Crank-Nicolson type and still achieve reasonably good accuracy. The reasons why it is desirable to do so will become evident when we consider two space dimensions.

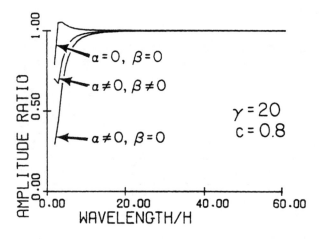

Figure 9.9 Amplitude ratio NAF/AAF for c = 0.8.

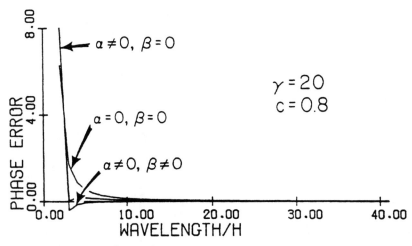

Figure 9.10 Phase error for c = 0.8.

Example 9.7

Let us reexamine the purely convective flow of Example 9.1 with c = 0.9. This time we use our Petrov-Galerkin formulation. Results when $\alpha \neq 0$ and $\beta = 0$ are shown in Fig. 9.11a; results for $\alpha \neq 0$ and $\beta \neq 0$ are shown in Fig. 9.11b. The figures show the initial conditions and the calculated numerical approximation after the perturbation has traveled two and four wavelengths. Because c < 1, some damping must be present in the calculation. The solution in Fig. 9.11b shows the least reduction in amplitude and the least amount of dispersion. Table 9.1 gives the maximum absolute errors; notice that these errors become zero if c = 1. ∎

Table 9.1 Maximum absolute error in the approximation of a convecting Gaussian wave form for c = 0.9

	t = 2.073	t = 4.055
$\alpha = \beta = 0$	0.472	0.520
$\alpha \neq 0 \quad \beta = 0$	0.414	0.465
$\alpha \neq 0 \quad \beta \neq 0$	0.129	0.198

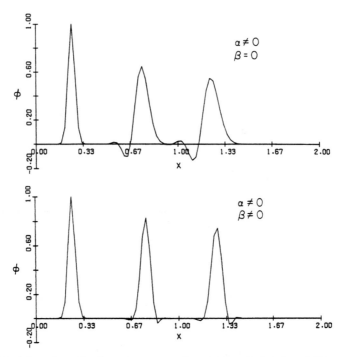

Figure 9.11 Advection of a Gaussian wave when (a) $\alpha \neq 0$, $\beta = 0$ and (b) $\alpha \neq 0$, $\beta \neq 0$.

Example 9.8

We now revisit Example 9.2 and solve it using our Petrov-Galerkin method. The solutions are for $\gamma = 20$ and $c = 0.8$. Figures 9.12a, 9.12b, and 9.12c show the numerical approximations for $\alpha = \beta = 0$; $\alpha \neq 0$, $\beta = 0$; and $\alpha \neq 0$, $\beta \neq 0$, respectively. The accuracy of the latter is very clear. Table 9.2 shows the maximum absolute errors. It is interesting that these errors have decreased from time $t = 2.0$ to $t = 4.0$. Observe that the solution for $\alpha = \beta = 0$ is much more accurate than that obtained in Example 9.2 using bilinear space-time weighting functions.

Table 9.2 Maximum absolute errors for transport and diffusion of a Gaussian wave at $\gamma = 20$ and $c = 0.8$

	$t = 2.05$	$t = 4.05$
$\alpha = 0, \beta = 0$	0.15	0.11
$\alpha \neq 0, \beta = 0$	0.062	0.052
$\alpha \neq 0, \beta \neq 0$	0.03	0.018

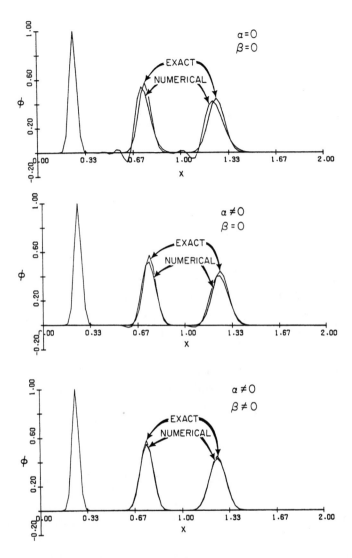

Figure 9.12 Advection and diffusion of a Gaussian wave when (a) $\alpha = \beta = 0$, (b) $\alpha \neq 0$, $\beta = 0$, and (c) $\alpha \neq 0$, $\beta \neq 0$.

Example 9.9

We now solve an example involving a variable diffusion coefficient $D(x)$ and a source term $S(x,t)$; the convective velocity is kept constant. The above functions are given by

$$D(x) = a^2 e^{x^2} \tag{9.41}$$

and

$$S(x,t) = 2a\left[1 + 2x\{x - u(t+1)\} - \frac{2}{a}\{x - u(t+1)^2\}\right]e^{x^2 - [x - u(+1)]^2 / a} \tag{9.42}$$

The analytical solution is

$$\phi(x,t) = e^{-[x - u(t+1)]^2 / a} \tag{9.43}$$

Figure 9.13 shows the solution at times t = 2.0 and 4.0 obtained from u = 0.3, a = 0.00359, and c = 0.8. In this problem the parameter γ varies over the range 10 < γ < 580. We can observe how the solution improves as the Petrov-Galerkin perturbations are added to the weighting functions. ■

As a final comment, recall that the order of convergence was predicted to be $O(h^3)$ in Eq. (9.36), when α ≠ 0 and β = 0. To illustrate this, we solve the equation of Example 9.2 with γ = 20 for five different meshes, keeping c fixed at 0.9. The maximum relative errors at time t = 2.07 are given in Table 9.3.

Table 9.3 Maximum error in the solution of Eq. (9.1) with s = 0 and c = 0.9 for various mesh sizes at t = 2.07

h	Max absolute error	Max relative error
0.06250	0.149	26.2
0.05000	0.077	14.0
0.04167	0.046	8.7
0.025	0.012	2.2
0.00125	0.002	0.3

A log-log plot of the relative error versus the mesh size is shown in Fig. 9.14. The slope of the linear fit is 2.7. This is the actual convergence rate, which agrees well with the predicted value of 3.

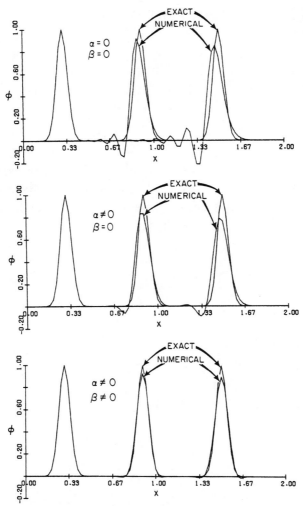

Figure 9.13 Advection-diffusion of a Gaussian wave with variable diffusivity and variable source terms: (a) $\alpha = \beta = 0$, (b) $\alpha \neq 0$, $\beta = 0$, and (c) $\alpha \neq 0$, $\beta \neq 0$.

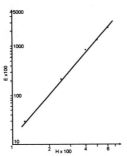

Figure 9.14 Convergence rate of the Petrov-Galerkin method with $\alpha \neq 0$ and $\beta \neq 0$.

9.4 MULTIDIMENSIONAL TIME-DEPENDENT CONVECTION-DIFFUSION

We now extend the Petrov-Galerkin method of the previous section to more than one space dimension. We look at the solution of

$$\frac{\partial \phi}{\partial t} + \mathbf{V} \cdot \nabla \phi = \nabla D \nabla \phi + S \qquad (9.44)$$

To obtain the weighting functions, we follow the same procedure used in Section 8.4, introducing anisotropic diffusion and dispersion in the direction of flow only. For constant velocity $\mathbf{V}$ the weighting functions can be expressed in the form (Exercise 9.31)

$$w_i(\mathbf{x}, t) = M_i + \frac{h}{2\|\mathbf{V}\|}\left(\alpha + \frac{\beta \Delta t}{2}\frac{\partial}{\partial t}\right)\mathbf{V} \cdot \nabla M_i \qquad (9.45)$$

where the functions M_i are quadratic in time with the time variation taking the form of Eq. (9.29).

The nodal numbering and coordinate system for a time + two space dimensions element is shown in Fig. 9.15. The shape functions are trilinear space-time functions given by

$$N_1(x,y,t) = L_1(x)L_1(y)L_1(t)$$
$$N_2(x,y,t) = L_2(x)L_1(y)L_1(t)$$
$$N_3(x,y,t) = L_2(x)L_2(y)L_1(t)$$
$$N_4(x,y,t) = L_1(x)L_2(y)L_1(t)$$
$$N_5(x,y,t) = L_1(x)L_1(y)L_2(t) \qquad (9.46)$$
$$N_6(x,y,t) = L_2(x)L_1(y)L_2(t)$$
$$N_7(x,y,t) = L_2(x)L_2(y)L_2(t)$$
$$N_8(x,y,t) = L_1(x)L_2(y)L_2(t)$$

where

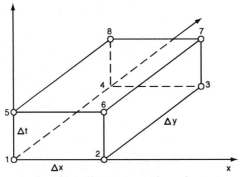

Figure 9.15 Nodal numbering for trilinear space-time element.

$$L_1(\alpha) = 1 - \frac{\alpha}{\Delta\alpha}$$

$$0 \le \alpha \le \Delta\alpha \qquad (9.47)$$

$$L_2(\alpha) = \frac{\alpha}{\Delta\alpha}$$

are the linear shape functions in each direction, $\alpha = x, y, t$.

The test functions are constructed using the quadratic variation in time $T(t)$ given in Eq. (9.29).

$$\begin{aligned}
M_5(x,y,t) &= L_1(x)L_1(y)T(t) \\
M_6(x,y,t) &= L_2(x)L_1(y)T(t) \\
M_7(x,y,t) &= L_2(x)L_2(y)T(t) \\
M_8(x,y,t) &= L_1(x)L_2(y)T(t)
\end{aligned} \qquad (9.48)$$

which, substituted into Eq. (9.45), gives the Petrov-Galerkin weights in two dimensions. The parameters α and β are obtained from

$$\alpha = \coth\frac{\overline{\gamma}}{2} - \frac{2}{\overline{\gamma}} \qquad (9.49)$$

and

$$\beta = \frac{\delta}{3} - \frac{2\alpha}{\overline{\gamma}\delta} \qquad (9.50)$$

with

$$\bar{\gamma} = \frac{\|\mathbf{V}\|\bar{h}}{D} \tag{9.51}$$

and

$$\delta = \frac{\|\mathbf{V}\|\Delta t}{\bar{h}} \tag{9.52}$$

where $\bar{h}$ is defined as in Section 8.4. When α and β are both different from zero, the stability limit in the two-dimensional case takes the form

$$\Delta t \le \frac{1}{\dfrac{|u|}{\Delta x} + \dfrac{|v|}{\Delta y}} \tag{9.53}$$

and we can define the Courant number as

$$c = \left(\frac{|u|}{\Delta x} + \frac{|v|}{\Delta y}\right)\Delta t \tag{9.54}$$

Example 9.10

We consider unidirectional flow, u = 0.25, v = 0.0, where the flow is parallel to the x-direction. The initial two-dimensional Gaussian perturbation is given by

$$\phi(x,y,0) = \exp\left\{-\left[(x-0.25)^2 + (y-0.25)^2\right]/0.00125\right\} \tag{9.55}$$

and is set equal to zero when $|\phi| < 10^{-10}$. The diffusion coefficient is D = .0003125, and we consider the region $0 \le x \le 1.0$, $0 \le y \le 0.5$. The boundary conditions are

$$\phi(0,y,t) = \phi(1,y,t) = 0$$

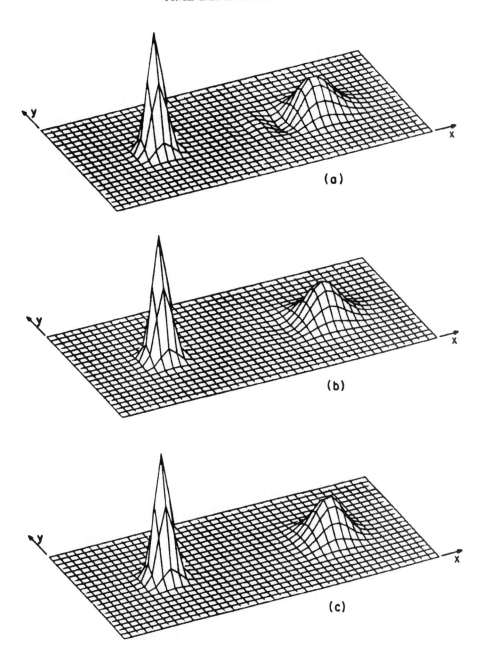

Figure 9.16 Convection and diffusion of a Gaussian wave parallel to the x-axis with c = 0.85, γ = 20: initial condition and results at t = 2.0.

$$\frac{\partial \phi}{\partial y}\bigg|_{y=0.0} = \frac{\partial \phi}{\partial y}\bigg|_{y=0.5} = 0$$

Figure 9.16 shows the initial condition and results calculated with a regular mesh of size $\Delta x = \Delta y = 0.025$, $c = 0.85$, and $\gamma = 20$. The left peak is the initial condition and the right one is the numerical solution at $t = 2.04$. The peak is traced over two full base diameters in the x-direction, defined as the region over which the initial perturbation is greater than zero. When $\alpha = \beta = 0$, using the Crank-Nicolson-Galerkin algorithm, the results show significant numerical dispersion, with the maximum amplitude $\phi_{max} = 0.297$. When $\alpha \neq 0$ and $\beta = 0$, the dispersion is greatly reduced and $\phi_{max} = 0.299$. The best solution is obtained when $\alpha \neq 0$ and $\beta \neq 0$ as expected; there is no dispersion and $\phi_{max} = 0.3104$. ∎

Example 9. 11

We now set $D = 0$; that is, we consider purely advective transport at a $25°$ angle to the x-axis with $u = 0.25$ and $v = 0.1166$. The initial condition is

$$\phi(x,y,0) = \exp\left\{-\left[(x-0.175)^2 + (y-0.175)^2\right]/0.00125\right\} \quad (9.56)$$

The Courant number is $c = 0.7332$ and $\gamma = 20$ with the same mesh as in the previous example. The boundary conditions at $x = 1.0$ and $y = 0.5$ are $\partial \phi / \partial n = 0$; the other two remain the same as before.

Figure 9.17 shows the initial condition and the results at $t = 1.3$. For the case $\alpha = \beta = 0$, we obtain $\phi_{max} = 0.72$ and large dispersive oscillation on the order of 61% of ϕ_{max}. For $\alpha \neq 0$ and $\beta = 0$, $\phi_{max} = 0.58$; when $\alpha \neq 0$ and $\beta \neq 0$, $\phi_{max} = 0.64$. In both cases there is some numerical dispersion on the order of 8% of ϕ_{max}. In this problem the initial Gaussian perturbation is described by six elements in each direction. A solution with twice the number of elements, $\Delta x = \Delta y = 0.0125$, is shown in Fig. 9.18 for $c = 0.91$. The improvement for the case $\alpha \neq 0$, $\beta \neq 0$ is dramatic, with $\phi_{max} = 0.91$ and numerical

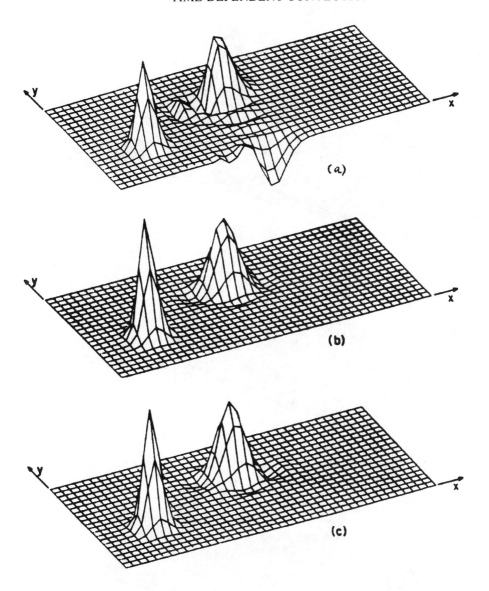

Figure 9.17 Convection of a Gaussian cone at 25° to the mesh with c = 0.73, L/h = 6 showing initial conditions and results at t = 1.3: (a) α = β = 0, (b) α ≠ 0, β = 0, and (c) α ≠ 0, β ≠ 0.

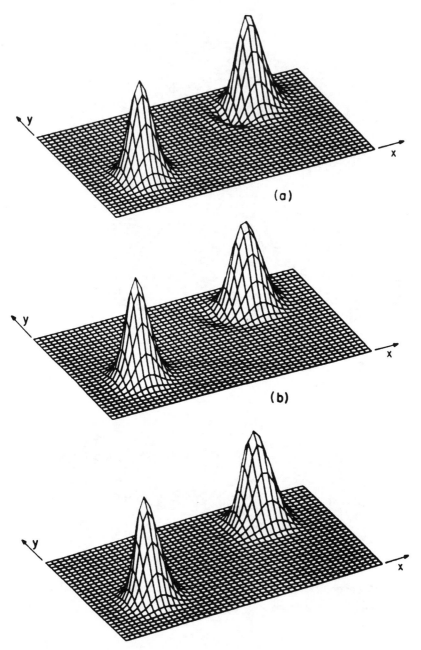

Figure 9.18 Convection of a Gaussian cone at 25° to the mesh with c = 0.73, L/h = 12 showing results at t = 1.3: (a) $\alpha = \beta = 0$, (b) $\alpha \neq 0$, $\beta = 0$, and (c) $\alpha \neq 0$, $\beta \neq 0$.

dispersion less than 1% of ϕ_{max}. The case $\alpha \neq 0$, $\beta = 0$ shows excessive damping, $\phi_{max} = 0.846$, and the amplitude of numerical dispersion is 4% of ϕ_{max}. The case $\alpha = \beta = 0$ shows better amplitude, with dispersion of the order of 8% of ϕ_{max}. ∎

REMARKS

1. The above example should illustrate how difficult it is to solve these problems numerically and how different algorithms may behave very differently with changes in the parameters that are not too large. Yu and Heinrich (1987) give more examples involving variable flow fields solved using the Petrov-Galerkin method.

2. The stiffness matrices for convection problems are non-symmetric, demanding large amounts of storage space. A practical way to avoid this is to treat the convective terms explicitly, thus keeping only the symmetric part of the matrix implicitly. Unfortunately, when $\alpha \neq 0$ and $\beta \neq 0$, this leads to an unconditionally unstable method. However, if $\beta = 0$, we can still compute with the stability condition $c \leq 1$. This makes the case $\alpha \neq 0$, $\beta = 0$ very practical in situations when the scheme behaves reasonably well, or when the added accuracy of $\beta \neq 0$ is not needed.

3. An important method developed for convective transport is the Taylor-Galerkin method (Donea, 1984). We will use the one-dimensional equation, Eq. (9.5), to explain how this method works. The idea is to approximate ϕ_t^n by a truncated Taylor series in time as

$$\phi_t^n = \frac{\phi^{n+1} - \phi^n}{\Delta t} - \frac{\Delta t}{2} \phi_{tt}^n - \frac{\Delta t^2}{6} \phi_{ttt}^n + O\left(\Delta t^3\right) \qquad (9.56)$$

where $\phi_t^n = \partial \phi / \partial t (x, t_n)$ with the space variable x continuous. From the differential equation, we have

$$\phi_t = -u\,\phi_x \qquad (9.57)$$

and differentiating this expression with respect to time, we have

$$\phi_{tt} = u^2\phi_{xx} \text{ and } \phi_{ttt} = u^2\phi_{txx} \tag{9.58}$$

From Eqs. (9.56) and (9.58), we can write

$$\frac{\phi^{n+1} - \phi^n}{\Delta t} - u^2 \frac{\Delta t^2}{6}\phi^n_{txx} + u\phi_x - u^2 \frac{\Delta t}{2}\phi^n_{txx} = 0 \tag{9.59}$$

This equation can now be discretized in time and space to yield a number of different algorithms. For example, replacing the time derivative in Eq. (9.59) by a forward difference leads to the Euler-Taylor-Galerkin form

$$\frac{\phi^{n+1} - \phi^n}{\Delta t} - u^2 \frac{\Delta t}{6}\left(\phi^{n+1}_{xx} - \phi^n_{xx}\right) + u\phi_x - u^2 \frac{\Delta t}{2}\phi^n_{xx} = 0 \tag{9.60}$$

which can be discretized in space using linear or higher order elements. Notice that the first-order equation, Eq. (9.57), was transformed into a second-order equation in space. Also notice that Eq. (9.60) will not approach the correct limit as the solution approaches steady state. There are some controversial aspects of these methods. However, the Taylor-Galerkin methods are accurate and simple to implement and have been widely used, particularly in the numerical approximation of solutions to the Euler equation for compressible flows. The extension to convection-diffusion equations requires Eq. (9.56) be replaced by (Donea et al.,1984)

$$\frac{\phi^{n+1} - \phi^n}{\Delta t} = \phi^n_t + \frac{1}{2}\Delta t\phi^n_{tt} + 0\left(\Delta t^2\right) \tag{9.61}$$

which is less accurate. This is made necessary because otherwise higher order spatial or time derivatives would be introduced. It is important to notice that the essential property of these algorithms is the introduction of stabilizing diffusion and dispersion terms of the same type as in our Petrov-Galerkin method. More properties of Taylor-Galerkin methods are treated in Exercises 9.32 to 9.38.

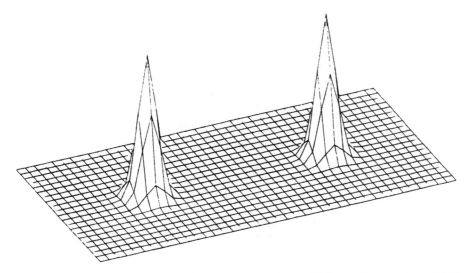

Figure 9.19 Advection of a Gaussian wave at c = 1, u = 0.25, v = 0, $\Delta x = \Delta y = 0.025$, $\alpha \neq 0$ and $\beta \neq 0$. Results at t = 2.0.

4. The Petrov-Galerkin method presented here possesses the so-called unit Courant-Friedrich-Levy property. This means that when the Courant number c = 1, the perturbations are undistorted as they propagate through the mesh. To illustrate this, the solution for pure convection of a Gaussian wave in the x-direction using $\alpha \neq 0$ and $\beta \neq 0$ is shown in Fig. 9.19 for c = 1. The initial condition is shown to the left, and the numerical solution at t = 2 on the right shows no errors.

5. The method of moments has also been used to treat transport by convection (Pepper and Long, 1978). The method calculates zeroth, first, and second moments of the concentration within the mesh and advects the concentration, maintaining conservation of the moments. Extension to convection with diffusion was done by Pepper and Baker (1980). These methods can be highly accurate but can become rather involved.

Many other methods have been proposed and continue to appear in the literature, including the variational approach of Idelsohn (1989) and the Euler-Lagrange approach of Varoglu and Finn (1980).

An interesting class of convection-diffusion algorithms in one dimension was introduced by Allen (1983) using cubic Hermite elements (Exercise 4.3), in which both the derivative and the function are calculated at each node. This has the potential to reduce numerical damping, given the extra information about the slope of the function at each node – at the expense of doubling the number of unknowns in the problem. Unfortunately, these algorithms ran into great difficulties when attempting to extend them to two dimensions and are prohibitive in three dimensions.

  6. The Petrov-Galerkin method derived in this chapter has been applied to the time-dependent Burgers equation by Hanshaw (1981). These results show that when sharp fronts develop that become smaller than the grid space, a small oscillation also appears just ahead of and behind the sharp front. To illustrate this, consider the solution of the time-dependent Burgers equation,

$$\frac{\partial u}{\partial t} + u\frac{\partial u}{\partial x} = \varepsilon\frac{\partial^2 u}{\partial x^2} \qquad 0 < x < 2.0 \qquad (9.62)$$

with initial condition

$$u(x,0) = \begin{cases} \cos\left[2\pi(x-0.25)\right] & 0 \le x \le 0.5 \\ \\ 0 & \text{otherwise} \end{cases}$$

and with boundary conditions $u(0,t) = u(2,t) = 0$. The results for $\Delta x = \Delta t = 0.025$ and $\varepsilon = 0.01$ are shown in Fig. 9.20 with the initial

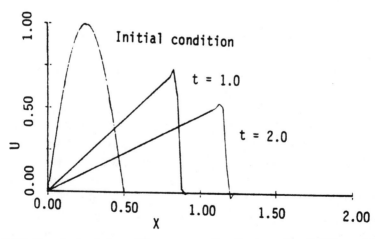

Figure 9.20 Transient solution of Burgers equation for $\Delta x = \Delta t = 0.025$, $\varepsilon = 0.01$.

condition at t = 0 and the calculated solutions at times t = 1.0 and t = 2.0. We observe slight oscillations, as indicated before, which are due to the inability of the mesh to resolve such sharp fronts. This is not a big problem unless the overshoot and undershoot cannot be tolerated, e.g. in chemical reactions, where their presence can be disastrous. Otherwise, these perturbations will not grow, as can be seen in Fig. 9.20. The desire to eliminate these inaccuracies has led to the development of the so-called discontinuity capturing Petrov-Galerkin methods by, e.g., Hughes et al. (1986) and Galeao and Dutra do Carmo (1988), already mentioned in Chapter 8. These methods, however, cannot be applied to one-dimensional problems because they are designed to introduce an additional diffusion in the direction orthogonal to the gradient of the dependent variable. Another effort to eliminate residual oscillations is the concept of a universal limiter introduced by Leonard (1991), which has been extensively used in the finite differences literature.

9.5 CLOSURE

The solution of time-dependent equations with convection can be very difficult because a continuous function must be transported using a discrete mesh. Conservation of the amplitude and moving speed of the different wave components must be achieved accurately for the numerical methods to be effective. We have constructed a Petrov-Galerkin method that is a natural extension of the method for steady state problems. Even though it can be very accurate under certain conditions, it cannot eliminate damping and dispersion in all cases. However, the method is one of the best available, and in most practical applications, it will perform very well. As is always the case with difficult problems, a great variety of algorithms has been proposed over the years to deal with convection-diffusion problems. In this chapter, we have mentioned some of the more salient methods and described their most basic features. However, there are many methods that are not mentioned. The interested reader should be able to obtain information about most other methods from the mentioned references.

REFERENCES

Allen, M. B. (1983) "How Upstream Collocation Works," *Int. J. Numer. Methods Eng.*, Vol. 19, pp. 1753–1763.

Cardle, J. A. (1995) "A Modification of the Petrov-Galerkin Method for the Transient Convection-Diffusion Equation," *Int. J. Numer. Methods. Eng.*, Vol. 38, pp. 171–181.

Donea, J. (1984) "A Taylor-Galerkin Method for Convective Transport," *Int. J. Numer. Methods Eng.*, Vol. 20, pp. 101–119.

Donea, J., Giuliani, S., Laval, H. and Quartapelle, L. (1984) "Time-Accurate Solutions of Advection-Diffusion Problems by Finite-Elements." *Comp. Math. Appl. Mech. Engr.*, Vol. 45, pp. 123–145.

Galeao, A. C. and Dutra do Carmo, E. G. (1988) "A Consistent Approximate Upwind Petrov-Galerkin Method for Convection-Dominated Problems," *Comput. Methods Appl. Mech. Eng.*, Vol. 68, pp. 83–95.

Hanshaw, T. C. (1981) "A Petrov-Galerkin Finite Element Solution to the Time Dependent Burgers Equation," Masters Thesis, Dept. of Aerosp. and Mech. Eng., Univ. of Arizona, Tucson.

Hughes, T. J. R., Mallet, M. and Mizukami, A. (1986) "A New Finite Element Formulation for Computational Fluid Dynamics: II. Beyond SUPG," *Comput. Methods Appl. Mech. Eng.*, Vol. 54, pp. 341–355.

Idelsohn, S. R. (1989) "Upwind Techniques via Variational Principles," *Int. J. Numer. Methods Eng.*, Vol. 28, pp. 769–784.

Idelsohn, S. R., Heinrich, J. C. and Oñate, E. (1996) "Petrov-Galerkin Methods for the Transient Advective-Diffusive Equation with Sharp Gradients," *Int. J. Numer. Methods Eng.*, Vol. 39, pp. 1455–1473.

Lapidus, L and Pinder, G. F. (1982) *Numerical Solution of Partial Differential Equations in Science and Engineering.* New York: John Wiley and Sons.

Leonard, B. P. (1991) "The Ultimate Conservative Differencing Scheme Applied to Unsteady One-Dimensional Advection," *Comput. Methods Appl. Mech. Eng.*, Vol. 88, pp. 17–74.

Lick, W. J. (1989) *Difference Equations from Differential Equations.* Berlin: Springer-Verlag.

Pepper, D. W. and Baker, A. J. (1980) "A High-Order Accurate Numerical Algorithm for Three-Dimensional Transport Prediction," *Comput. Fluids*, Vol. 28, pp. 371–390.

Pepper, D. W. and Long, P. E. (1978) "A Comparison of Results Using Second-Order Moments with and without Width Correction to

Solve the Advection Equation," *J. Appl. Meteorol.*, Vol. 17, p. 228.

Perrochet, P. (1995) "Finite Hyper-Elements: A 4-D Geometrical Framework Using Covariant Bases and Metric Tensors," *Comm. Numer. Methods Eng.*, Vol. 11, pp. 525–534.

Richtmeyer, R. D. and Morton, K. W. (1957) *Differenace Methods for Initial-Value Problems*. New York: John Wiley and Sons.

Varoglu, E. and Finn, W. D. (1980) "Finite Elements Incorporating Characteristics for One-Dimensional Diffusion-Convection Equations," *J. Comput. Phys.*, Vol. 34, pp. 371–389.

Westerink, J. J. and Shea, D. (1989) "Consistent Higher Degree Petrov-Galerkin Methods for the Solution of the Transient Convection-Diffusion Equation," *Int. J. Numer. Methods Eng.*, Vol. 28, pp. 1077–1102.

Yu, C-C (1986) *Finite Element Analysis of Time-Dependent Convection-Diffusion Equations*, Ph.D. Dissertation, Dept. of Aerosp. and Mech. Eng., Univ. of Arizona, Tucson.

Yu, C-C and Heinrich, J. C. (1986) "Petrov-Galerkin Methods for the Time-Dependent Convective Transport Equation," *Int. J. Numer. Methods Eng.*, Vol. 23, pp. 883–901.

Yu, C-C and Heinrich, J. C. (1987) "Petrov-Galerkin Method for Multidimensional, Time-Dependent, Convective-Diffusion Equations," *Int. J. Methods Eng.*, Vol. 24, pp. 2201–2215.

EXERCISES

9.1   Find the element equations for the Crank-Nicolson-Galerkin method applied to the problem of Example 9.1.

9.2   Find the element equations for the space-time element Galerkin method used in Example 9.2.

9.3   Show that the resulting algorithm in Exercise 9.2 is equivalent to the semidiscrete Galerkin method combined with the θ method for $\theta = 2/3$.

9.4   Consider the diffusion equation

$$\frac{\partial u}{\partial t} = \alpha \frac{\partial^2 u}{\partial x^2} \qquad \begin{array}{l} 0 < x < \pi \\ t > 0 \end{array}$$

with initial condition $u(x,0) = f(x)$ and boundary conditions $u(0,t) = u(\pi,t) = 0$. Show that the solution can be written as

$$u(x,t) = \sum_{-\infty}^{\infty} a_k e^{k(ix - k\alpha t)}$$

where

$$a_k = \frac{1}{2\pi} \int_{-\pi}^{\pi} f(x) e^{-ikx} dx$$

9.5   Using linear elements, a regular mesh of size h, and the θ method with a lumped mass matrix and $\theta = 0$, show that the Euler-Galerkin method produces the difference equation

$$u_j^{n+1} = \left(\frac{\alpha \Delta t}{\Delta x^2}\right)\left(u_{j-1}^n - 2u_j^n + u_{j+1}^n\right)$$

9.6   Show that if a_k is given as in Exercise 9.4 and the boundary conditions are applied, Eq. (9.10) gives the exact solution to the difference equation in Exercise 9.5.

9.7   Find ξ for the difference equation in Exercise 9.5.

9.8   Use Eq. (7.11) to find the Crank-Nicolson-Galerkin element equation Eq. (9.5) as given in Eq. (9.13).

9.9   Derive Eq. (9.14) from Eq. (9.13).

9.10   Calculate the solution to Example 9.3 using $c = 1$.

9.11   Find ξ for the Crank-Nicolson-Galerkin method using a lumped mass matrix, and determine how much damping is associated with the numerical scheme.

9.12   Find the element equations for the Petrov-Galerkin discretization of Eq. (9.5) using the backward-implicit θ method in time $\theta = 1$. Use it to obtain the difference equation, Eq. (9.17).

9.13   Show that the algorithm in Example 9.4 with $\theta = 0$ yields the difference equation, Eq. (9.20), and find the numerical amplification factor.

9.14   Perform a von Neumann stability analysis of the algorithm used in Example 9.4.

9.15   Consider the heat equation

$$\frac{\partial u}{\partial t} = \alpha \left(\frac{\partial^2 u}{\partial x^2} + \frac{\partial^2 u}{\partial y^2} \right)$$

Discretize this equation using the Euler-Galerkin method ($\theta = 0$) with lumped mass matrix. Find the stability condition using von Neumann's method.

9.16   Plot the damping versus L/h for the algorithm of Example 9.5 with $c = 0.25, 0.5$, and 0.75.

9.17   Plot the phase angle versus L/h for the algorithm of Example 9.4 with $c = 0.25, 0.5$, and 1.0.

9.18   Repeat Exercises 9.16 and 9.17 using the consistent mass matrix in the discretization.

9.19   Find the phase error for the Crank-Nicholson-Galerkin method of Example 9.3 and plot it for $c = 0.25, 0.5$, and 0.75.

9.20   Show that when space-time elements are used, only the weighting functions corresponding to nodes 3 and 4 in Fig. 9.2 are needed to define the algorithm.

9.21   Construct a Petrov-Galerkin scheme for the diffusion equation $\partial\phi/\partial t = D\partial^2\phi/\partial x^2$ using space-time bilinear elements and weight functions of the form $N_i(x)T(t)$, where $N_i(x)$ are the linear shape functions and $T(t)$ is given by

$$T(t) = \begin{cases} 2t/\Delta t & 0 \le t \le \Delta t/2 \\ 2(1-t/\Delta t) & \dfrac{\Delta t}{2} \le t \le \Delta t \end{cases}$$

Show that the resulting algorithm is identical to the Crank-Nicolson-Galerkin scheme.

9.22   Repeat Exercise 9.21 for the advective equation, Eq. (9.5).

9.23   Write the Galerkin weighted residuals form of Eq. (9.30) and integrate the higher order derivatives by parts to find the weights in Eq. (9.31).

9.24   Using the shape and weighting functions given in Eqs. (9.32) and (9.33), construct the Petrov-Galerkin weighted residuals form of Eq. (9.1) (with $S = 0$) and derive Eq. (9.34).

9.25   If we set the coefficient of ϕ_{txxx} to zero in Eq. (9.36), we obtain

$$\frac{\alpha}{12}c^3 + \frac{1}{3\gamma}c^2 = \frac{1}{6\gamma} + \frac{\alpha}{12} - \frac{\alpha}{\gamma^2}$$

Show that this equation always has a real root such that $0 < c < 1$ and that when $\gamma \to 0$, $c \to 0$, and when $\gamma \to \infty$, $c \to 1$.

9.26   Show that if $\beta = 0$ in Eq. (9.36), third-order accuracy in space may be recovered choosing $c = \sqrt{6\alpha/\gamma}$.

9.27   Show that when $\partial\phi / \partial t \rightarrow 0$, Eq. (9.34) reduces to Eq. (8.19).

9.28   Find the difference equations for the Petrov-Galerkin method applied to Eq. (9.1) using the bilinear weight functions

$$w_1(x,t) = \left(1 - \frac{x}{h}\right)\left(1 - \frac{t}{\Delta t}\right) - \frac{\alpha}{2}\left(1 - \frac{t}{\Delta t}\right) - \frac{\beta h}{4}\left(1 - \frac{x}{h}\right)$$

$$w_2(x,t) = \frac{x}{h}\left(1 - \frac{t}{\Delta t}\right) - \frac{\alpha}{2}\left(1 - \frac{t}{\Delta t}\right) - \frac{\beta h}{4}\left(\frac{x}{h}\right)$$

9.29   In Exercise 9.28 the optimal values of α and β are given by (Yu and Heinrich, 1986)

$$\alpha = \frac{6}{\gamma c}\left(\coth\frac{\gamma}{2} - \frac{2}{\gamma}\right)$$

$$\beta = \left(1 - \frac{6}{\gamma c}\right)\left(\coth\frac{\gamma}{2} - \frac{2}{\gamma}\right)$$

Find the amplification factor , and show that these algorithms are always overdamped for any value of α and β.

9.30   Find the amplification factor ξ for Eq. (9.34).

9.31   Derive the form of the weighting functions Eq. (9.45), following the same steps used in Section 8.4 but adding balancing diffusion and dispersion.

9.32   Find the difference equations for the Euler-Taylor-Galerkin method using linear shape functions in one dimension. Find the numerical amplification factor, and show that the scheme is stable if and only if $c \leq 1$.

9.33   Using the approximation

$$\frac{\phi^{n+1} - \phi^n}{\Delta t} = \frac{1}{2}\left(\phi_t^n + \phi_t^{n+1}\right) + u^2 \frac{\Delta t}{4}\left(\phi_{xx}^n - \phi_{xx}^{n+1}\right) + u^2 \frac{\Delta t^2}{12}\left(\phi_{txx}^n + \phi_{txx}^{n+1}\right)$$

and

$$\frac{1}{2}\left(\phi_t^n + \phi_t^{n+1}\right) = \frac{\phi^{n+1} - \phi^n}{\Delta t}$$

find the Crank-Nicolson-Taylor-Galerkin formulation

$$\frac{u^{n+1} - u^n}{\Delta t} - u^2 \frac{\Delta t}{6}\left(u_{xx}^{n+1} - u_{xx}^n\right) + \frac{1}{2}u\left(u_x^n - u_x^{n+1}\right) - u^2 \frac{\Delta t}{4}\left(u_{xx}^n - u_{xx}^{n+1}\right)$$

9.34   In Exercise 9.33, discretize in space using linear elements, and show that the algorithm is unconditionally stable.

9.35.   Using Eq. (9.61), derive the Taylor-Galerkin formulation for the convection-diffusion equation, Eq. (9.1), when $S = 0$.

9.36.   Extend the Taylor-Galerkin formulation to treat the nonlinear time-dependent Burgers equation

$$\frac{\partial u}{\partial t} + u \frac{\partial u}{\partial x} = \varepsilon \frac{\partial^2 u}{\partial x^2}$$

9.37   Extend the Taylor-Galerkin formulation to the two-dimensional purely convective equation, Eq. (9.44), when $D = S = 0$.

9.38.   Extend the Taylor-Galerkin method to the two-dimensional convection-diffusion equation, Eq. (9.44).

TEN

VISCOUS INCOMPRESSIBLE FLUID FLOW

10.1 OVERVIEW

The finite element discretization concepts introduced in the previous chapters will now be applied to the solution of the Navier-Stokes equations for viscous incompressible flows in two dimensions. We will discuss the three types of algorithms most commonly used in the solution of the equations, namely, the mixed formulation, the fractional step method, and the penalty formulation. The latter will be treated in more detail, as it is the basis for very practical procedures for the solution of the Navier-Stokes equations. We will face a new difficulty because the incompressibility condition is a constraint on the space of possible solutions and is related in a nonexplicit form to the pressure field, which in this case, is a dynamic variable. Mathematically, this will be expressed as a consistency condition known as the Ladyzhenkaya, Babuska, and Brezzi (LBB) condition, which established the compatibility of the velocity and pressure spaces when the Navier-Stokes equations are discretized directly.

We must be aware of the fact that a very large number of algorithms have been proposed for the solution of these equations, and only one such algorithm will be studied in detail here. However, this particular scheme has been extensively tested and applied to many practical problems in which it either outperforms or is as good as other methods, while being easy to use.

We will finish the chapter with a variety of examples of application to practical problems and the extension to three-dimensional calculations.

10.2 BASIC FORM OF THE NAVIER-STOKES EQUATIONS

The two-dimensional Navier-Stokes equations for laminar, viscous, incompressible flow consist of the equilibrium or momentum equations written as (refer to Chapter 2)

$$\rho\left(\frac{\partial u}{\partial t} + u\frac{\partial u}{\partial x} + v\frac{\partial u}{\partial y}\right) = -\frac{\partial p}{\partial x} + \frac{\partial}{\partial x}\left(\mu\frac{\partial u}{\partial x}\right) + \frac{\partial}{\partial y}\left(\mu\frac{\partial u}{\partial y}\right) + \rho B_x \quad (10.1)$$

and

$$\rho\left(\frac{\partial v}{\partial t} + u\frac{\partial v}{\partial x} + v\frac{\partial v}{\partial y}\right) = -\frac{\partial p}{\partial y} + \frac{\partial}{\partial x}\left(\mu\frac{\partial v}{\partial x}\right) + \frac{\partial}{\partial y}\left(\mu\frac{\partial v}{\partial y}\right) + \rho B_y \quad (10.2)$$

In Eqs. (10.1) and (10.2), u and v are velocity components in the x and y directions, respectively, ρ is density, μ is kinematic viscosity, p is perssure, and B_x and B_y are the components of the body forces.

Furthermore, the incompressibility constraint is given by the continuity equation

$$\frac{\partial u}{\partial x} + \frac{\partial v}{\partial y} = 0 \qquad (10.3)$$

In this chapter we will concern ourselves with flows at velocities that are small in comparison with the speed of sound waves propagating through the mediums, so that we can assume that large changes in pressure cause only small changes in density. Therefore we will assume that $\rho = \rho_0$ is always constant, where ρ_0 is the reference density. Later on we will consider flows that are driven by differences in density caused by changes in temperatures or the concentration of some species in the fluid. In these cases, the density changes will be accounted for through the body forces.

The Navier-Stokes equations are also often written in terms of the stress tensor σ_{ij}. Using indicial notation, they take the form

$$\rho \frac{Du_i}{Dt} = \frac{\partial \sigma_{ij}}{\partial x_j} + \rho B_i \tag{10.4}$$

where for a Newtonian fluid, the stress is given by the constitutive relation

$$\sigma_{ij} = -p\delta_{ij} + \mu\left(\frac{\partial u_i}{\partial x_j} + \frac{\partial u_j}{\partial x_i}\right) \tag{10.5}$$

This form has been used mainly because boundary conditions at open boundaries can be interpreted in terms of the components of the stress normal to the boundary. However, as we will see in Section 10.4, it can also lead to incorrect treatment of outflow boundary conditions. Therefore we will work with the momentum equations written in the form of Eqs. (10.1) and (10.2), and we will refer to them as the velocity form of the Navier-Stokes equations. The expression for the momentum equations using Eqs. (10.4) and (10.5) will be termed the stress form of the Navier-Stokes equations.

To introduce these concepts in an easier fashion, let us consider a specific example that contains many of the different features found in a typical flow problem.

Example 10.1

A benchmark problem in the numerical solution of the Navier-Stokes equations consists in calculating the flow over a backward facing step, as shown in Fig. 10.1. A fully developed flow is imposed along the inlet boundary, $x = 0$, $H/2 \le y \le H$; along the solid boundaries the fluid must move at the same velocity as the walls, which in this case are stationary; this is called the no-slip boundary condition. These are all boundary conditions of the Dirichlet type; therefore the segment Γ_1 of the boundary is composed of

$$\Gamma_1 = \{y = 0, L_1 \le x \le L_2\} \cup \{x = L_1, 0 \le y \le H/2\} \cup \{y = H/2, 0 \le x \le L_1\}$$
$$\cup \{x = 0, H/2 \le y \le H\} \cup \{y = H, 0 \le x \le L_2\}$$

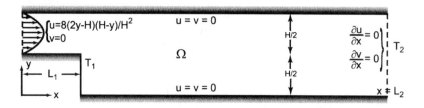

Figure 10.1 Flow over a backward facing step.

If $L_2 - L_1$ is large enough, at $x = L_2$ the flow should be fully developed. Therefore the gradients of the velocity components in the direction normal to the boundary must vanish. This is a Neumann boundary condition and constitutes the portion Γ_2 of the boundary, i.e.,

$$\Gamma_2 = \{x = L_2, 0 \le y \le H\}$$

In this particular case we will assume that no body forces are present. Therefore $B_x = B_y = 0$. Clearly, we could also have imposed the condition $v = 0$ at $x = L_2$, which would leave us with a Neumann condition for the first momentum equation and a Dirichlet condition for the second. This is perfectly acceptable but rarely leads to better approximations and more often will introduce inaccuracies in the calculations by modifying the flow field at the exit – if fully developed conditions have not been reached yet. ∎

REMARKS

1. Boundary conditions of the mixed type are rarely encountered when modeling fluid flow. However, if they should occur, they can be imposed in the same way as in Section 3.7.2. The boundary conditions that apply to flow with free surfaces are dealt with in Section 10.4.

2. In general, no boundary conditions are required for the pressure, except a reference value, which can be provided at any point in the domain. In some cases a known pressure may be the force driving the flow, in which case it can be readily applied at the corresponding boundary.

3. In some calculations it is necessary, or convenient, to impose cyclic boundary conditions along open boundaries. This can be achieved

in different ways, depending on the particular problem. We will discuss some of these in the examples.

4. Symmetry conditions also occur often in flow calculations and can be readily applied as a combination of a Dirichlet condition on one of the velocity components and a Neumann condition on the other velocity component. For example, a symmetric expansion such as the one shown in Fig. 10.2 can be modeled over one half of the domain using the boundary conditions $v = \partial u /\partial y = 0$ along the symmetry line.

The weak form of the boundary–value problem can be stated as follows: Find functions u* and v* in $H^1(\Omega)$ that satisfy the Dirichlet boundary conditions on Γ_1, and a function p* in $L^2(\Omega)$ such that

$$\int_\Omega \left\{ U\rho\left(\frac{\partial u^*}{\partial t} + u^*\frac{\partial u^*}{\partial x} + v^*\frac{\partial u^*}{\partial y}\right) - \frac{\partial U}{\partial x}p^* + \mu\left(\frac{\partial U}{\partial x}\frac{\partial u^*}{\partial x} + \frac{\partial U}{\partial y}\frac{\partial u^*}{\partial y}\right) + U\rho B_x \right\} d\Omega$$

$$+ \int_{\Gamma_2} U\left[\left(p^* + \mu\frac{\partial u^*}{\partial x}\right)n_x + \mu\frac{\partial u^*}{\partial y}n_y\right] d\Gamma = 0$$

(10.6)

$$\int_\Omega \left\{ V\rho\left(\frac{\partial v^*}{\partial t} + u^*\frac{\partial v^*}{\partial x} + v^*\frac{\partial v^*}{\partial y}\right) - \frac{\partial V}{\partial x}p^* + \mu\left(\frac{\partial V}{\partial x}\frac{\partial v^*}{\partial x} + \frac{\partial V}{\partial y}\frac{\partial v^*}{\partial y}\right) + V\rho B_x \right\} d\Omega$$

$$+ \int_{\Gamma_2} V\left[\mu\frac{\partial v^*}{\partial x}n_x + \left(-p^* + \mu\frac{\partial v^*}{\partial y}\right)n_y\right] d\Gamma = 0$$

(10.7)

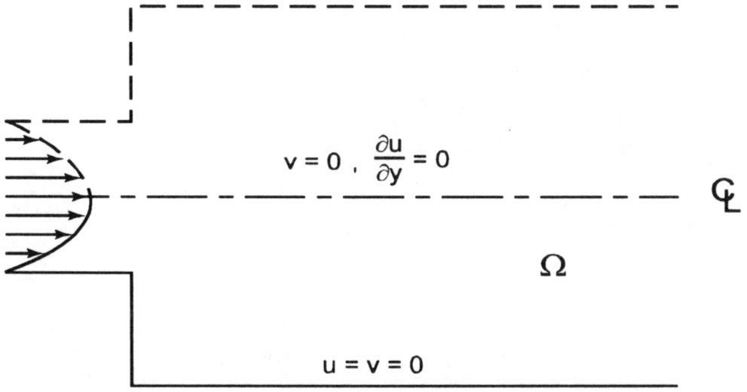

Figure 10.2 Symmetry boundary conditions in a divergent channel.

and

$$\int_{\Omega} Q \left(\frac{\partial u^*}{\partial x} + \frac{\partial v^*}{\partial y} \right) d\Omega = 0 \tag{10.8}$$

for all weighting pairs of functions U and V in $H_0^1(\Omega)$ and all functions Q in the space S that will be defined below.

REMARKS

1. This is called a "mixed variational formulation." Notice that we have not yet reached the point of a finite element discretization; this will come later. The mathematical theory behind this formulation is extremely tedious and is not within the scope of this book. The reader interested in the mathematical theory of finite element solution to the Navier-Stokes equations, and the particular aspects of existence and uniqueness of the solution to the above variational problem, is referred to the works of Ladyzhenskaya (1969), Teman (1979), Girault and Raviart (1979), and Carey and Oden (1986).

2. The pressure is sought in a space different from that for the velocity components, therefore the name "mixed" formulation. In fact, the pressure must be found over a subspace of $L^2(\Omega)$ referred to as "$L^2(\Omega)$ modulo constants" and denoted by $S = L^2(\Omega/\Re)$. The reason for this is that the pressure is only determined up to an additional constant; hence two functions that differ only by a real constant must be treated as the same.

3. Detailed mathematical analysis is only possible for the linear Stokes equations, that is,

$$\rho \frac{\partial u}{\partial t} = -\frac{\partial p}{\partial x} + \mu \left(\frac{\partial^2 u}{\partial x^2} + \frac{\partial^2 u}{\partial y2} \right) + B_x \tag{10.9}$$

$$\rho \frac{\partial v}{\partial t} = -\frac{\partial p}{\partial y} + \mu \left(\frac{\partial^2 v}{\partial x^2} + \frac{\partial^2 v}{\partial y2} \right) + B_y \tag{10.10}$$

plus the continuity equation, Eq. (10.3). The existence and uniqueness of solutions to these equations require the existence of a constant $\beta > 0$ such that

$$\text{SUP}_{\substack{\mathbf{V} \in H^1(\Omega) \\ \mathbf{V} \neq 0}} \frac{\left| \int_\Omega q \left(\frac{\partial u}{\partial x} + \frac{\partial v}{\partial y} \right) d\Omega \right|}{\| \mathbf{V} \|_1} \leq \beta \| q \|_0 \tag{10.11}$$

for all $q \in S$. This relation guarantees that the velocity and pressure spaces are "consistent" and is known as the LBB condition after Ladyzhenskaya (1969), Babuska and Aziz (1972) and Brezzi (1974).

The LBB condition is obviously satisfied for the Stokes equations. However, when we discretize the domain using finite elements to approximate the velocity and the pressure, it is easy to violate this condition in the chosen finite element subspace. Therefore we must check that the condition is satisfied for the finite element spaces. This has been done for many of the most commonly used spaces. The results are presented by Carey and Oden (1986) and will be discussed in the following sections.

10.3 CONSTANT-DENSITY FLOWS IN TWO DIMENSIONS

We turn our attention to the solution of Eqs. (10.1)-(10.3) when the body force terms B_x and B_y are zero. This will allow us to discuss the discretization process without the complications that arise in stratified flows in the presence of open boundaries, which will be dealt with in Section 10.4.

10.3.1 Mixed Formulation

Let us assume constant viscosity and rewrite Eqs. (10.1) and (10.2) in nondimensional form,

$$\frac{\partial u}{\partial t} + u \frac{\partial u}{\partial x} + v \frac{\partial u}{\partial y} = -\frac{\partial p}{\partial x} + \frac{1}{R_e} \left(\frac{\partial^2 u}{\partial x^2} + \frac{\partial^2 u}{\partial y^2} \right) \tag{10.12}$$

$$\frac{\partial v}{\partial t} + u\frac{\partial v}{\partial x} + v\frac{\partial v}{\partial y} = -\frac{\partial p}{\partial x} + \frac{1}{R_e}\left(\frac{\partial^2 v}{\partial x^2} + \frac{\partial^2 v}{\partial y^2}\right) \qquad (10.13)$$

where the Reynolds number Re is defined by

$$Re = \frac{\overline{U}\rho_0 L}{\mu} \qquad (10.14)$$

In Eq. (10.14), $\overline{U}$ is the characteristic velocity for the particular problem, and L is the characteristic length scale. The nondimensional variables are defined as

$$x = \frac{x'}{L} \quad y = \frac{y'}{L} \quad u = \frac{u'}{\overline{U}} \quad v = \frac{v'}{\overline{U}} \quad p = \frac{p'}{\rho_0 \overline{U}^2} \quad t = \frac{t'}{\left(L/\overline{U}\right)};$$

where the prime denotes a dimensional variable. The reference pressure is therefore $p_o = \rho_o \overline{U}^2$, and the reference time is $\tau = L/\overline{U}$.

The dependent variables u, v, and p are approximated using shape functions as

$$u(x,y,t) = \sum_j N_j(x,y)u_j(t) \qquad (10.15)$$

$$v(x,y,t) = \sum_j N_j(x,y)v_j(t) \qquad (10.16)$$

and

$$p(x,y,t) = \sum_k M_k(x,y)p_k(t) \qquad (10.17)$$

In order to satisfy the discrete LBB condition we must interpolate the pressure with a lower order polynomial than the velocity components. Figure 10.3 shows the simplest and most often used combination of such mixed interpolation. The combinations 1, 3 and 5 involve discontinuous pressure fields. The evidence from numerical experiments indicates that

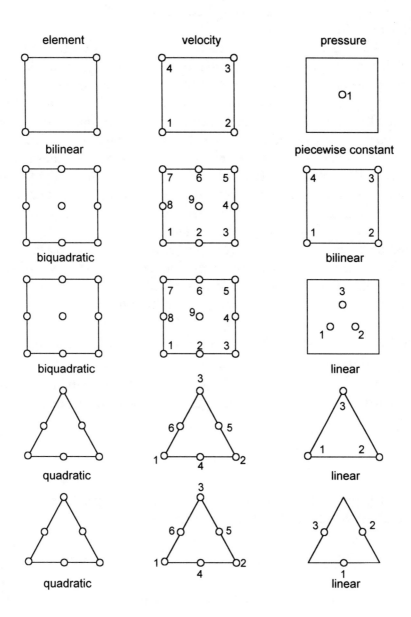

Figure 10.3 Elements for mixed interpolation of velocity and pressure.

these approximations to the pressure field result in improved accuracy over those combinations using continuous C^o pressures, as in combination 2 and 4.

The reader may be surprised that the interpolation of the velocity using linear triangles is not considered in Fig. 10.3. In fact, the linear velocity-constant pressure combination does not satisfy the LBB condition. However, the situation can be remedied if a constant pressure is defined over more that one linear triangle. This opens the door to many possibilities of other combinations in which the velocity and pressure are interpolated on different meshes, where the mesh for the velocity field corresponds to a refinement of the mesh for the pressure. These combinations can become cumbersome and limit the flexibility of the mesh generation. We will consider only schemes with velocity and pressure defined over a single mesh, as in Fig. 10.3. A more detailed discussion of mixed elements from the point of view of satisfaction of the LBB condition is given by Carey and Oden (1986). We will also defer the discussion of the rates of convergence. First, let us look at the implementation of the method using a Galerkin formulation.

Substituting Eqs. (10.15)–(10.17) into the weak formulation given by Eqs. (10.6)–(10.8), and setting the weighting functions U, V, and P equal to the shape functions, N_i, N_i, and M_i, respectively, yields the discrete Galerkin equations. For one element in which velocity is interpolated to n-noded elements and the pressure using m points, we obtain

$$\mathbf{M}\dot{\mathbf{d}} + \mathbf{K}\,\mathbf{d} + \mathbf{F} = \mathbf{0} \qquad (10.18)$$

The dimension of the system is $(2n + m) \times (2n + m)$. To simplify the notation, the degrees of freedom are ordered so that

$$\mathbf{d} = \left[u_1 u_2 \ldots u_n v_1 v_2 \ldots v_n p_1 \ldots p_m\right]^T = \begin{bmatrix} \mathbf{u} \\ \mathbf{v} \\ \mathbf{p} \end{bmatrix} \qquad (10.19)$$

For the bilinear velocity-constant pressure and biquadratic velocity-linear pressure elements we have

$$\mathbf{d} = \begin{bmatrix} u_1 u_2 u_3 u_4 v_1 v_2 v_3 v_4 p_1 \end{bmatrix}^T$$

and

$$\mathbf{d} = \begin{bmatrix} u_1 u_2 \ldots u_9 v_1 v_2 \ldots v_9 p_1 p_2 p_3 \end{bmatrix}^T$$

respectively.

The mass matrix **M** is defined by

$$\mathbf{M} = \begin{bmatrix} \mathbf{A} & \mathbf{0} & \mathbf{0} \\ \mathbf{0} & \mathbf{A} & \mathbf{0} \\ \mathbf{0} & \mathbf{0} & \mathbf{0} \end{bmatrix} \qquad (10.20)$$

where **A** is an n x n mass matrix given by

$$\begin{bmatrix} a_{ij} \end{bmatrix} = \begin{bmatrix} \int_{\Omega} N_i N_j \, d\Omega \end{bmatrix} \qquad (10.21)$$

The stiffness matrix **K** contains only the linear part of the operator, i.e., the viscous terms, the pressure terms, and the continuity constraint, and is given by

$$\mathbf{K} = \begin{bmatrix} \mathbf{B} & \mathbf{0} & \mathbf{C}_x \\ \mathbf{0} & \mathbf{B} & \mathbf{C}_y \\ -\mathbf{C}_x^T & -\mathbf{C}_y^T & \mathbf{0} \end{bmatrix} \qquad (10.22)$$

Here **B** is an n x n matrix defined by

$$\left[b_{ij} \right] = \left[\int_{\Omega} \frac{1}{Re} \left(\frac{\partial N_i}{\partial x} \frac{\partial N_j}{\partial x} + \frac{\partial N_i}{\partial y} \frac{\partial N_j}{\partial y} \right) d\Omega \right] \tag{10.23}$$

and corresponds to the Laplacian operator. The matrices $\mathbf{C}_x$ and $\mathbf{C}_y$ are both n x m and have the form

$$\left[(c_x)_{ik} \right] = \left[-\int_{\Omega} \frac{\partial N_i}{\partial x} M_k d\Omega \right] \tag{10.24}$$

and

$$\left[(c_y)_{ik} \right] = \left[-\int_{\Omega} \frac{\partial N_i}{\partial y} M_k d\Omega \right] \tag{10.25}$$

The function $\mathbf{F}$ is nonlinear, contains the convective terms, and is given by

$$\mathbf{F} = \begin{bmatrix} \mathbf{F}_x \\ \mathbf{F}_y \\ \mathbf{0} \end{bmatrix} \tag{10.26}$$

with the n x 1 vectors F_x and F_y defined by

$$\left[(f_x)_i \right] = \left[\int_{\Omega} N_i \left\{ \left(\sum_j N_j u_j \right) \left(\sum_k \frac{\partial N_k}{\partial x} u_k \right) + \left(\sum_j N_j v_j \right) \left(\sum_k \frac{\partial N_k}{\partial y} u_k \right) \right\} d\Omega \right] \tag{10.27}$$

and

$$\left[(f_y)_i \right] = \left[\int_{\Omega} N_i \left\{ \left(\sum_j N_j u_j \right) \left(\sum_k \frac{\partial N_k}{\partial x} v_k \right) + \left(\sum_j N_j v_j \right) \left(\sum_k \frac{\partial N_k}{\partial y} v_k \right) \right\} d\Omega \right] \tag{10.28}$$

The line integrals in Eqs. (10.6) and (10.7) will almost always vanish, except for the terms containing the pressure, which must be retained at open boundaries when the body forces are significant. We will discuss these in Section 10.4; here we will simply assume them to be zero.

Finally, note that in Eq. (10.18), nonhomogeneous Dirichlet conditions such as the developed inflow in Example 10.1 have not yet been applied. These will generate a constant (or time dependent) right-hand side vector. Let us now look at a very simple but important example – the case of a developed plane Poiseuille flow in a channel.

Example 10.2

One of the few flows for which an analytical solution to the Navier-Stokes equations is available is the Poiseuille flow between parallel plates. Figure 10.4 shows a region of height H and length 5H nondimensionalized using the characteristic length L = H. The solution is given by

$$u = 4\,u_m\,y\,(1-y)$$

$$v = 0$$

$$\frac{\partial p}{\partial x} = \frac{8\,u_m}{R_e}$$

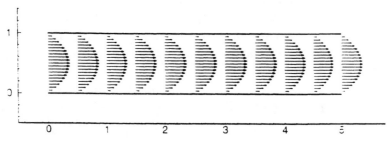

Figure 10.4 Velocity in a steady Poiseuille flow.

where u_m is the maximum velocity at the center of the channel. A finite element solution for the velocity field is also shown in Fig. 10.4. This was obtained using bilinear elements for the velocity and piecewise constant pressure in a regular mesh of size $\Delta x = 0.1$ and $\Delta y = 0.05$ at Re $= 100$.

The boundary conditions were as follows: developed flow at the entry, $x = 0$; no slip, $u = v = 0$ at the walls, $y = 0$, $y = 1$; and $p + \mu \partial u / \partial x = \partial v / \partial x = 0$ at the outflow boundary. In Fig. 10.4 we observe that the initial profile simply repeats itself as we move downstream, indicating a very accurate solution. The calculated pressure is also quite accurate, but this is discussed in Section 10.4. ∎

REMARKS

1. To calculate steady state flows, we may use a direct nonlinear iteration as described in Section 6.3.2, or we may use a time-dependent solution and calculate until the solution no longer changes in time.

The second approach is preferred when physical instabilities may occur that can change the mode of circulation. A direct iterative solution will normally not be able to branch off to a new mode that is physically more stable. This will be illustrated clearly when we discuss stratified flows. When a time-dependent algorithm is chosen, there are two basic ways to deal with the nonlinear terms.

(i) One method is treating the convective terms explicitly, that is, evaluating the nonlinear terms at each time step using the latest computed values and assembling them into the forcing term in the right-hand-side. This introduces a stability condition of the same form as given in Eq. (9.53), which means that the Courant number must be less than 1 for every element in the mesh. If the evolution of the flow field in time is important, this does not constitute a restriction. For accurate calculation of the time evolution, we strongly recommend that this condition be satisfied, even if an unconditionally stable method is used. Therefore this method is appropriate for problems in which the time dependency is important.

(ii) A second method is using an unconditionally stable time integrating algorithm combined with a Newton-Raphson iteration within each time step. This allows for the use of larger time steps in order to achieve steady state more rapidly. However, it can become expensive if the time evolution must be calculated accurately. Therefore this method offers an alternative to a direct Newton-Raphson iteration to compute steady state solutions. It has the advantage that the effect of high Rayleigh numbers, which makes solutions difficult to obtain with a direct iteration, can be counteracted by choosing a smaller time step at the beginning of the calculation. This time step can be increased as steady state is approached. Example 10.2 was solved using this method with a backward-implicit ($\theta = 1$) time stepping scheme.

2. At high Re the flow becomes highly convective, and the use of a Petrov-Galerkin formulation is necessary.

Unfortunately, for nonlinear problems our understanding of the effect of Petrov-Galerkin methods in the numerical calculations is limited. For example, experience shows that in problems involving body forces, the consistent weighting of those terms leads to unstable algorithms. For these and other reasons, discussed in Chapter 9, we prefer to apply the Petrov-Galerkin weights only to the convective terms. The results of extensive numerical calculations, some of which will be discussed here, show that this is sufficient to produce stable and accurate algorithms.

When time-dependent algorithms, such as discussed in points (i) and (ii) above, and bilinear elements are used, it suffices to weight the convective terms in the right-hand-side with the weighting functions given in Eq. (8.57), i.e.,

$$w_i = N_i + \frac{\alpha \overline{h}}{2|\mathbf{V}|}\left(u\frac{\partial N_i}{\partial x} + v\frac{\partial N_i}{\partial y}\right)$$

where $\alpha = \coth \gamma/2 - 2/\gamma$ as before and γ is now the cell Reynolds number given by

$$\gamma = \text{Re}|\mathbf{V}|\overline{h} \tag{10.22}$$

Referring to the results of Chapter 9, Section 9.3.2, improved accuracy will be achieved using a second-order time-stepping scheme. However, for many applications a cheaper first-order method is sufficient.

(ii) If a Newton-Raphson iteration is used in the algorithm, the convective term in the tangent matrix takes the form

$$u^k \frac{\partial \Delta u}{\partial x} + \frac{\partial u^k}{\partial x} \Delta u + v^k \frac{\partial \Delta u}{\partial y} + \frac{\partial u}{\partial y} \Delta v^k$$

in the momentum equation in the x-direction, and a similar expression is obtained for the convective terms in the y-direction. In this case, a straight application of our Petrov-Galerkin weighting functions will not be correct. In general, it may lead to a non-converging algorithm due to the presence of the second and fourth terms in the above expression if the time step is not sufficiently small. A method involving a second correction of the type proposed by Harari and Hughes (1994) or Idelsohn et al. (1995), combined with adequate mesh refinement, may then be necessary.

We will concentrate on time-dependent algorithms only, for which the Petrov-Galerkin methods defined in Chapter 9 are sufficient to achieve stability and the amount of balancing diffusion that must be added can be calculated unambiguously. However, we must remind the reader that a wide variety of methods that add diffusion to stabilize the algorithms have been proposed in the literature, most of which can be very effective. Even though no such thing as a best method exists, the algorithms described here have been extensively tested and are simple to use, providing a very useful tool for the practitioner.

Example 10.3

The steady state solution to flow over a backward facing step, shown in Fig. 10.1, is examined for Re = 900, based on the entry length H/2. A mesh of 3000 rectangular bilinear elements is used with piecewise constant pressure, in a mixed formulation. The mesh is irregularly spaced for better resolution close to the solid boundaries and at the entry. Referring to Fig. 10.1, $L_1 = 3H$ and $L_2 = 19H$. The solution was obtained using a backward-implicit time-stepping scheme combined with a Newton-Raphson iteration within each time step and Petrov-Galerkin weighting of the convective terms. The steady state streamfunction is shown in Fig. 10.5a. A second recirculation cell is observed at the top wall that was not present at the low Re flow. The pressure field is shown in Fig. 10.5b. The reader should examine these results for physical consistency. ∎

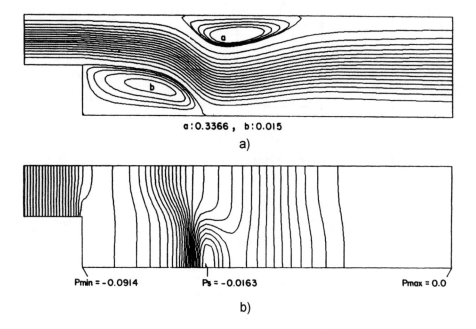

a : 0.3366 , b : 0.015

a)

Pmin = -0.0914   Ps = -0.0163   Pmax = 0.0

b)

Figure 10.5 Flow over a backward facing step at Re = 900: (a) streamfunction contours and (b) pressure contours.

REMARKS

1. Flow over a backward facing step has become one of the benchmark problems to test numerical models (Gartling, 1990; Blackwell and Pepper, 1992). Another very popular problem to test methods is the cavity-driven flow introduced by Burggraf (1966). There exists a great amount of information and solutions to this problem in the literature.

2. The pressure calculated in the program is piecewise constant. In some cases, particularly if the data are not smooth, the calculated pressure field oscillates from one element to the next in what is usually called a "checkerboard" mode. The oscillations occur because the discretization space of the pressure admits functions that are not constant, but for which the discrete divergence operator vanishes. We can express this as follows: there exists a piecewise constant function $p^*(x,y) \neq$ const., such that

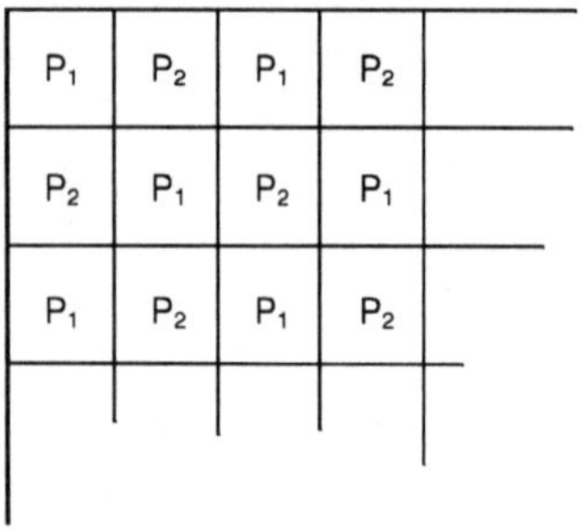

P$_1$	P$_2$	P$_1$	P$_2$
P$_2$	P$_1$	P$_2$	P$_1$
P$_1$	P$_2$	P$_1$	P$_2$

Figure 10.6 Checkerboard pressure patterns.

$$\int_{\Omega} p^* \left[\sum_j \frac{\partial N_i}{\partial x} u_j + \sum_j \frac{\partial N_i}{\partial y} v_j \right] d\Omega = 0 \qquad (10.23)$$

It can be shown (Exercise 10.6) that for the bilinear velocity–constant pressure mixed formulation over a square mesh, a pressure field that varies in a checkerboard pattern as shown in Fig. 10.6 satisfies Eq. (10.23). The pressure output is oscillatory whenever the checkerboard mode is excited, and it must be filtered out to obtain the actual pressure distribution.

An example when the checkerboard mode is excited is provided by flow over a backward facing step at Re = 100. Figure 10.7 shows the calculated pressures along the top wall. These are clearly oscillatory; also the smoothed pressure and the pressure calculated using a Poisson equation derived from the momentum equations are shown; the latter method is discussed in Section 10.4.1.

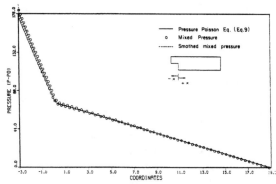

Figure 10.7 Calculated pressure along the top wall for flow over a backward facing step at Re =100.

3. The appearance of the checkerboard pressure mode should not be a cause for alarm. The smoothed pressure at the nodes can be readily obtained using a least squares fit with bilinear elements as follows.

Let $P(x,y) = \Sigma_j N_j(x,y)P_j$ denote the bilinear interpolation of the pressure, where P_j are the values of pressure at the nodes, and $p(x,y) = \Sigma_k M_k(x,y)p_k$ denote the piecewise constant calculated pressure. The functions M_k are such that

$$M_k = \begin{cases} 1 & \text{in element k} \\ 0 & \text{in all other elements} \end{cases}$$

and p_k is the value of the pressure over element k. We now construct the functional

$$J = \int_\Omega (P - p)^2 \, d\Omega \tag{10.24}$$

and minimize it with respect to the values P_i. The conditions for a minimum are

$$\frac{\partial J}{\partial P_i} = 2 \int_\Omega N_i \left[\sum_j N_j P_j - \sum_k M_k p_k \right] d\Omega = 0$$

which results in a system of equations of the form

$$\mathbf{A} \, \mathbf{P} = \mathbf{B} \tag{10.25}$$

for the values P_i, where $\mathbf{A}$ is the usual mass matrix

$$[a_{ij}] = \int_\Omega N_i \, N_j \, d\Omega \tag{10.26}$$

and right-hand side vector $\mathbf{B}$ is given by

$$[b_i] = \int_\Omega N_i \sum_k M_k p_k \, d\Omega \qquad (10.27)$$

The method can clearly be extended to other interpolation spaces; in particular, to the case of biquadratic velocities – bilinear pressure, which also exhibit a checkerboard mode.

Using mass lumping, this procedure simplifies to a simple weighted average of the pressure over rectangular meshes (Exercise 10.7). However, in the above form it is more accurate and applicable to isoparametric elements.

10.3.2 Fractional Step Method

The method of fractional steps has been used extensively, especially in finite difference discretizations. Some selected references are Chorin (1967), Yanenko (1971), Schneider et al (1978), Donea et al. (1982), and Quartapelle (1993). The method is designed for the solution of the time-dependent equations and consists in performing the solution in two steps as follows.

First, we split the velocity components into two parts as

$$u = u^* + u'$$

$$v = v^* + v' \qquad (10.28)$$

where u^* and v^* satisfy the momentum equations without the pressure terms, i.e., using the vector notation for conciseness,

$$\frac{\partial u^*}{\partial t} = -\mathbf{V} \cdot \nabla u + \frac{1}{Re} \nabla^2 u \qquad (10.29)$$

$$\frac{\partial v^*}{\partial t} = -\mathbf{V} \cdot \nabla v + \frac{1}{Re} \nabla^2 v \qquad (10.30)$$

Using an explicit forward Euler scheme to solve Eqs. (10.29) and (10.30), one obtains

$$\frac{u_{n+1}^* - u^n}{\Delta t} = -\mathbf{V}^n \cdot \nabla u^n + \frac{1}{Re} \nabla^2 u^n \qquad (10.31)$$

$$\frac{v_{n+1}^* - v^n}{\Delta t} = -\mathbf{V}^n \cdot \nabla v^n + \frac{1}{Re} \nabla^2 v^n \qquad (10.32)$$

The velocity components calculated form Eqs. (10.31), and (10.32) will not satisfy the continuity equation. Therefore we need to compute corrections u′ and v′ with the help of the pressure field. To do this, we must first find the pressure field. Differentiating Eq. (10.12) with respect to x and Eq. (10.13) with respect to y and adding, the time derivative terms vanish by virtue of the continuity equation, and we can write

$$\nabla^2 p = \frac{\partial}{\partial x}\left(-\mathbf{V} \cdot \nabla u + \frac{1}{Re} \nabla^2 u\right) + \frac{\partial}{\partial y}\left(-\mathbf{V} \cdot \nabla v + \frac{1}{Re} \nabla^2 v\right) \qquad 10.33)$$

Substituting Eqs. (10.31) and (10.32) into Eq. (10.33), we obtain

$$\nabla^2 p^{n+1} = \frac{1}{\Delta t}\left(\frac{\partial u_{n+1}^*}{\partial x} + \frac{\partial v_{n+1}^*}{\partial y}\right) \qquad (10.34)$$

which must be solved for the pressure using any of our already known methods. Once p^{n+1} is known, from the complete discretized equations, we get

$$\frac{u^{n+1} - u^n}{\Delta t} = -\mathbf{V}^n \cdot \nabla u^n + \frac{1}{Re} \nabla^2 u^n - \frac{\partial p^{n+1}}{\partial x} \qquad (10.35)$$

$$\frac{v^{n+1} - v^n}{\Delta t} = -\mathbf{V}^n \cdot \nabla v^n + \frac{1}{Re} \nabla^2 v^n - \frac{\partial p^{n+1}}{\partial y} \qquad (10.36)$$

Substituting Eqs. (10.28), (10.31), and (10.32) we obtain

$$u'_{n+1} = -\Delta t \frac{\partial p^{n+1}}{\partial x}$$

$$(10.37)$$

$$v'_{n+1} = -\Delta t \frac{\partial p^{n+1}}{\partial y}$$

The new velocity components $u^{n+1} = u^*_{n+1} + u'_{n+1}$ and $v^{n+1} = v^*_{n+1} + v'_{n+1}$ satisfy Eqs. (10.35) and (10.36) as well as the incompressibility condition.

It is not hard to imagine that many variations of the present methodology can be obtained once this basic form is understood. Furthermore, in this formulation the pressure never appears implicitly when calculating the velocity components. In some algorithms this means that the LBB condition does not need to be explicitly satisfied, and in particular, the pressure may be interpolated using the same shape functions as for the velocity components. This has led some authors to call them "equal-order interpolation" schemes.

The advantage of this method is that only the solution of the Poisson equation for the pressure must be solved implicitly. The disadvantage is that the time step can become extremely small to satisfy stability, and calculations become prohibitively lengthy. The implementation of the method using bilinear finite elements does not pose particular difficulties and we address some of the aspects in the exercises.

10.3.3 Penalty Function Formulation

We now turn to a very practical class of algorithms, which are also the ones used in the software included with this book. The penalty function methods can be derived directly from the Stokes viscosity law (Fukumori and Wake, 1991); recall Eq. (2.9) now written in the form

$$p = p_s - \left(\overline{\mu} + \frac{2}{3}\mu\right)\nabla \cdot \mathbf{V} \qquad (10.38)$$

where p_s denotes the thermodynamic or static component of the pressure, p is the mean pressure and $\overline{\mu}$ is the second coefficient of viscosity.

The second viscosity $\bar{\mu}$ is still a controversial quantity. Stokes hypothesized that p and $\bar{p}$ must be equal and hence $\bar{\mu} = -2/3\ \mu$. This has been shown experimentally to be true for some gases; however, experiments with liquids have shown that λ is a positive quantity much larger than μ (White, 1974). The coefficient

$$\lambda = \left(\bar{\mu} + \frac{2}{3}\mu \right)$$ (10.39)

represents the bulk viscosity of the fluid. If the fluid is incompressible, large changes in pressure produce very small changes in the volume, and therefore λ must be a large number. Furthermore, as the fluid becomes perfectly incompressible, λ must tend to infinity.

The basic idea of the penalty method consists in expressing the pressure through the pseudoconstitutive relation

$$p = p_s - \lambda \nabla \cdot \mathbf{V}$$ (10.40)

in which λ is a large number. We will refer to this as the modified penalty formulation (for reasons explained later). Equation (10.40) is then substituted into the momentum equations,

$$\frac{\partial u}{\partial t} + \mathbf{V} \cdot \nabla u = -\frac{\partial p_s}{\partial x} + \lambda \frac{\partial}{\partial x}(\nabla \cdot \mathbf{V}) + \frac{1}{Re}\nabla^2 u$$ (10.41)

$$\frac{\partial v}{\partial t} + \mathbf{V} \cdot \nabla v = -\frac{\partial p_s}{\partial y} + \lambda \frac{\partial}{\partial y}(\nabla \cdot \mathbf{V}) + \frac{1}{Re}\nabla^2 v$$ (10.42)

and the continuity equation is no longer necessary. The discretizations of Eqs. (10.41) and (10.42) result in the solution of a discrete system of equations that involve only the velocity degrees of freedom; which is about 15% smaller for bilinear elements. Furthermore, there is no occurrence of zero diagonal elements of the final linear matrices, although the system will remain ill conditioned due to the large value of the penalty parameter. Hence direct equation solvers are required.

REMARKS

1. For constant density flows, the static component p_s in Eq. (10.40) is eliminated through a redefinition of the pressure, so that the hydrostatic pressure due to gravity is canceled out. For this reason, the body force terms are not written in Eqs. (10.12) and (10.13) even though one of these directions might represent the vertical direction aligned with gravity. In this case, the penalty formulation may be written as

$$p = -\lambda \, \nabla \cdot \mathbf{V} \qquad (10.43)$$

which is the standard form found in most references. However, this form of the penalty method is incorrect when applied to stratified flows and flows with free surfaces, as is discussed in the next section.

2. We will derive the discretized equations in the more general context of stratified flows in the presence of body forces, but first, we analyze the properties of the penalty method. To simplify this presentation, assume that we are interested in finding steady state solutions to the Stokes equations, i.e.,

$$\mu \left(\frac{\partial^2 u}{\partial x^2} + \frac{\partial^2 u}{\partial y^2} \right) = \frac{\partial p}{\partial x} \qquad (10.44)$$

$$\mu \left(\frac{\partial^2 v}{\partial x^2} + \frac{\partial^2 v}{\partial y^2} \right) = \frac{\partial p}{\partial y} \qquad (10.45)$$

that satisfy the incompressibility condition, Eq. (10.3). The Galerkin penalty function formulation of Eqs. (10.44) and (10.45) yields (Exercise 10.9).

$$\left[\int_\Omega \mu \left(\frac{\partial N_i}{\partial x} \frac{\partial N_j}{\partial x} + \frac{\partial N_i}{\partial y} \frac{\partial N_j}{\partial y} \right) u_j \, d\Omega \right] + \left[\int_\Omega \lambda \frac{\partial N_i}{\partial x} \left(\frac{\partial N_j}{\partial x} u_j + \frac{\partial N_j}{\partial y} v_j \right) d\Omega \right] = 0$$

$$(10.46)$$

$$\left[\int_\Omega \mu \left(\frac{\partial N_i}{\partial x}\frac{\partial N_j}{\partial x} + \frac{\partial N_i}{\partial y}\frac{\partial N_j}{\partial y}\right) u_j\, d\Omega\right] + \left[\int_\Omega \lambda \frac{\partial N_i}{\partial y}\left(\frac{\partial N_j}{\partial x} u_j + \frac{\partial N_j}{\partial y} v_j\right) d\Omega\right] = 0$$

$$10.47)$$

where we assumed that all line integrals vanish, since they are irrelevant to this discussion. The final system of linear equations can be written as

$$\left[\mu K_1 + \lambda K_2\right] d = F \qquad (10.48)$$

where the vector F is generated from the boundary conditions and, in this case, the vector d contains only velocity degrees of freedom. Let us assume now that the matrix K_2 is nonsingular and that λ is increased more and more in an effort to satisfy incompressibility better. Because μ and K_1 remain constant as λ increases, they become negligible, and the solution to Eq. (10.48) can be written as

$$d = \frac{1}{\lambda} K_2^{-1} F \qquad (10.49)$$

Because K_2 and F are also constant, we note that $d \to 0$ as $\lambda \to \infty$. This phenomenon is called "locking" and is directly related to satisfaction (or the lack thereof) of the LBB condition. In this case, the space S of the discrete pressures is not compatible with the space of divergence-free velocities, and therefore the only element of that space is $V = 0$.

We must guarantee that the penalty matrix K_2 is singular. This is achieved using selective reduced integration of the penalty term and a quadrature rule with degree of precision lower than required to guarantee an optimal convergence rate in the matrix K_1. Mathematical techniques to analyze elements to be used with the reduced integration penalty term have been developed by Oden (1982), Oden et al. (1982), Carey and Krishnan (1982), and Idelsohn et al. (1995). The analyses also yield an estimate of the parameter β in Eq. (10.11); the results are shown in Fig. 10.8. It is clear from this figure that when full integration is used in the penalty term (cases 1, 3 and 6), the method always locks. The most commonly used elements are cases 2 and 7, both of which are labeled as prone to produce an "oscillatory pressure." This means that these elements may exhibit checkerboarding in the pressure, based on the calculated estimates of the constant β. However, these estimates are not optimal. It has been shown by Carey and Krishnan (1982, 1984) that in the penalty method, both bilinear velocities with one-point reduced integration of the penalty term, and biquadratic (nine node) velocities with 2 x 2 reduced Gauss integration, converge at optimal rates, provided the problem data are smooth. If the data lack regularity, the pressure calculated using these elements can be oscillatory if the checkerboard modes are excited. In these cases, the simple smoothing procedure described in Section 10.3.1 can be used to eliminate the oscillations.

3. The mathematical aspects of the penalty method have been analyzed in great detail throughout the years. Particularly noteworthy is the equivalence theorem of Malkus and Hughes (1978). They proved that the solutions obtained using the mixed method with bilinear elements and constant pressure, and those obtained with the penalty method using the bilinear element and one-point reduced integration of the penalty, are the same. Convergence of the solutions for viscous incompressible flows and error estimates for the approximations can only be obtained for the linear case of Stokes flow (Teman, 1979; Carey and Krishnan, 1982). For the nonlinear Navier-Stokes equations, this can only be done under the assumption of very low Re. Otherwise, uniqueness of the solution cannot be guaranteed (Carey and Krishnan, 1984). A review of the penalty method by Heinrich and Vionnet (1995a,b) gives an extensive list of references on the history, theory and practice of the method.

4. To select the value of the penalty parameter λ, we must take into consideration the specific form in which the Navier-Stokes equations are written, and the word length in the floating point calculations. The first of these considerations indicates that it is desirable to formulate the equations using nondimensional forms that best balance the coefficients of the viscous and convective terms, as well as other terms that will be introduced later. The second implies that the penalty

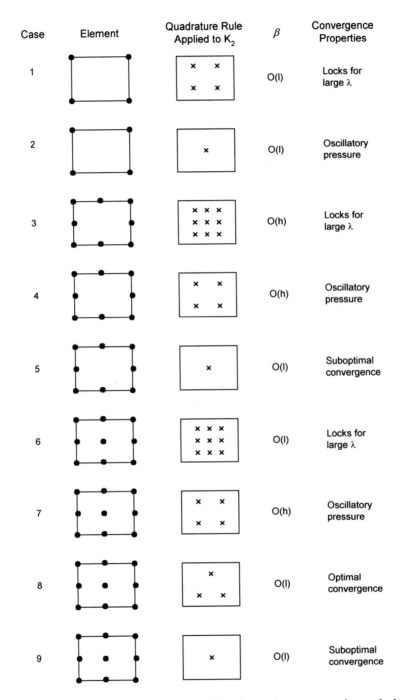

Figure 10.8. Analysis of the LBB condition for various commonly used elements (h denotes the mesh size parameter).

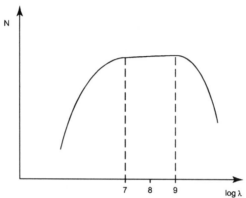

Figure 10.9 Schematic of accuracy versus penalty parameter.

parameter should be chosen to be approximately $O(10^{k/2})$, where k is the word length in the decimal system.

Computational experience shows that penalty calculations must be performed using double-precision 64 bit words. Under these circumstances, a penalty parameter λ between 10^7 and 10^9 will be adequate in most practical situations. The value of λ is problem independent, provided that the range of the governing parameters does not vary too much. Bercovier and Engelman (1979) and Carey and Krishnan (1982) have studied the effect of the penalty parameter on the solution. A schematic of the typical behavior of the solution in terms of the number N of digits of accuracy vs. $\log\lambda$ is shown in Fig. 10.9.

Example 10.4

To illustrate the effect of the magnitude of the penalty parameter λ, let us go back to the plane Poiseuille flow in a channel at Re = 100. Figure 10.10 shows the penalty solution for values of the penalty parameter $\lambda = 10^2$, 10^7, and 10^{15}. In Figure 10.8a, $\lambda = 10^2$, and we see that a loss of mass occurs as we move downstream along the channel. For $\lambda = 10^7$ in Fig. 10.10b, incompressibility is effectively imposed. In Fig. 10.10c, for $\lambda = 10^{15}$, the solution is meaningless because we do not have enough precision in the calculation.

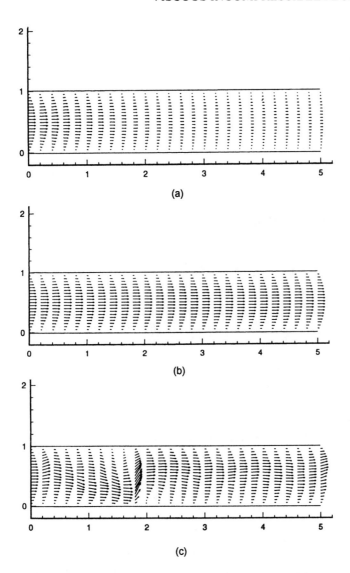

Figure 10.10 Poiseuille flow in a channel at Re = 100: (a) velocity profiles for $\lambda = 10^2$, (b) velocity profiles for $\lambda = 10^7$, and (c) velocity profiles for $\lambda = 10^{15}$.

Figure 10.11 shows the ratio of the total fluid mass leaving the computational domain at x = 5, to the total fluid mass entering the computational domain at x = 0. We observe that mass is accurately conserved for values $\lambda \geq 10^5$. Also notice that the solution in Fig. 10.10c, although inaccurate, does satisfy mass conservation. ∎

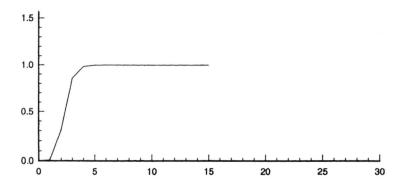

Figure 10.11 Mass flux ratio versus log λ for Poiseuille flow in a channel at Re = 100.

5. There are two more forms of the penalty formulation that are useful and yield similar algorithms. The first is based on allowing small density perturbations about the reference state ρ_0, i.e., we assume that the fluid is barotropic and satisfies an equation of state of the form

$$p - p_0 = \left(\frac{\partial p}{\partial \rho}\right)_0 (\rho - \rho_0) \tag{10.50}$$

Here p_0 is the reference pressure, and by definition $(\partial p / \partial \rho)_0 \equiv c_0^2$, where c_0 is the speed of sound at the reference state. We now use the continuity equation for a compressible flow, Eq. (2.5),

$$\frac{\partial \rho}{\partial t} + \mathbf{V} \cdot \nabla \rho + \rho \nabla \cdot \mathbf{V} = 0 \tag{10.51}$$

and assume that the flow is only slightly compressible, so that

$$\rho' = \rho - \rho_0 \ll \rho_0 \tag{10.52}$$

and ρ' can be neglected when compared to ρ_0. Substituting into Eq. (10.51), we obtain

$$\frac{\partial p}{\partial t} + \mathbf{V} \cdot \nabla p + \rho_0\, c_0^2\, \nabla \cdot \mathbf{V} = 0 \tag{10.53}$$

If c_0 is large in comparison with the characteristic velocity of the flow (low Mach number flow), it can be shown from the analytical solutions for acoustic waves of small amplitude that the term $\mathbf{V} \cdot \nabla p$ can be neglected (see, e.g., Malvern, 1969). We can then write

$$\frac{\partial p}{\partial t} = -\rho_0 \, c_0^2 \, \nabla \cdot \mathbf{V} \tag{10.54}$$

The time derivative of the pressure can be approximated using a simple backward difference,

$$p^{n+1} = p^n - \lambda \, \nabla \cdot \mathbf{V} \tag{10.55}$$

where $\lambda = \rho_0 c_0^2 \Delta t$. For incompressible flows, λ can be a very large number. For example, the speed of sound in water is approximately 1500 m/s. If we think of slowly moving air, on the other hand, we must nondimensionalize Eq. (10.55) in terms of the reference Mach number, which should be very small. In practice, λ can be viewed simply as the penalty parameter in Eq. (10.55). This formulation has been used extensively in the calculation of steady and time-dependent flows (Dyne and Heinrich, 1993).

Notice that in the above algorithm incompressibility is satisfied exactly for steady state solutions. Another algorithm has been devised to be able to calculate steady state solutions using smaller penalty parameters (Zienkiewicz et al. 1985) which is easier to explain using the time-independent Stokes equations, Eqs. (10.44) and (10.45). The mixed formulation using bilinear velocities and piecewise constant pressures results in the system of linear equations

$$\begin{bmatrix} \mathbf{S} & \mathbf{Q} \\ -\mathbf{Q}^T & \mathbf{0} \end{bmatrix} \begin{bmatrix} \mathbf{w} \\ \mathbf{p} \end{bmatrix} = \begin{bmatrix} \mathbf{F} \\ \mathbf{0} \end{bmatrix} \tag{10.56}$$

where, using the notation of Eqs. (10.19) and (10.22),

$$
S = \begin{bmatrix} B & 0 \\ 0 & B \end{bmatrix} \quad Q = \begin{bmatrix} C_X \\ C_y \end{bmatrix} \quad w = \begin{bmatrix} u \\ v \end{bmatrix}
$$

and F is the right-hand side vector resulting from the applied boundary conditions. To each side of Eq. (10.56) we add the vector

$$
\frac{1}{\lambda} \begin{bmatrix} 0 \\ p \end{bmatrix}
$$

and set up the following iteration:

$$
\begin{bmatrix} S & Q \\ -Q^T & \dfrac{-1}{\lambda} I \end{bmatrix} \begin{bmatrix} w^{n+1} \\ p^{n+1} \end{bmatrix} = \begin{bmatrix} F \\ -\dfrac{1}{\lambda} p^n \end{bmatrix}
\tag{10.57}
$$

Here the second set of equations has the form

$$
p^{n+1} = p^n - \lambda Q^T w^{n+1}
\tag{10.58}
$$

which is the discrete form of Eq. (10.55), only with a different interpretation for λ. Replacing Eq. (10.58) in the first set of equations results in

$$
\left[S - \lambda Q Q^T \right] w^{n+1} = F - Q p^n
\tag{10.59}
$$

Notice that this algorithm has the same form as the one described previously, except that the penalty parameter λ has a different interpretation. This iterative scheme is therefore equivalent to introducing a false time, or an artificial time evolution, to reach steady state.

10.4   STRATIFIED FLOWS

There is a large variety of important flows that are incompressible but in which changes in density due to temperature or species concentration must be taken into account because they provide the driving forces for the fluid motion. The standard way to model these kinds of flows is to use the Boussinesq approximation. This approximation consists in keeping the density constant, equal to the reference density in the total derivative of the velocity, i.e., the left-hand side of Eqs. (10.1) and (10.2), but allowing for a variable density in the body force, which is given by an equation of state of the form.[*]

$$\rho = \rho_0 \left[1 + \beta_T (T - T_0) + \beta_1 (c_1 - c_{1_0}) + \beta_2 (c_2 - c_{2_0}) + \cdots \right] \quad (10.60)$$

where T is temperature and c_i denotes concentration of species i; β_T is the coefficient of thermal expansion and β_i is the coefficient of volumetric expansion associated with species i. The subscript "0" denotes the values at the reference state.

We will consider only a temperature-dependent flow with equation of state $\rho = \rho_0[1 + \beta_T(T-T_0)]$. For examples of applications involving species concentrations as well, we refer the reader to Heinrich (1984) and Felicelli et al. (1996). We will assume that the gravity vector points in the negative y-direction. The governing equations are then written as

$$\frac{\partial u}{\partial x} + \frac{\partial v}{\partial y} = 0 \quad (10.61)$$

$$\frac{\partial u}{\partial t} + u\frac{\partial u}{\partial x} + v\frac{\partial u}{\partial y} = -\frac{1}{\rho_0}\frac{\partial p}{\partial x} + v\left(\frac{\partial^2 u}{\partial x^2} + \frac{\partial^2 u}{\partial y^2}\right) \quad (10.62)$$

$$\frac{\partial v}{\partial t} + u\frac{\partial v}{\partial x} + v\frac{\partial v}{\partial y} = -\frac{1}{\rho_0}\frac{\partial p}{\partial y} + v\left(\frac{\partial^2 v}{\partial x^2} + \frac{\partial^2 v}{\partial y^2}\right) - g\left[1 + \beta_T(T-T_0)\right] \quad (10.63)$$

---

[*] In Eq. (10.60) we are assuming only a linear dependence on temperature in each species. In some applications more accurate nonlinear equations of state may be required. This is the case when dealing with seawater (Fofonoff, 1962).

$$\frac{\partial T}{\partial t} + u\frac{\partial T}{\partial x} + v\frac{\partial T}{\partial y} = D_T\left(\frac{\partial^2 T}{\partial x^2} + \frac{\partial^2 T}{\partial y^2}\right) \qquad (10.64)$$

where $v = \mu/\rho_0$ is the kinematic viscosity. Note that the body force vector is

$$\frac{\rho}{\rho_0}\begin{bmatrix} B_x \\ B_y \end{bmatrix} = \frac{\rho}{\rho_0}\begin{bmatrix} 0 \\ -g \end{bmatrix}$$

In what follows we will use the penalty method exclusively.

10.4.1 Finite Element Approximations

We use bilinear elements to interpolate both velocity components and the temperature; extension to other elements should be evident. We construct a penalty Petrov-Galerkin formulation of the equations of motion and energy in which the Petrov-Galerkin weights are applied only to the convective terms. We do this for the following three reasons.

1. As discussed in Section 7.2.4, the solution of time-dependent problems using a consistent mass matrix may oscillate during the first few time steps, unless an unreasonably small time step is used. This has been found to be the case with the energy equation, Eq. (10.64), in many calculations. The problem is that the oscillatory temperatures enter the right-hand side in the momentum equations and the whole calculation may become unstable. To avoid these instabilities, we choose to use mass lumping. Mass lumping eliminates the effect of the Petrov-Galerkin weighting in the mass matrix, which can be constructed using the unperturbed shape functions as weighting functions.

2. A similar stability problem has been observed to occur in cases when the body force term is weighted using the Petrov-Galerkin functions. The reasons for the unstable behavior have not been fully investigated. To avoid this problem, we also will not use the perturbed Petrov-Galerkin weights in the body force terms.

3. In the viscous terms, only the cross derivatives of the Petrov-Galerkin weighting are nonzero. Numerical experiments have shown no discernible differences in results between formulations using the Galerkin and the Petrov-Galerkin weights. Furthermore, it becomes

significantly more expensive to use the Petrov-Galerkin weights in the diffusion terms because we introduce a new nonlinearity in the equations. Hence we use the Galerkin shape functions here too.

To obtain the weighted residuals form of Eqs. (10.62) – (10.64), first we separate the static component of the pressure using Eq. (10.40), rewritten as

$$p = p_s + P \qquad\qquad (10.65)$$

where p_s denotes the component of the pressure that is always in equilibrium with the conservative gravity field and P is the term called "modified pressure" (Bachelor, 1967), or the dynamic portion of the pressure. In the penalty formulation, we have

$$P = -\lambda \nabla \cdot \mathbf{V} \qquad\qquad (10.66)$$

The weighted residuals equations written for a general body force term will be (Exercise 10.10)

$$\int_\Omega \left[U\left(\frac{\partial u}{\partial t} + u\frac{\partial u}{\partial x} + v\frac{\partial u}{\partial y}\right) + v\left(\frac{\partial U}{\partial x}\frac{\partial u}{\partial x} + \frac{\partial U}{\partial y}\frac{\partial u}{\partial y}\right)\right] d\Omega - \oint_\Omega \frac{1}{\rho_0}\frac{\partial U}{\partial x}P\,d\Omega$$

$$= \int_\Omega U\left(\frac{\rho}{\rho_0}\beta_x - \frac{\partial p_s}{\partial x}\right)d\Omega + \int_\Gamma V\left[\left(p_s - p + v\frac{\partial u}{\partial x}\right)n_x + v\frac{\partial u}{\partial y}n_y\right]d\Gamma$$

$$(10.67)$$

$$\int_\Omega \left[V\left(\frac{\partial v}{\partial t} + u\frac{\partial v}{\partial x} + v\frac{\partial v}{\partial y}\right) + v\left(\frac{\partial V}{\partial x}\frac{\partial v}{\partial x} + \frac{\partial V}{\partial y}\frac{\partial v}{\partial y}\right)\right] d\Omega - \oint_\Omega \frac{1}{\rho_0}\frac{\partial V}{\partial y}P\,d\Omega$$

$$= \int_\Omega V\left(\frac{\rho}{\rho_0}\beta_y - \frac{\partial p_s}{\partial y}\right)d\Omega + \int_\Gamma V\left[v\frac{\partial v}{\partial x}n_x + \left(p_s - p + v\frac{\partial v}{\partial y}\right)n_y\right]d\Gamma$$

$$(10.68)$$

$$\int_{\Omega} \left[W\left(\frac{\partial T}{\partial t} + u\frac{\partial T}{\partial x} + v\frac{\partial T}{\partial y} \right) + D_T\left(\frac{\partial W}{\partial x}\frac{\partial T}{\partial x} + \frac{\partial W}{\partial y}\frac{\partial T}{\partial y} \right) \right] d\Omega \tag{10.69}$$

$$= \int_{\Omega} D_T W\left(\frac{\partial T}{\partial x}n_x + \frac{\partial T}{\partial y}n_y \right) d\Gamma$$

where the notation $\hat{R}$ is used to denote reduced integration and U, V, and W are used to denote the weighting functions for each of the equations. In this formulation, U and V are always the same, equal to the bilinear shape functions, except in the convective term, where we add a Petrov-Galerkin perturbation. The function W is the bilinear shape function except when applied to the convective terms, where a Petrov-Galerkin perturbation different from the one used in U and V is added. The perturbations differ only in the magnitude of the parameter α because of the difference between v and D_T.

The importance of the pressure splitting used in Eqs. (10.67) and (10.68) will become evident as we discuss some examples in the next section.

Example 10.5

We first consider a simple stratified steady state plane Poiseuille flow. This is the same as in Example 10.2 with the addition of a linear, stable temperature gradient. Equations (10.61) through (10.64) are nondimensionalized using the same reference values as in Section 10.3.1. The nondimensional temperature is obtained from

$$T = \left(T' - T_0 \right) / \left| \Delta T \right|$$

where $\left| \Delta T \right|$ is the maximum temperature difference in the fluid. The resulting nondimensional parameters are Re, defined in Eq. (10.14), the Peclet number, Pe, and the Grashof number, Gr, defined by

$$Pe = \frac{\overline{U}L}{D_T} \tag{10.70}$$

$$Gr = \frac{\beta_T g |\Delta T| L^3}{v^2} \qquad (10.71)$$

Setting $u_m = 1$, the exact solution is given by (Exercise 10.11)

$$u = 4y(y-1)$$
$$v = 0$$
$$T = y$$

and

$$p = \frac{8(5-x)}{Re} + \frac{G_r}{Re^2}(y^2 - 1)$$

The expression for the pressure is obtained by setting $p(5,1) = 0$. The length of the computational domain is 5H, as before, and the same mesh, with $\Delta x = 0.1$ and $\Delta y = 0.05$, is used.

The finite element solution has been calculated using the standard penalty formulation given by Eq. (10.65). The natural boundary conditions stemming from Eqs. (10.67) and (10.68), i.e.,

$$p + \frac{1}{Re} \frac{\partial u}{\partial x} = 0 \qquad (10.72)$$

$$\frac{1}{Re} \frac{\partial v}{\partial x} = 0 \qquad (10.73)$$

and zero heat flux in the x-direction, $\partial T / \partial x = 0$ are imposed at the outflow boundary, $x = 5$.

The solution is shown in Fig. 10.12, for the case Re = 100, Pe = 100, and Gr = 100, and is clearly erroneous. By the way, the same solution will result if we use a mixed formulation. Also notice that we did not have this kind of problem when we discussed the solution of the isothermal case in Example 10.2, which was also performed using the standard penalty method where $p = P$.

To understand what is happening, first consider the fluid to be at rest, so that the pressure gradient in the y-direction is balanced by the gravitational forces. Equation (10.63) become

$$\frac{1}{\rho_0}\frac{\partial p'}{\partial y'} = -g\left[1+\beta_T\left(T'-T_0\right)\right] \tag{10.74}$$

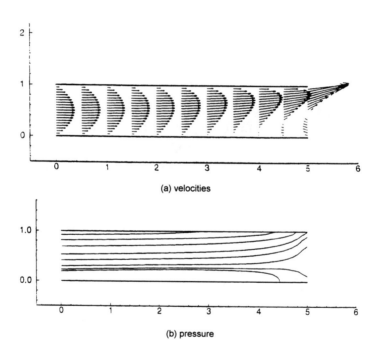

(a) velocities

(b) pressure

Figure 10.12 Finite element approximation to the stratified Poiseuille flow using the standard penalty formulation of Eq. (10.43) and the natural boundary condition of Eqs. (10.72) and (10.73): (a) flow field and (b) calculated pressure.

where the prime denotes dimensional quantities. The first term on the right-hand side corresponds to the hydrostatic pressure at the reference density and is always eliminated by redefining the pressure gradient as $\partial p'/\partial y - \rho_0 g$. For this reason, it is usually eliminated from Eq. (10.63), and we will also do this from now on. Therefore we can write

$$\frac{1}{\rho_0} \frac{\partial p'_s}{\partial y'} = -g\beta_T (T' - T_0) \qquad (10.75)$$

In nondimensional form, integrating and applying the condition $p(5,1) = 0$, we obtain

$$p_s = \frac{Gr}{Re^2} (y^2 - 1) \qquad (10.76)$$

so the dynamic pressure constitutes the first term of the analytical solution

$$P = \frac{8}{Re} (5 - x) \qquad (10.77)$$

In the numerical solution, shown in Fig. 10.12b, we observe that the standard penalty method is trying to approximate the total pressure, $p = p_s + P$. However, it runs into an insurmountable difficulty at the open boundary, $x = 5$, where the natural boundary condition, Eq. (10.72), is imposed weakly.

Because Re is relatively large, and $\partial u/\partial x$ should be zero, the pressure is being forced to be constant along the exit boundary, which clearly contradicts Eq. (10.76). As a result, the standard penalty method cannot describe the quadratic behavior of the static pressure and a solution different from the correct one is obtained.

In the case just discussed, the terms

$$-\int_\Omega V \frac{\partial p_s}{\partial y} d\Omega \qquad \int_\Gamma V p_s n_y d\Gamma$$

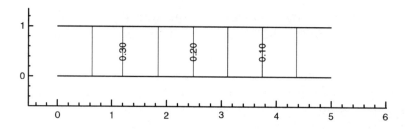

Figure 10.13 Modified pressure in a stratified Poiseuille flow calculated using the penalty formulation of Eq. (10.65). Numerical solution denoted by the dotted line and the exact solution by the solid line.

were omitted in the calculation and p was set equal to P in the standard penalty formulation. If we now set $p = p_s + P$, and evaluate the above integrals using Eq. (10.76) for p_s, an extremely accurate solution is obtained. The velocity field is the same as in Fig. 10.4, and the calculated pressure P is the modified pressure shown in Fig. 10.13. The pressure gradient $\partial p / \partial x$ so calculated is 0.07999, compared to the exact gradient of 0.08, giving a relative error of only $1.74 \times 10^{-4}\%$. ∎

REMARKS

1. The above example serves as an illustration of how important the proper treatment of the static pressure can be and how we must be aware of the pressure in a calculation. Evidently, the splitting used in Example 10.5 is only possible if a static solution such as given by Eq. (10.74) is available. This is the case only for a very limited number of flows constrained between parallel boundaries, but which are of great practical importance. In more general situations these solutions cannot be obtained if, for example, one of the walls is not vertical or horizontal (Philips, 1970). However, the pressure distribution will only differ from a purely static distribution over a thin boundary layer along the tilted walls. Therefore we may still separate the pressure components using Eq. (10.74) to calculate the static component, being aware that the modified pressure calculated near the tilted walls will be slightly perturbed. As we will see in the next example, this still offers great computational advantages.

2. Another important class of flows in which the static component of pressure plays a significant role is flows in rotational

systems, such as Couette flows, that are subject to strong inertial forces. We will give an example of this type of flow when we discuss open boundary conditions in more detail in the next section.

The present discussion applies only to the steady state case. It is easy to see that during a transient calculation the conditions at the outlet boundary may not be correct until steady state is approached – unless the static pressure is calculated and updated at every time step, which can be impractical. This will also be discussed in the next section.

Example 10.6

This example involves Rayleigh-Bernard convection in a rectangular container uniformly heated below and cooled from above. The vertical walls are adiabatic, and no slip is prescribed for the velocity over the entire boundary. The lower wall has nondimensional temperature $T_L = 0$, and the upper wall is set at $T_u = -1$. An appropriate nondimensionalization is obtained using the heat diffusion D_T to define the time and velocity scales as $\tau = L^2/D_T$ and $\overline{U} = D_T/L$ respectively, and the reference pressure as $p_o = \rho_o D_T^2/L_2$. The temperature is left as before. The nondimensional governing equations, whose derivation is left to Exercise 10.13, are given in Section 6.3.3, Eqs. (6.68)–(6.71). The nondimensional parameters are the Prandtl number Pr and the Rayleigh number Ra, defined in Eqs. (6.72) and (6.73), respectively. Notice that Ra = Gr Pr.

The computational domain consists of an enclosure of height H and length 2H discretized with a mesh of 40 by 20 uniform bilinear elements. The Navier–Stokes and energy equations accept a quiescent solution in this case, which is given by $u = v = 0$, $T = -y$ and $p_s = -Ra$ Pr $y^2/2$. Figure 10.14 shows the steady state solution for Ra = 20,250 and Pr = 2.5. These values were chosen for comparison with the results of Argyris et al. (1984). Figures 10.14a and 10.14d show the temperature and flow field, respectively. Figures 10.14b and 10.14c show the modified pressure and the total pressure. Figure 10.14b clearly shows how the fluid motion is driven by opposing pressure gradients along the bottom and top walls. On the other hand, we observe that even though these lateral pressure gradients can still be observed in Fig. 10.14c, the

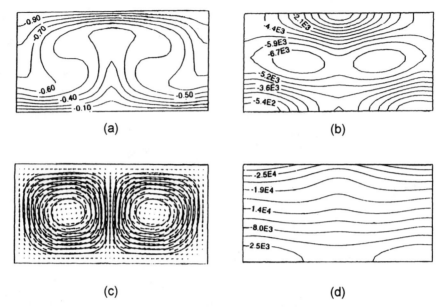

Figure 10.14 Steady state calculation of Rayleigh-Benard convection for (a) temperature, (b) modified pressure **P**, (c) total pressure p, and (d) flow field.

total pressure is completely dominated by the static component. This is shown more dramatically in Fig. 10.15, where a three-dimensional plot of the modified pressure P and the total pressure p is shown. ∎

If the standard penalty method defined by Eq. (10.43), or the modified pressure form of Eq. (10.65), is used, the same solution is obtained. The absence of an open boundary allows the standard penalty to converge to the correct solution, calculating the total pressure. However, convergence is much slower than if the static component is removed and the gradients are the dominant part of the forces. In this case, this results in 73% more CPU time required by the standard penalty method to obtain a steady state solution to within the same tolerance, using a fully implicit algorithm and a Newton-Raphson iteration. Furthermore, in those cases where analytical solutions are available, such as Poiseuille, Couette, or Hagen-Poiseuille flows (Lai, et al., 1974), calculations with the modified pressure consistently show significantly better accuracy. A final remark pertains to free surface flows. We show in Section 10.4.3 that using modified pressure is crucial to correctly approximating free surface flows.

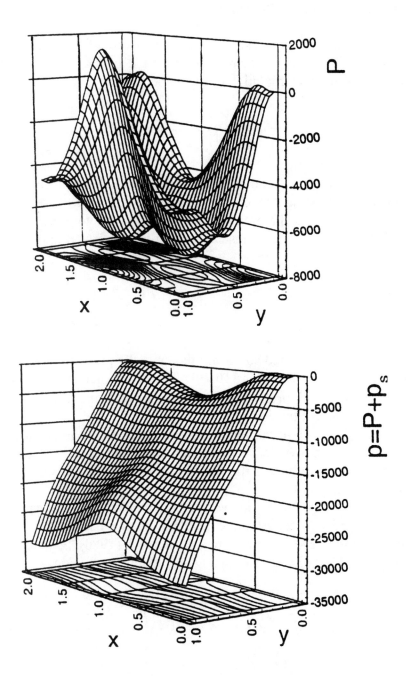

Figure 10.15 Three-dimensional pressure surfaces corresponding to cases in Figs. 10.12b and 10.12c.

10.4.2 Calculation of the Pressure

In penalty calculations the pressure P may be recovered from Eq. (10.66) regardless of whether P represents the modified pressure or the total pressure. For bilinear interpolation of the velocities we have

$$P^e = -\frac{\lambda}{A^e} \iint_e \left(\frac{\partial u}{\partial x} + \frac{\partial v}{\partial y} \right) de \qquad (10.78)$$

where the superscript e denotes the restriction to the element under consideration and A^e is the element area. In this case, a one-point reduced integration quadrature formula is employed to evaluate Eq. (10.78). If Eq. (10.55) is used instead, a similar expression is obtained, i.e.,

$$\left(P^e\right)^{n+1} = \left(P^e\right)^n - \frac{\lambda}{A^e} \iint_e \left(\frac{\partial u}{\partial x} + \frac{\partial v}{\partial y} \right) de \qquad (10.79)$$

In the calculations presented so far in this book, the pressures have been obtained from Eqs. (10.78) or (10.79). These are piecewise constant pressures – the nodal pressures are obtained with the least squares procedure explained at the end of Section 10.3.1.

If Eq. (10.79) is used, the pressures may become unbounded in time. It is therefore convenient to normalize them at each time step. This can be accomplished either using the largest calculated pressure or the pressure at a fixed point.

An alternative to the above method to calculate the pressure, which may yield a more accurate solution, is the use of a pressure Poisson equation introduced by Sohn and Heinrich (1990). Starting from Eqs. (10.62) and (10.63), differentiating and adding, we obtain

$$\frac{1}{\rho_0} \left(\frac{\partial^2 p}{\partial x^2} + \frac{\partial^2 p}{\partial y^2} \right) = -\frac{\partial}{\partial x} \left(u \frac{\partial u}{\partial x} + v \frac{\partial u}{\partial y} \right) - \frac{\partial}{\partial y} \left(u \frac{\partial v}{\partial x} + v \frac{\partial v}{\partial y} \right)$$

$$+ \nu \left[\frac{\partial}{\partial x} \left(\frac{\partial^2 u}{\partial x^2} + \frac{\partial^2 u}{\partial y^2} \right) + \frac{\partial}{\partial y} \left(\frac{\partial^2 v}{\partial x^2} + \frac{\partial^2 v}{\partial y^2} \right) \right] - g\beta \frac{\partial T}{\partial y} \qquad (10.80)$$

The viscous terms on the right-hand side of Eq. (10.80) are normally eliminated using the continuity equation and written as

$$
\frac{\partial^2 p}{\partial x^2} + \frac{\partial^2 p}{\partial y^2} = -\rho_0 \left[\frac{\partial}{\partial x}\left(u\frac{\partial u}{\partial x} + v\frac{\partial u}{\partial y} \right) + \frac{\partial}{\partial y}\left(u\frac{\partial v}{\partial x} + v\frac{\partial v}{\partial y} \right) - g\beta\frac{\partial T}{\partial y} \right] \quad (10.81)
$$

Denoting the weighting functions as w, the weak form becomes

$$
\int_\Omega \left(\frac{\partial w}{\partial x}\frac{\partial p}{\partial x} + \frac{\partial w}{\partial y}\frac{\partial p}{\partial y} \right) d\Omega
$$

$$
= -\rho_0 \int_\Omega \left[\frac{\partial w}{\partial x}\left(u\frac{\partial u}{\partial x} + v\frac{\partial u}{\partial y} \right) + \frac{\partial w}{\partial y}\left(u\frac{\partial v}{\partial x} + v\frac{\partial v}{\partial y} \right) - wg\beta\frac{\partial T}{\partial y} \right] d\Omega
$$

$$
+ \int_\Gamma w \left\{ \left[\rho_0\left(u\frac{\partial u}{\partial x} + v\frac{\partial u}{\partial y} \right) + \frac{\partial p}{\partial x} \right]n_x + \left[\rho_0\left(u\frac{\partial v}{\partial x} + v\frac{\partial v}{\partial y} \right) + \frac{\partial p}{\partial y} \right]n_y \right\} d\Gamma
$$

$$
(10.82)
$$

Equation (10.82) has two disadvantages: it requires a boundary condition on the gradient of p along the entire boundary, which is not readily available; and the evaluation of the convective terms along open boundaries where the derivatives of the velocity cannot be calculated accurately. It is better to work with Eq. (10.80). The weak form is

$$
\int_\Omega \left(\frac{\partial w}{\partial x}\frac{\partial p}{\partial x} + \frac{\partial w}{\partial y}\frac{\partial p}{\partial y} \right) d\Omega
$$

$$
= -\rho_0 \int_\Omega \left[\frac{\partial w}{\partial x}\left(u\frac{\partial u}{\partial x} + v\frac{\partial u}{\partial y} \right) + \frac{\partial w}{\partial y}\left(u\frac{\partial v}{\partial x} + v\frac{\partial v}{\partial y} \right) - wg\beta\frac{\partial T}{\partial y} \right] d\Omega
$$

$$
+ \rho_0 \nu \int_\Gamma \left[\frac{\partial w}{\partial x}\left(\frac{\partial^2 u}{\partial x^2} + \frac{\partial^2 u}{\partial y^2} \right) + \frac{\partial w}{\partial y}\left(\frac{\partial^2 v}{\partial x^2} + \frac{\partial^2 v}{\partial y^2} \right) \right] d\Omega
$$

$$
- \int_\Gamma w \left(\frac{\partial u}{\partial t}n_x + \frac{\partial v}{\partial t}n_y \right) d\Gamma
$$

$$
(10.83)
$$

This equation does not require specification of the pressure gradients along the wall – we only need to specify the pressure at one point in the domain to determine the solution uniquely. Moreover, the line integral vanishes at all solid walls; it is only nonzero along open boundaries in time-dependent flows, or if the flow is excited by the time-varying motion of a wall. In either case its evaluation is straightforward. The only difficulty arises from the presence of the second derivatives of the velocities in the viscous terms. However, in the computational domain Ω, these can be evaluated accurately utilizing a least squares method as described in Section 10.3.1. We will not elaborate further on this method; the details and a variety of numerical examples are discussed by Sohn and Heinrich (1990).

10.4.3 Open Boundaries

Many of the difficulties associated with open boundaries and the static pressure that we have discussed so far stem from the fact that we are using low-order interpolation functions for velocities and piecewise constant pressures. Because of this we must apply Gauss's theorem to the pressure gradients in the momentum equations in order to shift the derivatives to the weighting functions. This gives rise to the line integral of the pressure along the open boundaries as in Eqs. (10.6), (10.7) and Eqs. (10.67), (10.68). If higher order elements such as quadratic velocities and nonconforming linear pressure are used (case 3 in Fig. 10.3), there is no need to integrate the pressure gradients, and some of the difficulties can be avoided. However, many formulations are written using the tensor form of Eq. (10.41), and the Gauss theorem applied to the whole stress tensor σ given by Eq. (10.5), which includes the pressure.

      The line integral along the open boundaries that results from the weak formulation is normally eliminated by invoking a "natural" boundary condition. In Examples 10.2, 10.3, and 10.5, the boundary conditions applied at the outflow boundary are given by Eqs. (10.72) and (10.73) or their equivalents. The first of these makes a statement about the pressure, which was not a problem in the isothermal flows because the hydrostatic component of the pressure has been subtracted out and the modified pressure is constant at the exit. This is consistent with the rest of the problem. However, in a stratified flow, we had to first make sure that only the modified pressure was being represented by the penalty term in order to obtain meaningful solutions.

The issue of boundary conditions at outflow boundaries has been treated by many authors (see, e.g., Sani and Gresho, 1994), although in a rather ad hoc way. An analysis performed by Heinrich et al. (1996) points out that we should not apply boundary conditions on the pressure; only a reference pressure must be prescribed at some point in the domain. Therefore the difficulties at the open boundaries can also be eliminated by removing the pressure from the boundary condition given by Eq. (10.72), so that only

$$\frac{\partial u}{\partial x} = \frac{\partial v}{\partial x} = 0$$

are used. To achieve this, the line integrals

$$\int_{\Gamma} U p n_x \, d\Gamma \qquad \int_{\Gamma} V p n_y \, d\Gamma$$

in Eqs. (10.6) and (10.7) must be discretized and retained in the weighted residuals formulation. If a mixed formulation is being used, these integrals will contribute to the matrices $\mathbf{C}_x$ and $\mathbf{C}_y$ in Eq. (10.22), but no contribution will be added to $\mathbf{C}_x^T$ and $\mathbf{C}_y^T$ in the last row. Thus $\mathbf{K}$ will become an asymmetric matrix even in the case of Stokes flow.

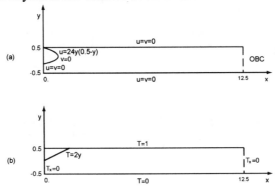

Figure 10.16 Domain and boundary conditions for stratified flow over a backward facing step at Re = Pe = 800 and Gr = 3.6 x 10^5.

If a penalty formulation is used, retrieving the integrals is only possible in a time-dependent formulation where the integrals are evaluated explicitly on the right-hand side. To try to include them implicitly by means of the penalty term will result in an ill-conditioned stiffness matrix. We will give an example of such an algorithm in the next section.

Example 10.7

We return to flow over a backward facing step, only this time the flow is stably stratified. The domain and boundary conditions are shown in Fig. 10.16. A fully developed flow is prescribed at the step wall and the parameter values are $Re = Pe = 800$ and $Gr = 3.6 \times 10^5$. This particular case was proposed by Leone (1990). The simulation used a regular mesh of bilinear elements of size $\Delta x = 0.0625$ and $\Delta y = 0.03125$. Figure 10.17 shows that when the standard penalty method is used and the

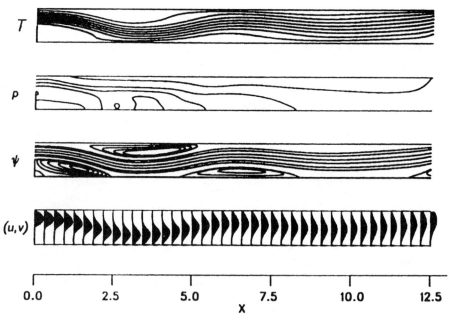

Figure 10.17 Solution to the stratified flow over a backward facing step using the standard penalty method.

pressure is not removed at the outflow boundary, the same problem observed in Fig. 10.12 is evident in this case. On the other hand, Fig. 10.18 shows the solution when the pressure line integrals are retained in the weak formulation, as discussed above. The effect is the same as when the static pressure p_s is separated and only the modified pressure P calculated; both modified pressure algorithms give very similar solutions. In this case we have $p_s = Gr(y^2 + y - 0.75)/2Re^2$ at x = 12.5. In Fig. 10.18, both the modified pressure and the total pressure are shown. The first clearly shows the regions of flow reattachment, while it is evident from the second that the flow is dominated by the hydrostatic pressure distribution. In Fig. 10.19, we show a comparison of the total pressure distribution along the bottom wall obtained with the standard penalty, retaining the line integrals, modified pressure penalty, and the pressure Poisson equation.

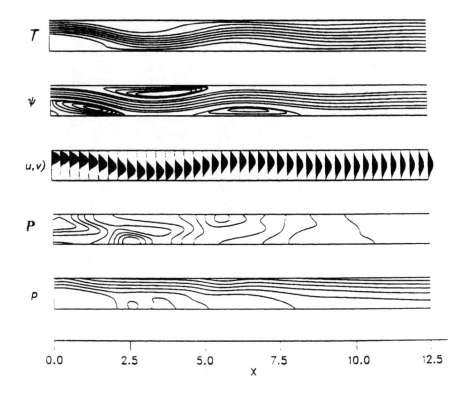

Figure 10.18 Solution to the stratified flow over a backward facing step using the modified pressure penalty method.

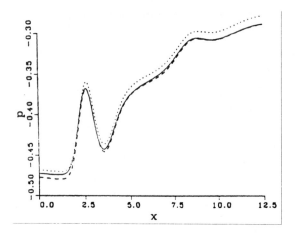

Figure 10.19 Total pressure along the bottom wall for stratified flow over a backward facing step. The standard penalty method is the dotted line, the modified penalty method is the dashed line, and the pressure Poisson equation is the solid line.

To close this example and show how effective it can be to retain the pressure line integrals in the formulation, Fig. 10.20 shows the results of a calculation on a domain truncated at $x = 7$, which cuts the third recirculation almost in the middle. The agreement with the previous calculation is remarkable. A more detailed comparison of the two can be found in the work by Heinrich and Vionnet (1995a, b). ■

For a similar benchmark problem involving heat fluxes along channel boundaries that has been solved with a variety of existing numerical models, the interested reader should look at the book by Blackwell and Pepper (1992). Among the solutions represented, there is one obtained with the present penalty method.

10.5 FREE SURFACE FLOWS

Consider now flow of a fluid with a free surface that can be subjected to small-amplitude gravity waves. We will assume that the flow is isothermal, since the basic concepts are not affected by density changes of the type discussed at the beginning of this chapter. The fluid layer will be assumed to be two-dimensional with a reference depth h_0, as shown in Fig. 10.21. The deviation of the depth from the reference depth is denoted by h, so that the position of the free surface $S(x, y, t)$ is

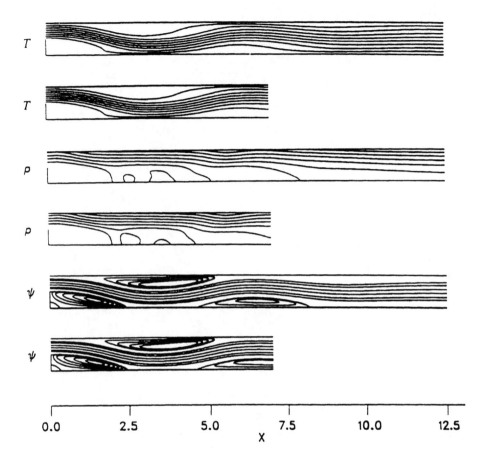

Figure 10.20 Solution to the stratified flow over a backward facing step on a shorter domain of length 7x using modified penalty methods.

$$S = y - h_0 - h = 0 \qquad (10.84)$$

Let us rewrite the governing equations of motion as

$$\nabla \cdot \mathbf{V} = 0 \qquad (10.85)$$

and

$$\frac{D\mathbf{u}}{Dt} = -\frac{1}{\rho_0}\nabla p + \nu\nabla^2\mathbf{u} + \mathbf{B} \qquad (10.86)$$

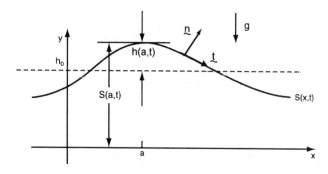

Figure 10.21 Free surface flow in two dimensions.

The boundary conditions at the free surface express continuity of stresses and are given by

$$\sigma_{ij}\,n_j = \left(-p_a + 2\sigma\,G\right)n_i \qquad \text{on} \quad S \qquad (10.87)$$

and

$$\sigma_{ij}\,t_j = 0 \qquad \text{on} \quad S \qquad (10.88)$$

where p_a is the pressure at the interface, e.g., atmospheric pressure at a water-air interface, and is assumed to be constant; σ is the surface tension; G the mean surface curvature; $\mathbf{n}$ the unit outward normal vector and $\mathbf{t}$ the unit tangent vector, as shown in Fig. 10.21.

We also need a kinematic condition to describe the surface motion, which is

$$\frac{DS}{Dt} = \frac{\partial S}{\partial t} + \mathbf{V} \cdot \nabla S = 0 \qquad (10.89)$$

Using Eq. (10.82), this expression becomes

$$\frac{\partial h}{\partial t} + u\frac{\partial h}{\partial x} - v = 0 \qquad (10.90)$$

Example 10.8

Consider flow over a broad-crested weir as shown in Fig. 10.22. In this problem, surface tension effects are small and are neglected, i.e., $\sigma = 0$

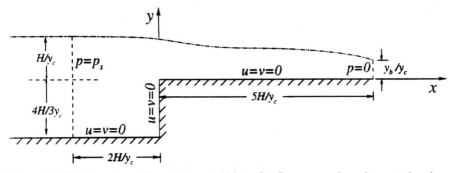

Figure 10.22 Domain and boundary conditions for flow over a broad–crested weir.

in Eq. (10.89). The shape of the free surface depends on the geometry of the weir as well as the free fall characteristics at the end of the hydraulic structure (Henderson, 1966). We choose $H/y_c = 2/3$, where H is the total head and y_c is called the critical depth. Using $\bar{U} = \sqrt{gy_c}$, the velocity of infinitesimal gravity waves as the characteristic velocity, and y_c as the characteristic length, the relevant nondimensional numbers are the Froude number Fr, defined as

$$\mathrm{Fr} = \frac{U^2}{gy_c}$$

and the Reynolds number, which in this case is given by

$$\mathrm{Re} = \sqrt{\frac{8g}{27}}\, H^{3/2} / v$$

The static pressure p_s becomes

$$p_s = \rho g (h + h_0 - y) \tag{10.91}$$

Hence, the pressure gradients are

$$\frac{\partial p_s}{\partial x} = \rho g \frac{\partial h}{\partial x} \qquad \frac{\partial p_s}{\partial y} = -\rho g \tag{10.92}$$

Steady state results for $F_r = 1$ and $Re = 100$ are shown in Fig. 10.23. The pressure boundary conditions are hydrostatic pressure at the inflow boundary and ambient pressure along the free surface. The head H/y_c was held constant at the inflow boundary. ∎

REMARKS

1. This kind of simulation cannot be performed correctly unless the modified pressure formulation is used and the static component given by Eq. (10.92) is explicitly incorporated in the weak formulation, Eqs. (10.67) and (10.68). From Fig. 10.23d, we see that the absolute pressure will exhibit a strong deviation from the hydrostatic conditions, especially around the upper left edge of the weir and in the region just upstream of the brink section. Furthermore, the boundary conditions in the open boundaries are very easy to implement.

2. The solution of free surface problems becomes complicated by the fact that the location of the free surface must be calculated as part of the problem. Figure 10.23a shows the location of the free surface at the initiation of the calculations. At every time step the location of the free surface is recalculated so as to satisfy dynamic equilibrium at the end of each time step. Here we have followed the basic idea of Rushak (1980) to modify the mesh.

3. There is much literature on free surface flows in which surface tension is important. Finite element algorithms for such flows have been proposed by, among others, Nickell et al. (1974), Kistler and Scriven (1984) and Chippada et al. (1995). Capillary flows are of extreme interest in space sciences, since in the absence of gravity, capillary effects become dominant. However, this is a vast subject, and we will refrain from discussing it further. Suffice it to say that the numerical techniques we have discussed are also applicable to such flows.

10.6   CLOSURE

Finite element algorithms to solve the Navier-Stokes equations for viscous incompressible flows have been presented in this chapter. We have discussed both the theoretical and practical difficulties faced in their solution, with emphasis on the use of the penalty function method with low-order bilinear isoparametric elements. We also included a discussion on the pressure and the various issues that arise with its

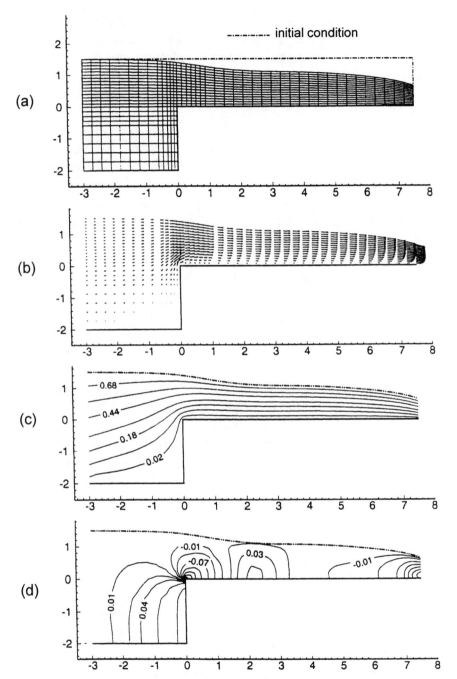

Figure 10.23 Calculated steady state flow over a broad-crested weir: (a) final mesh with free surface location, (b) velocity field, (c) streamfunction, and (d) modified pressure contours.

approximation in the presence of open and free boundaries. In the next chapter we will address the problem of generating finite element meshes when the geometry is complex, or when adapting the mesh is necessary to improve the solutions.

REFERENCES

Argyris, J. H., Doltsinis, J. St., Pimenta, P. M. and Wustenberg, H. (1984) "Natural Finite Element Techniques for Viscous Fluid Motion," *Comput. Methods Appl. Mech. Eng.*, Vol. 45, pp. 3–55.

Babuska, I. And Aziz, A. K. (1972) "Survey Lectures on the Mathematical Foundations of the Finite Element Method," pp. 5–359 in *The Mathematical Foundations of the Finite Elelemnt Method with Applications to Partial Differential Equations* (A. K. Aziz, editor). New York: Academic Press.

Batchelor, G. K. (1967) *An Introduction to Fluid Dynamics*. Cambridge, U. K.: Cambridge University Press.

Bercovier, M. and Engelman, M. (1979) "A Finite Element for the Numerical Solution of Viscous Incompressible Flow," *J. Comput. Phys.*, Vol. 30, pp. 181–201.

Blackwell, B. and Pepper, D. W. (editors) (1992) *Benchmark Problems for Heat Transfer Codes*. ASME Book No. G00709, HTD Vol. 222. New York: ASME Publication.

Brezzi, F. (1974) "On the Existence, Uniqueness and Approximation of Saddle Point Problems Arising from Lagrangian Multipliers," *Rev. Fr. Autom. Inf. Rech. Oper. Numer. Avel*, R 2, pp. 129–151.

Burggraf, O. R. (1966) "Analytical and Numerical Studies of the Structure of Steady Separated Flows," *J. Fluid Mech.*, Vol. 24, pp. 113–151.

Carey, G. F. and Krishnan (1982) "Penalty Approximation of Stokes Flow," *Comput. Methods Appl. Mech. Eng.*, Vol. 35, pp. 169–206.

Carey, G. F. and Krishnan (1984) "Penalty Finite Element Method for Navier-Stokes Equations," *Comput. Methods Appl. Mech. Eng.*, Vol. 42, pp. 183–224.

Carey, G. F. and Oden, J. T. (1986) *Finite Elements, Vol. VI, Fluid Mechanics*. Englewood Cliffs, N. J.: Prentice-Hall.

Chippada, S., Jue, T. C. and Ramaswamy, B. (1995) "Finite Element Simulation of Combined Buoyancy and Thermocapillary Driven

Convection in Open Cavities," *Int. J. Numer. Methods Eng.*, Vol. 38, pp. 335–351.

Chorin, A. J. (1967) "A Numerical Method for Solving Viscous Incompressible Flow Problems," *Comput. Phys.*, Vol. 2, pp. 12–26.

Donea, J., Giuliani, S. and Laval, H. (1982) "Finite Element Solutions of the Unsteady Navier-Stokes Equations by Fractional Step Method," *Comput. Methods Appl. Mech. Eng.*, Vol. 30, pp. 53–73.

Dyne, B. R. and Heinrich, J. C. (1993) "Physically Correct Penalty-Like Formulations for Accurate Pressure Calculation in Finite Element Algorithms of the Navier-Stokes Equations," *Int. J. Numer. Methods Eng.*, Vol. 36, pp. 3883–3902.

Felicelli. S., Poirier, D. R. and Heinrich, J. C. (1996) "Macrosegregation Patterns in Multicomponent Ni-Base Alloys," *J. Cryst. Growth* (to appear)

Fofonoff, N. P. (1962) "Physical Properties of Sea Water," pp. 3–30 in *The Sea* (M. N. Hill, editor). New York: Interscience.

Fukumori, E. and Wake, A. (1991) "The Linkage Between Penalty Function Method and Second Viscosity Applied to Navier-Stokes Equation," in *Computational Mechanics* (Y. K. Cheung, J. H. W. Lee, and A. Y. T. Leung, editors) A. A. Blke: Rotterdam.

Gartling, D. K. (1990) "A Test Problem for Outflow Boundary Conditions – Flow over a Backward Facing Step," *Int. J. Numer. Methods Fluids*, Vol. 11, pp. 957–967.

Girault, V. and Raviart, P. A. (1979) *Finite Element Approximations of the Navier-Stokes Equations.*, Lecture Series in Mathematics No. 749. Berlin: Springer-Verlag.

Harari, I. and Hughes, T. J. R. (1994) "Stabilized Finite Element Method for Steady Advection Diffusion with Production," *Comput. Methods Appl. Mech. Eng.*, Vol. 115, pp. 165–191.

Heinrich, J. C. (1984) "A Finite Element Model for Double-Diffusive Convection," *Int. J. Numer. Methods Eng.*, Vol. 20, pp. 447-464.

Heinrich, J. C., Idelsohn, S. R., Oñate, E. and Vionnet, C. A. (1996) "Boundary Conditions for Finite Element Simulations of Convective Flows with Artificial Boundaries." *Int. J. Num. Mech. Engr.*, Vol. 39, pp. 1053–1071.

Heinrich, J. C. and Vionnet, C. A. (1995a) "On Boundary Conditions for Unbounded Flows,"*Comm. Numer. Methods Eng.*, Vol. 11, pp. 179–185.

Heinrich, J. C. and Vionnet, C. A. (1995b) "The Penalty Method for the Navier–Stokes Equations," *Arch. of Comput. Methods Eng.*, Vol. 2, pp. 51–65.

Henderson, F. M. (1966) *Open Channel Flow*. New York: Macmillan.

Idelsohn, S. R., Nigro, N., Storti, M. A. and Buscaglia, G. (1996) "A Petrov-Galerkin Formulation for Advection-Reaction Diffusion Problems," *Comput. Methods Appl. Mech. Eng.*, Vol. 136, pp. 27–46.

Idelsohn, S., Storti, M. and Nigro, N. (1995) "Stability Analysis of Mixed Finite Element Formulations with Special Mention of Equal-Order Interpolate." *Int. J. Num. Meth. Fluids*, Vol. 20, pp 1003–1022.

Kistler, S. F. and Scriven, L. E. (1984) "Coating Flow Theory by Finite Element and Asymptotic Analysis of the Navier-Stokes System," *Int. J. Numer. Methods Fluids*, Vol. 4, pp. 207–229.

Ladyzhenskaya, O. A. (1969) *The Mathematical Theory of Viscous Incompressible Flows*, 2nd ed., New York: Gordon and Breach.

Lai, W. M., Rubin, D. and Krempl, E. (1974) *Introduction to Continuum Mechanics*. New York: Pergamon Press.

Leone, J. M. (1990) "Open Boundary Conditions Mini-symposium, Benchmark Solution: Stratified Flow over a Backward Facing Step," *Int. J. Numer. Methods Fluids*, Vol. 11, pp. 953–967.

Malkus, D. S. and Hughes, T. J. R. (1978) "Mixed Finite Element Methods – Reduced and Selective Integration Techniques: A Unification of Concepts, *Comput. Methods Appl. Mech. Eng.*, Vol. 15, pp. 63–81.

Nickell, R. E., Tanner, R. I. and Caswell, B. (1974) "The Solution of Viscous Incompressible Jets and Free-Surface Flows Using Finite Element Methods," *J. Fluid Mech.*, Vol. 65, pp. 189–206.

Oden, J. T. (1982) "RIP Methods for Stokesian Flows," pp. 305–318 in *Finite Elments in Fluids*, Vol. 4 (R. H. Gallagher, D. H. Norrie, J. T. Oden, and O. C. Zienkiewicz, editors). New York: Wiley.

Oden, J. T., Kikuchi, N. and Song, Y. J. (1982) "Penalty-Finite Element Methods for the Analysis of Stokesian Flows," *Comput. Methods Appl. Mech Eng.*, Vol. 31, pp. 183–224.

Phillips. O. M. (1970) "On Flows Induced by Diffusion in a Stably Stratified Fluid," *Deep Sea Res.*, Vol. 17, pp. 435–443.

Quartapelle, L. (1993) *Numerical Solution of the Navier-Stokes Equations*. Boston: Basel-Birkhauser Verlag.

Rushak, K. J. (1980) "A Method for Incorporating Free Boundaries with Surface Tension in Finite Element Fluid Flow Simulators," *Int. J. Numer. Methods. Eng.*, Vol. 15, pp. 639–648.

Sani, R. L. and Gresho, P. M. (1994) "Reviews and Remarks on the Open Boundary Condition Minisymposium," *Int. J. Numer. Methods Fluids*, Vol. 18, pp. 983–1008.

Schneider, G. E., Raithby, G. D. and Yovanovich, M. M. (1978) "Finite Element Solution Procedures for Solving the Incompressible Navier-Stokes Equations Using Equal Order Interpolation," *Numer. Heat Transfer*, Vol. 1, pp. 435–451.

Sohn, J. L. and Heinrich, J. C. (1990) "A Poisson Equation Formulation for Pressure Calculations in Penalty Finite Element Models for Viscous Incompressible Flows," *Int. J. Numer. Method Eng.*, Vol. 30, pp. 349–361.

Teman, R. (1979) *Navier-Stokes Equations: Theory and Numerical Analysis*. Amsterdam: North-Holland.

White, F. M. (1974) *Viscous Fluid Flow*. New York: McGraw-Hill.

Yanenko, N. N. (1971) *The Method of Fractional Steps*. Berlin: Springer-Verlag.

Zienkiewicz, O. C., Villotte, J. P., Toyoshima, S. and Nakazawa, S. (1985) "Iterative Method for Constrained and Mixed Approximation: An Inexpensive Improvement for F. E. M. Performance," *Comput. Methods Appl. Mech. Eng.*, Vol. 51, pp. 3–29.

EXERCISES

10.1   Using weighting functions U and V for the momentum equations and P for the continuity equation, find the weighted residuals formulation of Eqs. (10.1)–(10.3).

10.2   Calculate the element equations corresponding to Eq. (10.18) for the bilinear velocity – constant pressure element in $0 \le x \le h$, $0 \le y \le k$.

10.3   Use the enclosed computer program to reproduce the results of Example 10.2.

10.4   Discuss the results in Fig. 10.12, and determine if they are consistent with what you would expect from physical considerations.

10.5   Use the enclosed computer program to reproduce the results of Fig. 6.5 for flow over a backward facing step at Re = 100, and those in Example 10.4 at Re = 900. Discuss the differences encountered as Re increases.

10.6   Consider a square mesh of size h, and assemble Eq. (10.23) over the four elements containing the shape function N_i. Notice that it is only necessary to consider the shape function N_i one at a time. Assuming that other velocity components are zero, show that the pressure field in Fig. 10.5 satisfies Eq. (10.23). (A detailed analysis was given by Carey and Krishnan, 1982.)

10.7   Show that for a rectangular mesh of bilinear elements, if the matrix **A** is lumped in Eq. (10.25), the smoothed pressures at the nodes are given by

$$P_i = \frac{A_1 p_1 + A_2 p_2 + A_3 p_3 + A_4 p_4}{A_1 + A_2 + A_3 + A_4}$$

where A_i denotes the area of element i = 1,2,3,4, which are the elements containing node i, and p_i is the constant pressure over element i. Furthermore, if all elements are of the same size, the above reduces to

$$P_i = \frac{1}{4}(p_1 + p_2 + p_3 + p_4)$$

10.8   Following the description of the fractional step method, discretize all variables using bilinear rectangular elements.
    (a) Find the element equations corresponding to Eqs. (10.31) and (10.32) using mass lumping and Petrov-Galerkin weighting of the convective terms.

(b) Find the element equations for the pressure equation, Eq. (10.34).

(c) Find the element equations for the velocity corrections given in Eq. (10.37).

(d) Can you now put together a simple program and try it for the Poiseuille problem of Example 10.2?

10.9   Derive Eqs. (10.46) and (10.47).

10.10   Derive Eqs. (10.67) through (10.69).

10.11   Obtain the analytic solution to the stratified Poiseuille flow of Example 10.5.

10.12   Solve the isothermal Poiseuille flow of Example 10.2, for Re = 5, using the tensor form of the momentum equations given by Eqs. (10.4) and (10.5) with natural boundary conditions

$$-p + \frac{2}{R_e} \frac{\partial u}{\partial x} = 0$$

and

$$\frac{\partial u}{\partial y} + \frac{\partial v}{\partial x} = 0$$

at the outflow boundary x = 5. Can you explain what happens to the flow field at x = 5?

10.13   Starting from Eqs. (10.62) and (10.63), derive Eqs. (6.69) through (6.72) including the time derivative terms.

ELEVEN

MESH GENERATION

11.1 OVERVIEW

Mesh (or grid) generation establishes the locations of nodal (or grid) points, element connectivities, and specification of boundary values, i.e., prescribed Dirichlet and Neumann conditions for a specific problem. A physical problem domain (or region) must be *discretized* by the user, and a solution achieved using node points within either the physical or computational domain. Some numerical models require the use of structured (or ordered) meshes, which must be orthogonal. If the physical domain is rectangular, the mesh is easy to construct; if the domain is irregular, or highly distorted, a transformation must first be made to create a rectangular, computational domain. This type of procedure is common to finite difference (or finite volume) methods and accounts for the popular use of boundary-fitted coordinates (BFC); the physical problem is *globally* transformed to a rectangular geometry. The FEM requires no such constraint and operates by utilizing unstructured meshes, i.e., orthogonality is not necessary. Each element is individually treated by performing a *local* isoparametric transformation (from x,y to ξ, η space), calculations are made and stored, then the procedure is repeated for the next element, not necessarily in sequence.

The ability to create effective, reliable meshes is not as simple as it appears, generally requiring the user to create several meshes before a satisfactory solution is achieved. The *dues* one must pay in using finite element methods is this preprocessing, mesh generation step; once the mesh is established, it is easy for the finite element user to change boundary condition constraints and requirements.

In this chapter, we will discuss the types of meshes and generation schemes commonly employed, structured versus unstructured meshes, bandwidth and optimization techniques (node numbering), and briefly describe adaptation. There are numerous commercial mesh generation packages now available on the market; short descriptions of several of the more popular models are given.

11.2 INTRODUCTION

The history of grid solution problems can be traced back to two key periods. The first period goes back several hundred years to when mathematicians began interpreting grids as a set of points used to sample a function. This type of analysis prevailed until the introduction of the computer age. The second key period occurred in the mid–1950s with the introduction of finite element analysis for stiffness and deflection of airframes. In this period, the grid points became the corners, or vertices, of elements used to discretize a physical object (domain). The need for preprocessing, i.e., mesh generation, began in earnest and has advanced to our present state. Both early approaches have essentially merged today and have created major spin-off fields including CAD/CAM and computer graphics. One has only to view the latest science fiction movies to see the realism associated with the special effects. A good review of the history of grid generation is given by Oran and Boris (1987). Mesh generation with orthogonal meshes is discussed by Thompson et al. (1985). Finite element meshes are described in numerous texts; see Carey (1997), Szabo and Babuska (1991), George (1991), Segerlind (1984), Zienkiewicz (1977), and Akin (1993).

11.2.1 Types of Meshes

There are basically two types of meshes: structured and unstructured. A structured mesh consists of horizontal and vertical lines that cross at intersections called nodes. The intersection of lines must be orthogonal. This constraint is best achieved by discretizing a physical domain that is defined by square or rectangular boundaries. One can then create grid intersections or nodes. The intersection of lines must be orthogonal. This constraint is best achieved by discretizing a physical domain that is

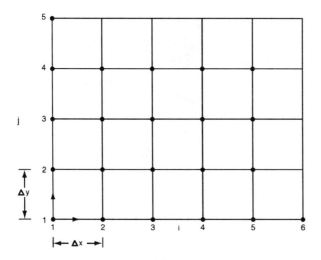

Figure 11.1 Structured mesh.

defined by square or rectangular boundaries. One can then create grid lines by inspection, and the physical domain becomes the computational domain as well. Much of the early numerical simulations were conducted on problems that were first reduced to rectangular physical systems of interest. Once the structured mesh is established, the nodes are then counted sequentially in order, e.g., top down and left to right, which is easily established using simple nested "do loops." An example of a 2-D structured mesh is shown in Fig. 11.1. Node (1,1) would be located at the lower left corner, with (i, j) denoting node locations within the mesh. If the nodal spacing is the same in each direction, i.e., $\Delta x = \Delta y = $ const, then the mesh is considered to be uniform in both x and y. When Δx and Δy vary between nodes, the mesh is nonuniform in x and y. Such meshes are commonly used when one wishes to create smaller grid spacings near surfaces (and hopefully, enhance accuracy), e.g., to capture the boundary layer. The problem with structured meshes occurs when the physical geometry is not regular. One must then resort to either ignoring curvature effects, utilizing transformations (typical of complex variables), or employing BFCs (which is used in most commercial finite volume codes). These procedures require *global* transformation of the physical *geometry* as well as governing equations.

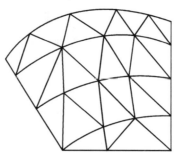

Figure 11.2 Unstuctured mesh.

In unstructured meshes, a 2-D physical domain is discretized by a set of seemingly randomly placed nodes that are connected to other nodes via triangular or quadrilaterally shaped subdomains, or elements. The most common types of elements are linear three-noded triangles or linear four-noded quadrilaterals (called bilinear quadrilaterals). Other popular types of elements include quadratic and cubic triangles and quadrilaterals. In three dimensions these would become tetrahedrals and hexahedrals, respectively. An example of a 2-D unstructured mesh employing three-noded triangular elements is shown in Fig. 11.2. The generation of unstructured meshes requires more thought and effort than structured meshes. In general, one puts more nodes (or elements) near surfaces and in regions where activity (or steep gradients) is likely to occur. Many times, the user must guess as to where the most nodes should be placed, ultimately necessitating the generation of a second mesh (and comparing solutions for accuracy). It is up to the user to specify the mesh density (number of nodes and elements), which is best achieved through experience. When using unstructured meshes, the need to globally transform the physical geometry to an orthogonal, rectangular computational domain is eliminated. The user discretizes the physical domain in situ and solves the governing equations in the physical coordinates of the problem[*] – a major advantage over structured meshes.

A relatively recent addition to the family of mesh generation techniques is the use of adaptation. The intent of adaptation is to begin with a coarse mesh, which can be easily and quickly generated.

---

[*]When using isoparametric elements, each element is locally transformed to ξ,η space, the calculations performed, and results stored, then returned to x, y physical space for the next element. Such operations do not require *global* transformation of the problem geometry or governing equations.

Adaptation criteria are established by the user, e.g., the first difference in density for compressible flow problems or concentration in species transport, which are calculated over each element in the domain. If the criteria are not met, the element (or elements) are refined, or subdivided; these subdivisions are continued as the solution progresses in time until a set number of adaptations have been reached, or the criteria are satisfied. In our experience, we have found that two to three adaptations are generally sufficient. Likewise, as the solution progresses, certain regions of the computational domain may no longer contain steep gradients, i.e., the flow may be smooth. In this case, elements within the region will unrefine and return to larger, coarser elements. Hence the process of adaptation involves both refining and unrefining groups of elements. It is well known that adaptation significantly improves solution accuracy and speed and eliminates the need to regenerate meshes typically required of more conventional finite element procedures.

Mention should also be made about meshless methods and Lagrangian techniques. Meshless methods are typically hybrid numerical techniques that incorporate some degree of analytical solution capability. An example of one such method is the employment of Gaussian puff/plume techniques to calculate concentration transport and diffusion. In Lagrangian techniques a particle, or "quantity" is advected from one location to another in time, i.e., $v \equiv dx/dt$ and $a \equiv dv/dt$. Lagrangian and analytical methods are usually tied back to a reference Eulerian mesh, primarily for establishing a basis for comparing results and graphical display. Lagrangian-Eulerian techniques have been developed over many years; Arbitrary Lagrangian-Eulerian (ALE) algorithms developed at Los Alamos National Laboratory (LANL) are discussed by Hirt and Amsden (1978). Several quasi-Lagrangian schemes are discussed by Long and Pepper (1981); the method of second moments (Fig, 11.3) is a high-order, quasi-Lagrangian method (see Pepper and Long, 1978).

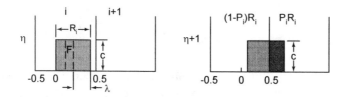

Figure 11.3 Method of Moments within an Eulerian mesh (R_i denotes the width of the distribution – the 2nd moment, F is the 1st moment, and C is the 0th moment).

11.2.2 Popular Mesh Generation Schemes

There are numerous ways in which a mesh can be generated (see Table 11.1); the choice of methodology is usually dictated by ease of use and familiarity with a preferred approach. Needless to say, there is no method yet developed that will automatically create the ideal mesh and appropriate boundary condition constraints without detailed user input and guidance. The following methods are the more common

Table 11.1 Mesh generation schemes

Type	Structured	unstructured
Manual	x	x
Semi–automatic	x	
Transport mapping	x	
Explicit PDE	x	
Overlapping/Deformation	x	x
Advancing–front		x
Combination	x	x

and popular approaches employed to create meshes (a more thorough discussion on these techniques is given by George (1991)

11.2.2.1 Manual Generation

The user defines the elements by their vertices and places the nodes at specific (user prescribed) locations. This approach is quick and suitable for physical domains with simple geometries that can be discretized with a coarse element array. In most instances, the preprocessing consists of merely creating a simple date file that can be read by the solver. Several of the exercise problems throughout this book can best be solved using this approach.

11.2.2.2 Semi-Automatic Generation

The user creates a simple mesh construct that is used to create a more complex mesh. This method is used when the domain is rectangular or cylindrical. Succeeding 2-D layers are used to create a three-dimensional mesh, e.g., groundwater modeling.

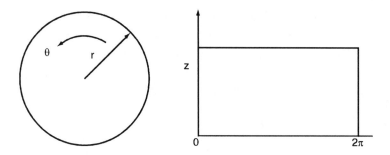

Figure 11.4 Transformation mapping: cylinder to rectangle.

11.2.2.3 Transport Mapping

This type of mesh generation corresponds to a domain with an elementary geometry, i.e., triangle, quadrilateral. The mapping function is predefined by the user. Transport mapping is usually performed when a physical geometry can be transformed to a rectangular computational domain in a global sense; e.g., a cylindrical geometry can be transformed to a rectangular geometry using $r = e^z$, as shown in Fig. 11.4. The use of complex variables employing Swartz-Christofel relations is typically employed. An example of a transformation from eccentric annuli to a rectangular domain is described by Pepper and Cooper (1980).

11.2.2.4 Explicit Solution of PDEs

A set of partial differential equations are solved which permits transformation from the physical domain to a rectangular computational domain, creating a set of structured meshes in the computational domain. The mesh is usually quadrilateral (in 2-D) or hexahedral (in 3-D) and permits finite difference (or volume) methods to be employed on an orthogonal mesh. A set of Poisson equations is established that permit attraction of grid lines to specified surfaces; the equations typically consist of second-order derivatives to account for the $(x,y) \rightarrow (\xi,\eta)$ transformation and are likewise solved using finite difference techniques. This technique, known as BFC, is commonly used in most finite volume commercial codes. The development and mathematical bases of the BFC are discussed by Thompson et al. (1985) and Warsi (1993). EAGLE is a 2-D and 3-D BFC mesh generator originally developed for the Air Force by Mississippi State University (see Thompson et al., 1985). A version of the EAGLE code, known as GRIDGEN (Pointwise, Inc., 1996), is a commercially available program for workstations; pcGridGen (Numerical Technologies, 1996) is a PC-based grid generator program that also simplifies the 2-D BFC approach.

11.2.2.5 Overlapping and Deformation

A physical domain, defined by its contour, is inclosed within a mesh composed of quadrilaterals or hexahedrals (boxes). The grid is then deformed in such a way that the real domain is accurately covered. The boxes of the mesh can be split to obtain elements of the final mesh. An example of an overlapping mesh would be the creation of an orthogonal mesh that is laid over an irregular shape. While simple to implement, the ability to accurately capture details at surfaces is inhibited. This is similar to aliasing in computer graphics, where curved surfaces are displayed in stair-step fashion. Deformation of meshes should always be considered with care. Optimum accuracy is best obtained when elements and meshes are not distorted.

A coarse mesh is first created, generally consisting of blocks assuming a simple geometry. A coarse partitioning is then conducted to subdivide the coarse mesh (or blocks). The interfaces of the blocks must be carefully defined to ensure proper connectivity and determine the partitioning required in succeeding blocks. A simple mesh may consist of a set of *macro-elements* (generally prescribed by eight-node serendipity quadrilaterals) used to define specific *regions* of the physical domain. These macro-elements are then subdivided into smaller elements. The density of elements is typically prescribed by the user within each macro-element; however, the user must ensure that element and nodal connectivity are consistent among the macro-element interfaces. This particular approach is used in the mesh generation code described by Pepper and Heinrich (1992). A simple 2-D mesh generator subroutine for rectangular domains is included within FLOW-2D.

Structured partitioning is commonly used in commercial finite element mesh generators, e.g. PATRAN (PDA Engineering, Mesa Verda, CA) and LAPCAD (LAPCAD Engineering (1996), San Diego, CA), as well as most finite element commercial codes with preprocessors (FIDAP, FLOTRAN, NEKTON, etc.). One of the most widely used mesh generators is PATRAN, which interfaces with most finite element (and some finite difference/volume) commercial codes; unfortunately, PATRAN is not available for the PC or Macintosh. The use of structured partitioning permits any arbitrarily shaped domain to be discretized; however, one may have to define a large number of initial blocks (i.e., regions, patches, or macro-elements). Such meshes should then be optimized (renumbered) to reduce bandwidth (which will be discussed later).

11.2.2.6 Advancing-front Method

In this technique, a polygonal approximation of the physical problem boundary is made in 2-D, or a polyhedral approximation in the form of triangular faces made for 3-D. A front is established using the initial contour. Internal elements are then created from the edge of the front. The front is then updated by adding or suppressing some edges. The process is repeated as long as the front is not empty. This technique is suitable for the creation of triangular or tetrahedral meshes; however, the method must be used with some care to ensure convergence of the process. A variation on this approach is the Delaunay-Voronoï technique (see Carey, 1997), whereby the mesh is created from points on the domain boundary utilizing Delaunay-Voronoï construction. In this approach, a mesh is created that consists of a convex hull of a set of points; angles within triangular elements are then kept to within prescribed limits (long, skinny triangular elements do not yield accurate solutions and may lead to divergence). It is also possible to generate a mesh from a set of points initially located within the domain. Delaunay-Voronoï generation is especially useful in computer graphics.

11.2.2.7 Combination

Meshes can also be generated by using combinations of the previously described approaches, as well as previously created meshes. This approach typically employs geometric transformations or topologic transformations (global refinement). Geometric properties of the domain are fully considered to mesh only useful parts; the mesh of the various parts, which can be derived from use of symmetry, rotation, and mirror imaging, can then be obtained using simple transformations. The final mesh is then created by connecting all the subregion meshes together. An example of this type of procedure at a relatively simple level is the multiple-grid or telescoping-grid techniques used in the past by the National Weather Service to make forecasts. A large, coarse grid is first created to calculate synoptic, global flow patterns; results from these simulations are then used as initial and boundary conditions for a finer mesoscale grid over a region. Subsequently, a fine microscale region on the order of kilometers is created to predict local effects. This type of mesh nesting is very effective when dealing with flow problems that occur over a wide range of spatial scales.

11.3   MESH GENERATION TECHNIQUES

It is critical that all the information about a problem be made available to the user before beginning creation of a mesh. This information includes problem geometry and range of distances, initial and boundary conditions (location of loads), type of problem to be solved, and recognition of simple geometric shapes' symmetry lines (to reduce unnecessary nodal calculations and problem size) and cut lines. The generation of a mesh is generally a trial-and-error technique; the more experience one gains, the less trial-and-error is required.

11.3.1 Structured Meshes

When the physical geometry is simple and an orthogonal computational mesh can be quickly generated, finite difference/volume methods are advantageous and quick. The FEM typically utilizes unstructured meshes, but is equally effective on structured meshes.

Using a FEM preprocessor to generate structured meshes is not very efficient; FEM preprocessors are generally slower in creating simple meshes when compared to finite difference/volume methods. However, there are occasions when a problem is readily amenable to a structured mesh approach, and simpler mesh generation approaches should be employed.

11.3.1.1 One Dimension

A one-dimensional mesh is trivial to make and primarily useful for academic purposes or to test algorithms. Both finite difference and finite element one-dimensional meshes are structured and sequentially ordered. Figure 11.5 shows a 1-D mesh for both methods. In the finite difference

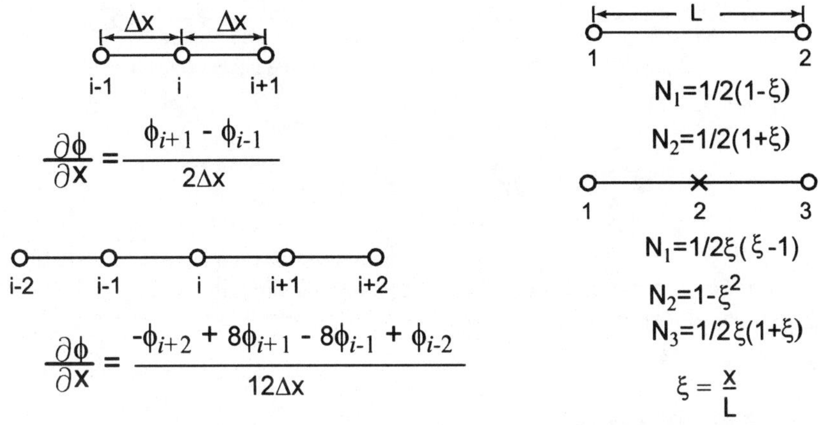

Figure a) central difference shows:

$$\frac{\partial \phi}{\partial x} = \frac{\phi_{i+1} - \phi_{i-1}}{2\Delta x}$$

$$\frac{\partial \phi}{\partial x} = \frac{-\phi_{i+2} + 8\phi_{i+1} - 8\phi_{i-1} + \phi_{i-2}}{12\Delta x}$$

a) central difference

Figure b) finite element shows:

$$N_1 = 1/2(1-\xi)$$

$$N_2 = 1/2(1+\xi)$$

$$N_1 = 1/2\xi(\xi-1)$$

$$N_2 = 1-\xi^2$$

$$N_3 = 1/2\xi(1+\xi)$$

$$\xi = \frac{x}{L}$$

b) finite element

Figure 11.5 One-dimensional mesh comparison between finite difference and finite element.

approach, more adjacent nodes are included in the gradient operator to enhance spatial accuracy, i.e., more terms are included in the Taylor series expansion. In the FEM, one begins with a linear, two-node element and progresses to higher order elements by including more internal nodes within the element. The text by Pepper and Heinrich (1992) includes a 1-D mesh generator for linear, quadratic, and cubic elements. Specifying the Δx intervals, it is easy to place node locations in the finite difference (or finite volume) scheme using $x_{i+1} = x_i + \Delta x$, where i is incremented from 1 to $n - 1$ nodes. The more interesting (and troublesome) mesh generation schemes begin at the 2-D level.

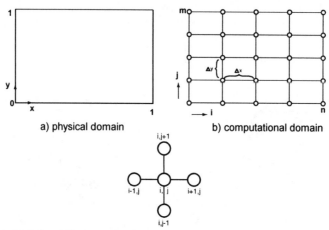

Figure 11.6 Finite difference mesh for 2-D rectangular domain for (a) physical domain and (b) computational domain.

11.3.1.2 Two Dimensions

Structured meshes in two dimensions are simple to make, providing the domain is rectangular. If the physical domain is also rectangular, the resulting computational domain is identical. For example, Fig. 11.6 shows a simple 2-D domain and resulting mesh. Utilizing pseudo code, the mesh is easily created with only a few lines of code

$$\text{do } i = 1, \, n - 1$$
$$x(i+1) = x(i) + \Delta x$$
$$\text{do } j = 1, \, m - 1$$
$$y(j+1) = y(j) + \Delta y$$
$$\text{end do}$$

Nonuniform grid spacing is likewise simple to create. For example, to increment a mesh so that the spacing between nodes is much finer near a surface, an algorithm like $x_{i+1} = x_i + \Delta x \cdot i / n$ can be used.

11.3.1.3 Three Dimensions

In order to create a three-dimensional, structured mesh, one need only add an additional "do loop" in the third direction, i.e.,

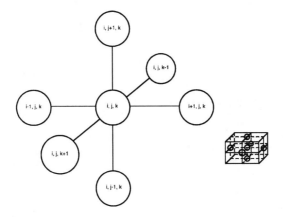

Figure 11.7 A 3-D structured mesh molecule.

$$\text{do } i = 1, \, n-1$$
$$x(i+1) = x(i) + \Delta x$$
$$\text{do } j = 1, \, m-1$$
$$y(j+1) = y(j) + \Delta y$$
$$\text{do } k = 1, \, \ell-1$$
$$z(k+1) = z(k) + \Delta z$$
$$\text{end do}$$

Notice that a node point and unknown variable can be located anywhere within the 3-D mesh by specifying the i, j, k values, e.g., $\phi(3,6,7)$ would locate the variable at i = 3, j = 6, and k = 7. A 3-D structured mesh molecule is shown in Fig. 11.7. Notice that it would take 8, eight-node trilinear hexahedral elements to access the nodes associated with the 3-D structured mesh (in this case we would be using the local element operations to create a global recursive relation). Refer to Pepper and Baker (1979) about the use of time splitting finite elements to create recursive relations.

11.3.1.4 Boundary Fitted Coordinates

As mentioned above, the major problem in using structured meshes is when one wishes to discretize irregular shapes. The most common and

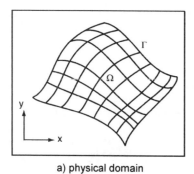

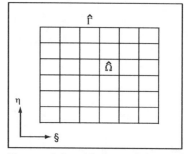

a) physical domain   b) computational (reference) domain

Figure 11.8 Boundary Fitted Coordinate mesh.

widely used method today is to employ a change of variables transformation, also known as BFC. This technique creates a transformation function that maps the physical domain onto a computational (or reference) domain that is orthogonal. The governing equations are transformed, and the solution computed on the reference domain mesh. This particular technique was developed in the 1970s. A comprehensive review of the BFC approaches is discussed by Thompson et al. (1985); also see Patankar (1980, 1991).

As an illustration of the BFC, Fig. 11.8 shows an irregular, 2-D physical domain that has been discretized using a non-orthogonal mesh, and the equivalent, orthogonal mesh in the reference (or computational) domain. Thus $(x,y) \rightarrow (\xi, \eta)$. Notice that in x,y space the mesh lines tend to follow the curvature of the boundaries, hence its name. Since the final mesh must be orthogonal, one actually begins with the reference domain in space. In 2-D this is a quadrilateral; in 3-D the mesh becomes a cube (or parallelepiped). This *canonically* constructed mesh is structured and permits use of i,j,k nodal indices. The reference domain mesh points map back to the mesh points and appropriate connectivity in the physical domain. The problem lies in the choice of the PDE relations to generate an acceptable mesh with the correct properties (density, boundary constraints, and orthogonality). There are basically two approaches one can use to develop these generating system equations:

1. A set of PDE relations is posed in terms of x,y (or x,y,z). These relations are then inverted, i.e., functions in terms of $x(\xi, \eta)$, or $x(\xi, \eta, \zeta)$, are computed instead of $\xi(x,y,z)$. A coupled system of equations is then solved iteratively to yield the desired result.

2. The PDEs are posed in terms of ξ,η,ζ with exact values of the boundary condition on the boundary contours. The PDEs are solved in the structured mesh domain and solutions for the values of $x(\xi,\eta,\zeta)$, $y(\xi,\eta,\zeta)$, and $z(\xi,\eta,\zeta)$ are used to construct the physical domain mesh. This method is a simplification of the other approach and is more empirical.

There are three methods employed to establish the PDE relations used in either method 1 or method 2 above. These are elliptic methods, hyperbolic methods, and parabolic methods, which conform to the three types of PDE relations. The most widely used approach is the elliptic method, which we will discuss at some length now.

Assume the following relations exist for a 2-D domain and we wish to follow method 1:

$$\frac{\partial^2 \xi}{\partial x^2} + \frac{\partial^2 \xi}{\partial y} = 0 \qquad (11.1)$$

$$\frac{\partial^2 \eta}{\partial x^2} + \frac{\partial^2 \eta}{\partial y^2} = 0 \qquad (11.2)$$

with appropriate boundary conditions. Likewise assuming solutions of Eqs. (11.1) and (11.2) are known, we can perform an inverse to obtain $x(\xi,\eta)$ and $y(\xi,\eta)$. Utilizing the chain rule,

$$dx = \frac{\partial x}{\partial \xi}\underbrace{\left[\frac{\partial \xi}{\partial x}dx + \frac{\partial \xi}{\partial y}dy\right]}_{d\xi} + \frac{\partial x}{\partial \eta}\underbrace{\left[\frac{\partial \eta}{\partial x}dx + \frac{\partial \eta}{\partial y}dy\right]}_{d\eta} \qquad (11.3)$$

and

$$dy = \frac{\partial y}{\partial \xi}\left[\frac{\partial \xi}{\partial x}dx + \frac{\partial \xi}{\partial y}dy\right] + \frac{\partial y}{\partial \eta}\left[\frac{\partial \eta}{\partial x}dx + \frac{\partial \eta}{\partial y}dy\right] \qquad (11.4)$$

Noting that combinations of the products of these derivatives yield values that are equal to 0 or 1, we can obtain the matrix expression

$$\begin{bmatrix} \dfrac{\partial x}{\partial \xi} & \dfrac{\partial x}{\partial \eta} \\ \dfrac{\partial y}{\partial \xi} & \dfrac{\partial y}{\partial \eta} \end{bmatrix} \begin{bmatrix} \dfrac{\partial \xi}{\partial x} & \dfrac{\partial \xi}{\partial y} \\ \dfrac{\partial \eta}{\partial x} & \dfrac{\partial \eta}{\partial y} \end{bmatrix} = \begin{bmatrix} 1 & 0 \\ 0 & 1 \end{bmatrix} \tag{11.5}$$

$$\underbrace{\phantom{\begin{bmatrix} \dfrac{\partial x}{\partial \xi} & \dfrac{\partial x}{\partial \eta} \\ \dfrac{\partial y}{\partial \xi} & \dfrac{\partial y}{\partial \eta} \end{bmatrix}}}_{\equiv M}$$

or

$$\begin{bmatrix} \dfrac{\partial \xi}{\partial x} & \dfrac{\partial \xi}{\partial y} \\ \dfrac{\partial \eta}{\partial x} & \dfrac{\partial \eta}{\partial y} \end{bmatrix} = \dfrac{1}{J} \begin{bmatrix} \dfrac{\partial y}{\partial \eta} & -\dfrac{\partial x}{\partial \eta} \\ -\dfrac{\partial y}{\partial \xi} & \dfrac{\partial x}{\partial \xi} \end{bmatrix} \tag{11.6}$$

$$\underbrace{\phantom{\dfrac{1}{J} \begin{bmatrix} \dfrac{\partial y}{\partial \eta} & -\dfrac{\partial x}{\partial \eta} \\ -\dfrac{\partial y}{\partial \xi} & \dfrac{\partial x}{\partial \xi} \end{bmatrix}}}_{\equiv M^{adj}}$$

where J is the familiar Jacobian, $J = \partial x/\partial \eta \cdot \partial y/\partial \eta - \partial x/\partial \eta \cdot \partial y/\partial \xi$. Utilizing Eq. (11.6), one can obtain the second derivative expressions for ξ and η, e.g.,

$$\dfrac{\partial^2 \xi}{\partial x^2} = \dfrac{\partial}{\partial \xi}\left[\dfrac{1}{J} \dfrac{\partial y}{\partial \eta} \right] \dfrac{\partial \xi}{\partial x} + \dfrac{\partial}{\partial \eta}\left[\dfrac{1}{J} \dfrac{\partial y}{\partial \eta} \right] \dfrac{\partial \eta}{\partial x} \tag{11.7}$$

or

$$\dfrac{\partial^2 \xi}{\partial x^2} = \dfrac{\partial}{\partial \xi}\left[\dfrac{1}{J} \dfrac{\partial y}{\partial \eta} \right] \dfrac{1}{J} \dfrac{\partial y}{\partial \eta} - \dfrac{\partial}{\partial \eta}\left[\dfrac{1}{J} \dfrac{\partial y}{\partial \eta} \right] \dfrac{1}{J} \dfrac{\partial y}{\partial \xi} \tag{11.8}$$

Hence

$$\dfrac{\partial^2 \xi}{\partial x^2} = \dfrac{1}{J^2}\left[\dfrac{\partial y}{\partial \eta} \dfrac{\partial^2 y}{\partial \eta \partial \xi} - \dfrac{\partial y}{\partial \xi} \dfrac{\partial^2 y}{\partial \eta^2} \right] + \dfrac{1}{J^3}\left[-\dfrac{\partial J}{\partial \xi}\left(\dfrac{\partial y}{\partial \eta} \right)^2 + \dfrac{\partial J}{\partial \eta} \dfrac{\partial y}{\partial \eta} \dfrac{\partial y}{\partial \xi} \right] \tag{11.9}$$

which obviously becomes somewhat complicated. Similar expressions are obtained for $\partial^2 \xi/\partial \eta^2$, $\partial^2 \eta/\partial x^2$ and $\partial^2 \eta/\partial y^2$. A little additional effort shows that x and y must satisfy the expressions

$$g_{11}\frac{\partial^2 x}{\partial \xi^2} + g_{22}\frac{\partial^2 x}{\partial \eta^2} + 2g_{12}\frac{\partial^2 x}{\partial \xi \partial \eta} = 0 \qquad (11.10)$$

$$g_{11}\frac{\partial^2 y}{\partial \xi^2} + g_{22}\frac{\partial^2 y}{\partial \eta^2} + 2g_{12}\frac{\partial^2 y}{\partial \xi \partial \eta} = 0 \qquad (11.11)$$

where

$$g_{ij} = \sum_{m=1}^{2} A_{mi}A_{mj} \qquad (11.12)$$

with $A_{mi} = (-1)^{i+m}M^{cofactor}$. The resultant equations are coupled and nonlinear and usually solved using an iterative technique. By imposing real boundary conditions, an initial guess is then used to start the solution procedure.

A variation of the above approach is to add attractions (P and Q) to Eqs. (11.1) and (11.2):

$$\frac{\partial^2 \xi}{\partial x^2} + \frac{\partial^2 \xi}{\partial y^2} = P \qquad (11.13)$$

$$\frac{\partial^2 \eta}{\partial x^2} + \frac{\partial^2 \eta}{\partial y^2} = Q \qquad (11.14)$$

The inverse system to be solved subsequently becomes

$$g_{11}\frac{\partial^2 x}{\partial \xi^2} + g_{22}\frac{\partial^2 x}{\partial \eta^2} + 2g_{12}\frac{\partial^2 x}{\partial \xi \partial \eta} + J^2\left(P\frac{\partial x}{\partial \xi} + Q\frac{\partial x}{\partial \eta}\right) = 0 \quad (11.15)$$

$$g_{11}\frac{\partial^2 y}{\partial \xi^2} + g_{22}\frac{\partial^2 y}{\partial \eta^2} + 2g_{12}\frac{\partial^2 y}{\partial \xi \partial \eta} + J^2\left(P\frac{\partial y}{\partial \xi} + Q\frac{\partial y}{\partial \eta}\right) = 0 \quad (11.16)$$

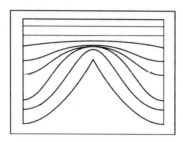

Figure 11.9 Attraction of BFC generated grids to a point.

When P > 0, points are attracted to the right; if P < 0, points are attracted to the left. Likewise, Q > 0 attracts points to the top, while Q < 0 draws them to the bottom. Expressions for P and Q can also be obtained to concentrate lines to a surface or attract lines to a point, as shown in Fig. 11.9. More detailed discussions on the expressions for P and Q are given by Thompson et al. (1985) and Warsi (1993).

In method 2, assume that the reference domain $\hat{\Omega}$ is a unit square with boundary $\hat{\Gamma}$, whereby

$$\hat{\Gamma} = \left\{ (\eta, \xi), (\eta = 0) \bigcup (\xi = 1) \bigcup (\eta = 1) \bigcup (\xi = 0) \right\} \quad (11.17)$$

and the system equations exist such that

$$\frac{\partial^2 x}{\partial \xi^2} + \frac{\partial^2 x}{\partial \eta^2} = 0 \qquad \left(\text{in } \hat{\Omega} \right) \qquad (11.18)$$

with

$$x(\xi, \eta) = x_\Gamma \qquad \left(\text{on } \hat{\Gamma} \right) \qquad (11.19)$$

and

$$\frac{\partial^2 y}{\partial \xi^2} + \frac{\partial^2 y}{\partial \eta^2} = 0 \qquad \left(\text{in } \hat{\Omega} \right) \qquad (11.20)$$

$$y(\xi, \eta) = y_\Gamma \qquad \left(\text{on } \hat{\Gamma} \right) \qquad (11.21)$$

Real values for the boundary conditions are imposed at points along $\hat{\Gamma}$, i.e., $x(\xi,\eta) \equiv x(x,y)$ and $y(\xi,\eta) \equiv y(x,y)$. The solutions of Eqs. (11.18)–(11.21) yield values for $x(\xi,\eta)$ and $y(\xi,\eta)$ at each nodal point in Ω. Thus the mesh in Ω is known with connectivity associated with the canonical mesh in $\hat{\Omega}$, and nodal locations determined by (x,y). Difficulties relating to nonlinearity and coupling common to method 1 disappear; however, the mesh may not be totally valid. It is possible to create node points outside the true physical domain $\hat{\Omega}$, especially when dealing with concave boundaries. Corrective procedures must be employed to overcome these errors.

Two other methods previously mentioned are hyperbolic and parabolic generation systems. For example, the relations

$$\frac{\partial x}{\partial \xi}\frac{\partial x}{\partial \eta} + \frac{\partial y}{\partial \xi}\frac{\partial y}{\partial \eta} = 0 \qquad (11.22)$$

$$\frac{\partial x}{\partial \xi}\frac{\partial y}{\partial \xi} - \frac{\partial x}{\partial \eta}\frac{\partial y}{\partial \xi} = f(\xi,\eta) \qquad (11.23)$$

produce a hyperbolic method that is controlled by the function $f(\xi,\eta)$, which operates on the system boundary. An orthogonal mesh is created; however, $f(\xi,\eta)$ is not always easy to establish. Likewise, it is difficult to establish the exterior boundary explicitly, and may create a mesh with low precision for the exteriors.

When dealing with arbitrary geometries, it is convenient to partition, or decompose, the region into several blocks having the desired topology, ultimately leading to a quadrilateral (or hexahedral) reference domain. Three of the most common types of decomposition geometries are the 0 types, C types, and H types, which are shown in Fig. 11.10. When these types of shapes cannot be employed, artificial cuts (cut lines) must be made, a not so trivial task in many applications to correctly ensure a reliable mesh. The use of C and H type decompositions are extremely useful and simpler to construct than unstructured finite element meshes (unless adaptation is employed).

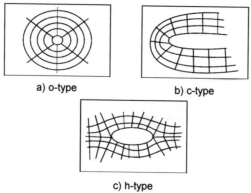

a) o-type
b) c-type

c) h-type

Figure 11.10 Types of decomposition.

11.3.2 Unstructured Meshes

The creation of unstructured meshes is generally relegated to the finite element approach. Recently, the combination of finite volume and finite element models have evolved that operate on unstructured meshes (typically triangles) (Fig. 11.11); however, the preprocessing involved in creating a mesh stems back to classical finite element methodology, including element connectivity. In such cases, a patch of elements with a common vertex node is generated to create a control volume. A transformation is then performed, and a simple integral expression established for ensuring conservation. The procedure entails some effort in bookkeeping and does not permit flexibility in using higher order elements. However, adaptation can be utilized. The interested reader can find detailed descriptions of the method in the work by Schneider and Raw (1987).

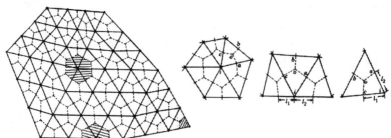

Figure 11.11 Finite volume – finite element mesh. (From *Handbook of Numerical Heat Transfer*, W. J. Minkowycz, E. M. Sparrow, G. E. Schneider, and R. H. Pletcher, Eds., Copyright 1988, John Wiley & Sons, Inc. Reprinted by permission of John Wiley & Sons, Inc.).

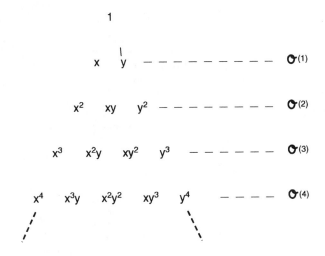

Figure 11.12 Pascal's triangle.

11.3.2.1 General Types of Elements

There are an unlimited number of elements available for the user; the basis of the element configuration lies in the choice of the degree of polynomial desired. Note that the shape function is also typically used to define the element shape; i.e., isoparametric elements are preferred. It is useful to recall Pascal's triangle to examine the number of terms occurring in a polynomial in x,y (see Fig. 11.12); a first-order polynomial requires three terms, a second-order requires six terms, etc.

11.3.2.2 One-Dimensional Elements

The most common element shapes are the linear, quadratic, and cubic forms; in fluid flow problems, one generally tries to use the *lowest* order element to reduce bandwidth and storage. The three most common one-dimensional elements are shown in Fig. 11.5b. In two and three dimensions the user need only to recall the one-dimensional element form and apply this to each edge (or leg) of the element. This technique produces a *serendipity* family of elements; if the user includes interior nodes, the family becomes Lagrangian (with basis in Lagrange polynomials), which can be found by simple products in each coordinate.

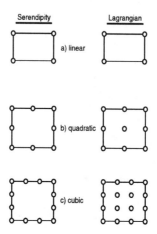

Figure 11.13 Standard 2-D quadrilateral element family.

11.3.2.3 Two-Dimensional Elements

Examples of the three types of quadrilateral elements are shown in Fig. 11.13 for Lagrangian and serendipity families in two dimensions; extension to three dimensions is straightforward. Dividing a quadrilateral Lagrangian element at its shortest diagonal distance produces two triangles, with similar properties and formulation, as shown in Fig. 11.14. The triangle has an advantage over quadrilaterals in approximating any boundary shape, and the fact that the number of nodes coincides exactly with the number of polynomial terms is

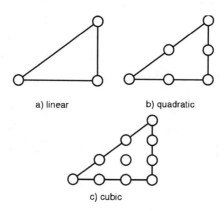

Figure 11.14 Triangular element family.

desirable. Most of the image generation in computer graphics is performed using triangles and tetrahedrals (when no solutions to nonlinear equations of motion are sought); likewise, structural mechanics problems are routinely solved using triangular and tetrahedral elements.

It is important to remember that elements generated within the physical domain are subsequently mapped onto the computational (or reference) domain on a local level (as opposed to global in BFC). Providing coordinate relationships exist, shape functions specified in local coordinates can be transformed to yield element properties in the global, physical space. This is achieved using isoparametric relationships, as previously discussed. Illustration of this mapping process is shown in Fig. 11.15 for the two-dimensional quadrilateral and triangle. Note that for a parametric transformation based on bilinear shape functions, no internal angle can be greater than $180°$, i.e., $\theta < 180°$. It is also necessary to ensure that the mid-side nodes are located within

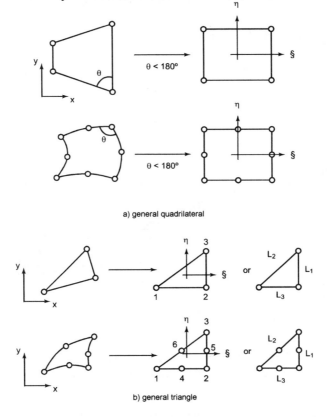

a) general quadrilateral

b) general triangle

Figure 11.15 Two-dimensional element mapping.

the middle third of the leg for serendipity-type approximations; such general rules are impractical for cubic functions.

When one elects to use higher order elements to discretize a problem domain, some care should be exercised in selecting the choice of element. For linearly distorted elements, the serendipity family of $O(4)$ or greater will yield quadratic convergence[*] (refer to Pascal's triangle, Fig. 11.12). The nine-noded Lagrangian element represents better Cartesian polynomials than the eight-noded serendipity element and is the preferred choice in modeling smooth solutions. When there is no distortion, both elements yield similar results. The most accurate 2-D quadratic element for fluid flow simulation is the nine-node Lagrangian; this has been amply demonstrated in numerous examples reported in the literature. In three dimensions the 27-noded element yields the most accurate convergence. Recall that the amount of computational storage and effort increases dramatically over the use of bilinear quadrilaterals. It is often preferable to use more, lower order elements.

11.3.2.4 Three-Dimensional Elements

Mesh generation in two dimensions can become troublesome but is not terribly difficult. Mesh generation in three dimensions is considerably

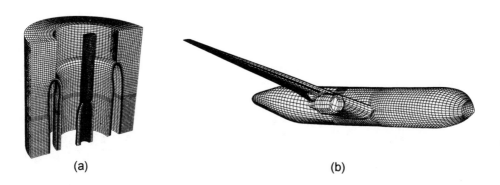

(a)   (b)

Figure 11.16 Three-dimensional element meshes: (a) cylindrical cross-section and (b) airplane surface (from XYZ Scientific Applications, Inc., Livermore, CA).

---

[*]Convergence rate is typically $O(h^{p+1})$, where h is the element size and p is the degree of polynomial expansion.

more complicated, generally requiring a great deal of trial-and-error (and patience) for the user when regions are complex. Such mesh generation practices are best left to users with experience in running reliable, multidimensional commercial packages, the most popular being PATRAN. In some instances, the user can create two-dimensional planes that can be discretized with two-dimensional elements. These planes are then simply connected in the third dimension; this practice is commonly done in atmospheric and groundwater modeling, and in environmental transport modeling situations where the physical domains are not overly distorted, e.g., HVAC and indoor air quality within rooms and buildings. In many engineering applications, such luxuries are not available, and an extensive three-dimensional discretization must be performed. Figure 11.16 shows example three-dimensional element meshes; it is not uncommon to spend many hours creating an acceptable element mesh for a complex geometry.

Examples of the three most common types of three-dimensional elements are shown in Fig. 11.17. Notice that the elements are extensions of the one- and two-dimensional element families previously discussed. To obtain Lagrangian forms of the serendipity elements shown in Fig. 11.17, one can add the appropriate interior nodes, as illustrated in the two-dimensional elements shown in Figs. 11.13 and 11.14. Triangular prisms are particularly useful in cylindrical geometries and serve as useful "filler" elements when used with hexahedrals.

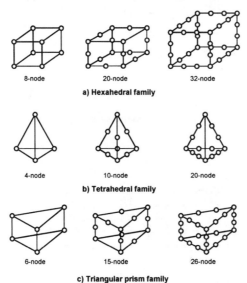

Figure 11.17 Families of 3-D serendipity elements.

11.3.3 Mesh Generation Guidelines

As one can see, it is considerably more difficult to create an algorithm to generate an unstructured mesh than a structured mesh. However, a finite element mesh must keep track of node location (x,y), element connectivity (the global nodes that make up the local nodes of an element), and boundary condition nodes and/or elements (Dirichlet and/or Neumann conditions). Hence the bookkeeping becomes considerably more involved.

There is no reason one should end up using severely distorted elements to discretize a domain. In fact, the interior of most domains can be meshed using non-distorted elements; as one approaches the boundaries, several slightly distorted elements can be constructed. Curved sides should only be employed on curved boundaries, and the curvature should be rather mild ($\leq 30°$ arc). When this is not possible, more elements should be utilized. An example of an irregular domain discretized with eight-noded quadrilaterals (macro-elements) is shown in Fig. 11.18. Notice that the interior of the region employs rectangular elements; elements at the boundary are curved and overly distorted in Fig. 11.18a. Refinement of the mesh in Fig. 11.18b has reduced the distortion, which may or may not be enough. If the solution is still inaccurate, further refinement would be required (especially along the boundary).

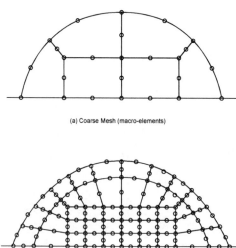

(a) Coarse Mesh (macro-elements)

(b) Enhanced Mesh

Figure 11.18 Discretization of a hemispherical domain using eight-noded quadratic quadrilaterals.

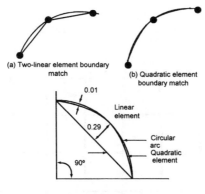

(a) Two-linear element boundary match

(b) Quadratic element boundary match

0.01

Linear element

0.29

Circular arc

Quadratic element

90°

(c) Boundary mismatch utilizing arc with r = 1.

Figure 11.19 Boundary curvature matching. (From *Finite Element Analysis*, D. S. Burnett, 1987, Addison-Wesley, with permission from Addison-Wesley).

If the physical domain has all boundary sides straight, with no internal curved surface (e.g., hole), any type of element will match the boundary exactly. Likewise, boundaries defined by higher degree polynomials can also be matched exactly with corresponding higher order elements. However, nonpolynomial curvature cannot be matched exactly by polynomial elements; hence the domain boundary becomes altered to the outer edge of the defining element. Suppose one wishes to use linear elements to prescribe a boundary, as shown in Fig. 11.19. In order to reduce the error associated with the area omissions, more linear elements are required. This leads to the decision by the user whether to increase the number of lower order elements or use a higher order element. The quadratic yields about 1% geometric error while the linear element produces about 29% error for a $90°$ arc (Burnett, 1987). In practice, preprocessing typically requires several refinements of the mesh, including boundary matching, before a suitable solution is achieved.

There may be instances when the user wishes to combine elements of different order (although this is discouraged in fluid flow related problems). Interelement continuity and connectivity must be maintained; otherwise a solution mismatch occurs (but convergence may be achieved). Each element side, or face, must match with its adjacent element. There are several ways to achieve this linking of two different order elements. The first approach utilizes *transition* elements, i.e., specialty elements that have an odd number of nodes prescribed on their boundaries. An example of this type of transition is shown in Fig. 11.20

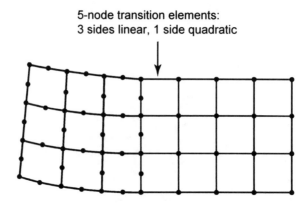

Figure 11.20 Quadratic to linear transition.

for a quadratic to linear transition. A second approach is to impose a linear constraint equation that constrains the higher order midside node to match the nodal solutions in the lower order element. This is shown in Fig. 11.21; this technique is commonly used when employing h-adaptation (to prevent using higher order elements). It is also possible to combine elements of different dimensions, a practice commonly performed in commercial finite element codes and preprocessors, as well as graphics routines in postprocessors.

A uniform mesh with identically sized elements is easy to generate, much like the finite difference mesh mentioned previously. In complicated regions and/or when the mesh must be refined in a particular region, a finer element density is generated. This can be accomplished in several ways. The first and simplest method is to use mesh gradation in one or more directions, as mentioned earlier. This is

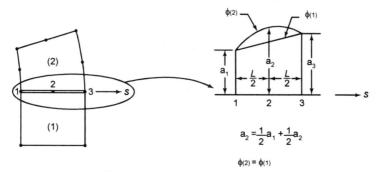

Figure 11.21 Quadratic to linear constraint.

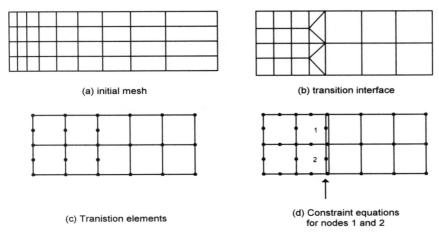

(a) initial mesh (b) transition interface

(c) Tranistion elements (d) Constraint equations for nodes 1 and 2

Figure 11.22 Mesh refinement.

illustrated in Fig. 11.22. It is best to make the gradation gradual.[*] The secondary way is to employ transition elements, as previously discussed, to transition from a finer region to a coarser region, as shown in Figs. 11.22b–11.22d. The meshes in Fig. 11.22a and 11.22b employ the same order elements; the use of transition with higher order elements keeps the element size constant. This is a form of p-adaptation, as shown in Fig. 11.22c and 11.22d. Figure 11.23 illustrates three ways in which a region in the lower left corner can be refined using the same type elements (Fig. 11.23a) and mixed elements (Figs. 11.23b and 11.23 c). The use of triangles as interfaces between different sized quadrilaterals is often used in h-adaptation (Ramakrishnam et al., 1990).

Elements should not be locally refined to the point that the smallest element is significantly smaller than the largest element, typically 10^2:1 (length) or 10^4:1 (area). Too large a difference can result in an ill-conditioned stiffness matrix and may produce inhibitively small time steps if using explicit marching methods. The more sophisticated (and usually expensive) commercial mesh generation codes, and some preprocessors used with commercial solvers, allow a very refined grid to

---

[*] One of the troublesome problems arising from the use of nonuniform meshes is the generation of numerical dispersion errors, i.e., reflections that occur as a solution leaves a fine mesh region and is transported to a coarser mesh; see Long and Pepper (1981).

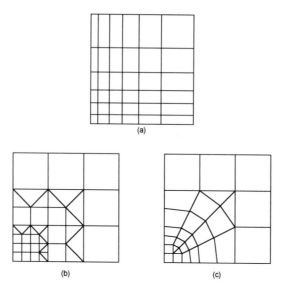

Figure 11.23 Local mesh refinement approaches.

be generated at the boundaries, utilizing nodal interpolation established from initial nodal values used to create the boundary contours. This procedure generates a uniformly fine mesh within the entire local region of interest and is relatively fast. However, the same error limitations still exist at the boundary; results from the refined mesh at or near the boundary should be examined for error and possibly discarded in the overall analysis of results. The refined mesh somewhat removed from the boundary should be suitable.

A problem dealing with the use of straight-sided linear elements on curved boundaries may occur in the vicinity of the boundary when solving second-order problems – the solution may converge to the wrong answer. This phenomenon can occur within a few elements of the boundary, producing very high error in the normal flux components. This problem is discussed by Babuska (1971), Zlamal (1973), and Krauthammer (1979). Curve-sided isoparametric elements (quadratic or higher order) do not exhibit this problem. Likewise, when one applies a flux to a curved boundary, and a straight-sided element is used for the discretization, the location and orientation of the boundary flux are altered, producing some error in the results. To reduce the error, one can use more elements (refined) near the boundary or shift to using higher order elements. A modeling error can also arise when a boundary flux is distributed over a small region, producing a need for refined elements near the load. If the user is not interested in the local behavior within the region, the distributed load can be approximated (or concentrated) at a node point at the centroid of the distribution (preferably a corner node if using higher order elements). The error is usually small when the distance is approximately five times the width of the original distributed load (Burnett, 1987).

11.4 BANDWIDTH

In 2-D and 3-D meshes, the node number pattern dictates the bandwidth of the assembled global matrix. Unless the user is employing an explicit marching scheme, the naturally implicit nature of the finite element method creates a banded, sparse matrix that may or may not be symmetric. Hence it behooves the user to minimize the bandwidth of the matrix to reduce storage and computer time. Finding the optimal minimum pattern can be difficult; however, any effort to achieve a near-optimal pattern is worth trying.

There are many algorithms available that automatically renumber a mesh to minimize its bandwidth; for frontal solvers, the wave front is minimized by renumbering the elements; the procedures are similar for nodal or elemental renumbering. The user must create the starting nodes, i.e., an initial mesh, which then gets reordered. These minimization routines are commonly used in most commercial finite element codes.

When dealing with symmetric matrices, one needs only to deal with the half-bandwidth; for asymmetric matrices, the full bandwidth must be considered (since the row-by-row bandwidth is not consistent). Symmetric matrices typically occur in structural and field type problems; asymmetric matrices arise in nonlinear problems, e.g. fluid flow. For simplicity, assume we have a symmetric matrix. The half-bandwidth over each element is calculated from the simple relation

$$B_{1/2}^{(e)} = N_{max}^{(e)} - N_{min}^{(e)} + 1 \tag{11.24}$$

where $N_{max}^{(e)}$ and $N_{min}^{(e)}$ are the largest and smallest node numbers,[*] respectively. The maximum $B_{1/2}$ value is determined as

---

[*]Or largest and smallest degrees of freedom (DOF).

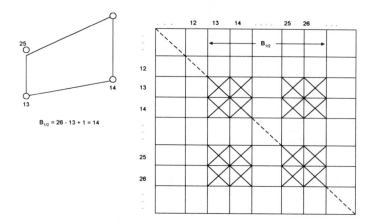

Figure 11.24 Half-bandwidth calculation for bilinear quadrilateral element and assembled stiffness matrix.

$$B_{1/2} = \max B_{1/2}^{(e)} \tag{11.25}$$

which then dictates the matrix bandwidth for the solver (including any unnecessary zero row values embedded within the band). To illustrate, Fig. 11.24 shows a four-noded, bilinear quadrilateral with global node numbers at the corner nodes. The half-bandwidth is $B_{1/2} = 26 - 13 + 1 = 14$, where $N_{max} = 26$ and $N_{min} = 13$. Thus the largest distance of a nonzero term from the main diagonal for this element is 14, including the diagonal, as shown in the assembled stiffness matrix.

Consider a simple rectangular plate consisting of eight triangular elements as shown in Fig. 11.25. The element connectivity and half-bandwidth per element are shown in Fig. 11.25b, and the resulting global stiffness matrix displayed in Fig. 11.25c. All nonzero values are shown by x's. Notice the apparent poor numbering sequence for the node numbers; the half-bandwidth varies from 3 to 9. The solution, however, is based on the maximum half-bandwidth of 9. What we would like to do is keep the node numbers as close together as possible. Thus one should number the nodes along paths containing the least number of nodes from one boundary to another, sequentially passing from each element to its adjacent element. As a rule of thumb, when the spacing among nodes is

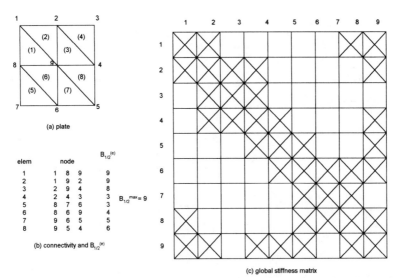

(a) plate

elem	node			$B_{1/2}^{(e)}$
1	1	8	9	9
2	1	9	2	9
3	2	9	4	8
4	2	4	3	3
5	8	7	6	3
6	8	6	9	4
7	9	6	5	5
8	9	5	4	6

$B_{1/2}^{max} = 9$

(b) connectivity and $B_{1/2}^{(e)}$

(c) global stiffness matrix

Figure 11.25 Half-bandwidth calculation for a plate discretized with eight triangular elements.

similar, one should number across the shortest dimension of the domain; however, if the nodal spacing is not similar, numbering across the shortest dimension may not yield the smallest half-bandwidth.

In Fig. 11.25c, notice that one can take advantage of the symmetry of the matrix by juggling the order of the global array to reduce the half-bandwidth. This is illustrated in Fig. 11.26. To reduce the half-bandwidth, the problem must be renumbered to minimize the differences among the related nodes. Figure 11.27 shows the renumbered nodes for the triangular elements of the plate and the resulting global matrix array. In this instance, the half-bandwidth is 5 in every element.

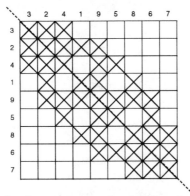

Figure 11.26 Final matrix after accounting for symmetry in the plate problem.

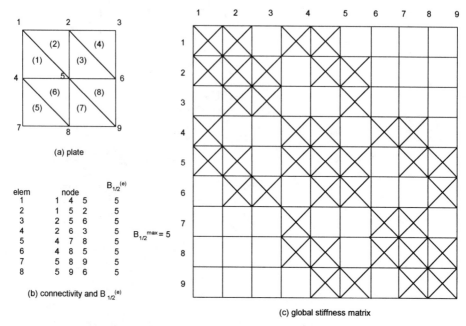

Figure 11.27 Reordered nodes for plate problem.

What happens if the nodes are renumbered sequentially in the vertical, beginning top-left and ending bottom-right? The challenge is to reduce (or eliminate) the zero values that lie between the maximum and minimum nonzero values within each row. What happens if more elements (and nodes) are added to discretize the domain (Exercise 11.1)? An example of a mesh using mixed elements and four different renumbering patterns is shown in Fig. 11.28, which is obtained from Burnett (1987). If one is using a frontal solver instead of a bandwidth solver, the procedures are similar; in this instance, the element numbers are renumbered to minimize the wave front.

11.4.1 Nodal Renumbering Schemes

Renumbering the nodes (or elements) of a mesh allows one to minimize the storage size required by the matrix solver and reduce the number of operations required by the final system, which ultimately reduces the CPU time. There are many methods that perform this renumbering operation, most of them automatic. The more popular schemes include

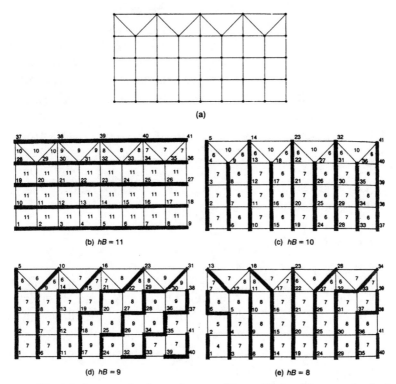

Figure 11.28 Mesh renumbering schemes. (From *Finite Element Analysis*, D. S. Burnett, 1987, Addison-Wesley, with permission from Addison-Wesley).

Gibbs' method (Gibbs et al., 1976), Grooms' method (Grooms, 1972), Lipton-Tarjan method (Lipton and Tarjan, 1979), Akhras-Dhatt method (Akhass and Dhatt, 1976), Frontal method, Element Colorization method (Melhem, 1987), and Nested Dissection method (George and Liu, 1978). A detailed discussion on the advantages and disadvantages of each method is given by Marro (1980) and George (1991). A more recent update on the application of renumbering schemes is given by Carey (1997). A brief discussion of each method follows.

11.4.1.1 Gibbs Method

There are three steps involved in using Gibbs' method. These steps consist of (1) searching for a good departure point, (2) optimization of the numbering descent, and (3) numbering using the Cuthill-McKee algorithm (Cuthill and McKee, 1969).

In step (1), a node is selected from the nodes on the contour of the boundary such that the degree of the node (i.e., the number of its neighbors) is minimal. Hopefully, a well-balanced distribution of nodes is obtained throughout the array. The last step is then implemented using the Cuthill-McKee algorithm.

This last step is the reverse of the renumbering and produces a profile width that should be globally better than the width before the reverse. Renumbering elements directly affects the nodal numbering connectivity, and should be done, if elected, before the nodal renumbering operation

11.4.1.2 Grooms' Method

Grooms' method is an iterative technique that starts from an initial numbering and the longest row of the matrix, and decreases the difference between the first column of this row and the diagonal. While the algorithm is effective, it can become time consuming.

11.4.1.3 Lipton-Tajan Method

The Lipton-Trajan method is based on the notion of a graph and a graph separation technique (a graph indicates the connections between nodes in terms of neighbors). The problem may be separated into subproblems of the same type but smaller in size. The interested reader should consult Lipton and Tarjan (1979) for details on employing this method.

11.4.1.4 Akhras and Dhatt Method

This particular algorithm, developed by Akhras and Dhatt (1976), uses an analysis of the optimal numbering of the regular mesh of a quadrilateral. Let S be the sum of the numbers of nodes within a set of node i's neighbors, denoted as V_i. Calculate the average of the numbers of the nodes, P, in V_i, and set the sum of the smallest and greatest number of nodes (S_p) in V_i. Since P and S_p increase, the goal is to construct a numbering that satisfies these properties in the domain. The algorithm consists of two phases. In the first phase the nodes are renumbered (one or more iterations) such that P and S_p increase. The second phase renumbers the result such that P and S increase (where the first phase can be restarted). While the technique is automatic and yields good results, the CPU can become lengthy.

11.4.1.5 Frontal Method

In the Frontal method the algorithm operates by radiating from a starting front. A set is reduced to one node (or collection of nodes) generally chosen on the contour of the domain. The renumbering depends strongly on the choice of the initial front, which must be selected by the user. Variations of the method exist and can be found in the literature (see George, 1991).

11.4.1.6 Element Colorization Method

The elements are renumbered by creating packets of elements such that any neighbor of an element in a given packet is not located in the same packet. The idea is similar to the four-color theorem. When only two packets are created, the method has the appearance of the Red-Black technique (Melhem, 1987), separating nodes into two disjoint sets. The goal of the method is to search for the neighbors of an element and separate them into various sets. This method is effective on vector computers.

11.4.1.7 Nested Dissection Method

The Nested Dissection method recursively splits the mesh into two sub-problems. Initially developed for structured meshes, the method has been extended to arbitrary meshes (see George and Liu, 1978).

11.4.2 Simple Bandwidth Reduction Algorithm

A simple algorithm is described that will renumber nodes so that the matrix bandwidth is reduced. The user can initially number the nodes of the problem in any manner. A list of pointers that relate the original numbering scheme to the new scheme is generated; these pointers are used again to refer back to the original numbering at output. The user does not have to be concerned about renumbering the original nodes to new nodal values, nor be concerned about bandwidth in numbering the nodes. This particular method was originally developed by Collins (1973).

To illustrate application of the algorithm, return to Fig. 11.25, which shows a plate discretized with eight triangular elements; note that the algorithm also handles four-node quadrilaterals. The half-bandwidth is 9. The first task is to generate the relationships among the nodes, and calculate the original matrix bandwidth. A pointer is set that contains information on the location of the beginning of the list of nodes related to the element node under consideration. Related nodes are identified and prior relationships examined. If the relationship does not exist, the array containing the set of pointers continues to be found. The width of the original matrix band is then calculated. Once these steps have been completed, the data are then passed to the renumbering phase of the algorithm. This is shown in Table 11.3 for the plate in Fig. 11.25a.

Table 11.3 Nodal relationships for plate problems

Node no	Related nodes
1	9,8,2
2	1,9,4,3
3	2,4
4	2,9,3,5
5	9,6,4
6	8,7,9,5
7	8,6
8	1,9,6,7
9	1,8,2,4,6,5

The origin of the new numbering system is located at each node in turn, so that the number of different numbering schemes attempted is equal to the number of nodes. As the new numbering scheme progresses, two arrays are established that are used as pointers to eliminate the need to search the old array. As each new numbering scheme is developed, these arrays are initialized to zero. After initializing for new node number 1, each node related to the new node number is assigned a new number. The related nodes are located in the initial array established at the beginning. If the related node has already been renumbered, then skip to the next related node. Once this step has been completed, a new number is assigned to the old number, and the difference between the

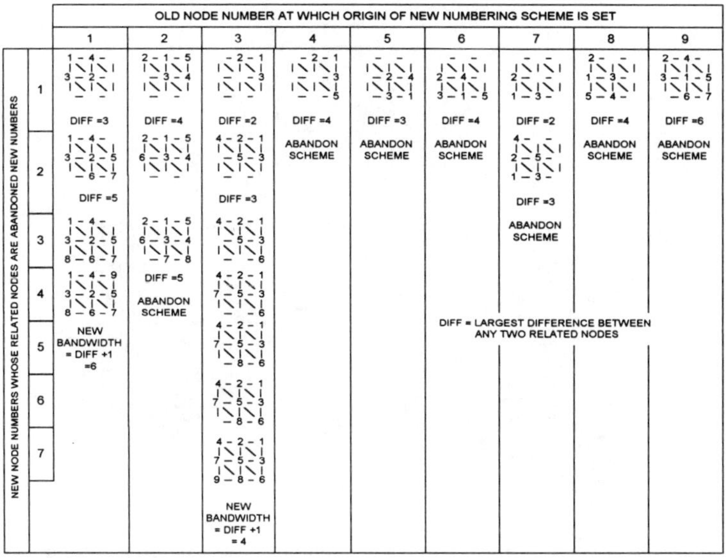

	OLD NODE NUMBER AT WHICH ORIGIN OF NEW NUMBERING SCHEME IS SET								
	1	2	3	4	5	6	7	8	9
1	DIFF =3	DIFF =4	DIFF =2	DIFF =4	DIFF =3	DIFF =4	DIFF =2	DIFF =4	DIFF =6
2	DIFF =5		DIFF =3	ABANDON SCHEME	ABANDON SCHEME	ABANDON SCHEME	DIFF =3	ABANDON SCHEME	ABANDON SCHEME
3							ABANDON SCHEME		
4		DIFF =5 ABANDON SCHEME							
5	NEW BANDWIDTH = DIFF +1 =6				DIFF = LARGEST DIFFERENCE BETWEEN ANY TWO RELATED NODES				
6									
7			NEW BANDWIDTH = DIFF +1 = 4						

(Left axis label: NEW NODE NUMBERS WHOSE RELATED NODES ARE ABANDONED NEW NUMBERS)

Figure 11.29 Renumbering attempts. (From *Int. J. Numer. Methods Eng.*, Vol. 6, R. J. Collins, Copyright 1973, John Wiley & Sons Limited. Reproduced by permission of John Wiley & Sons Limited).

new numbers of the related nodes checked. If this number is greater than the bandwidth obtained from previous numbering, the current scheme is abandoned, and a new one initiated. If the difference is the maximum yet encountered after the current numbering scheme, the value is retained. The process continues so that nodes related to the next new joint are assigned new numbers. At this point, a completely new numbering scheme has been created that is superior to the previous numbering schemes; the bandwidth is kept for comparison with future numbering schemes and the pointers for the new numbers are kept. An illustration of the renumbering sequence is shown in Fig. 11.29 for the nine-node plate problem (from Collins, 1973). The illustrations in the vertical columns represent the sequence of renumbering for a particular location of the new origin, reading from top to bottom. Table 11.4 shows the list of pointers to the new numbers (third column of Fig. 11.29). Figure 11.30 shows a simple 2-D mesh consisting of bilinear quadrilateral and triangular elements that has been renumbered using the algorithm (from Collins, 1973). Two passes of the scheme were required to renumber the nodes to the bandwidth shown in Fig. 11.30.

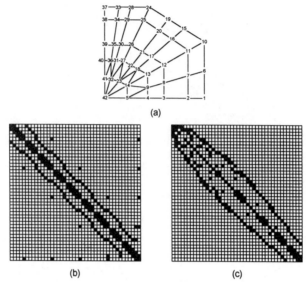

(a)

(b)   (c)

Figure 11.30 Simple 2-D mesh with quadrilateral and triangular elements: (a) configuration, (b) original matrix, and (c) final matrix. (From *Int. J. Numer. Methods Eng.*, Vol. 6, R. J. Collins, Copyright 1973, John Wiley & Sons Limited. Reproduced by permission of John Wiley & Sons Limited).

Table 11.4 New versus old node numbers.

Old node	New node
1	4
2	2
3	1
4	3
5	6
6	8
7	9
8	7
9	5

The algorithm is fairly simple, yet robust enough to be applicable to a wide range of analyses, e.g., structural analysis, fluid flow, heat transfer, or species transport meshes. The algorithm is also applicable to 3-D meshes, but some care should be exercised, as the scheme may not always produce a more efficient nodal numbering scheme (as the algorithm forces new numbering to take a direction that may not be optimal). Collins (1973) points out that "cartwheel" type problems (node at a hub is assigned a new number of low order) may be troublesome. The algorithm numbers nodes around the rim of the wheel–optimal node numbering is better produced when the hub is assigned a number equal to half the number of spokes, plus one. In such cases, the algorithm will not produce a bandwidth that is worse than the original bandwidth. It should be noted that the most popular renumbering scheme and bandwidth optimizer is the Cuthill-McKee algorithm and its variants; this scheme is found in most of the better commercial preprocessors and mesh generator codes. Although somewhat more complicated and lengthy than the algorithm presented here, the method is worth further investigation for those interested in optimal bandwidth reduction.

11.4.3 Delaunay Triangulation

Dirichlet (1850) first proposed a mechanism to connect an arbitrary set of points in space. For a given set of points, $\{P_k\}$, $k = 1,..., N_1$, the regions $\{V_k\}$, $k = 1, ..., N$, are the territories that can be assigned to each point P_k, such that V_k represents the space closer to P_k than to any other point in the set, i.e.,

$$V_k = \left\{ P_i : |p - P_i| < |p - P_j| \right\} \quad \forall_j \neq i$$

This type of geometric construction of tiles is known as Dirichlet tessellation, or a Voronoï (1908) diagram. This tesselation of a closed domain creates a set of nonoverlapping convex polyhedral, or Voronoï regions, which cover the whole domain. If all point pairs, which have some segment of a Voronoï boundary in common, are joined, a triangulation of the convex hull of the set of points is generated. This triangulation is known as Delaunay (1934) triangulation and is valid for n-dimensional space.

In two dimensions a line segment of the Voronoï diagram is equidistant from the two points it separates. Thus the vertices of the Voronoï diagram are equidistant from each of the three nodes that form.

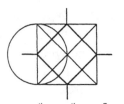

(a) Delaunay and Voronoi diagram showing some geometrical properties; points and Voronoi vertices P_i, $i=1,5$ and V_j, $j=1,4$, respectively. (From Weatherhill et al, 1995. Copyright (c) 1995 AIAA - Reprinted with permission).

(b) Delaunay triangulation and the associated Voronoi tessellation (Reprinted from *Comput. Fluids*, Vol. 20, No. 2, C. A. Hall, J. C. Cavendish, and W. H. Frey, "The Dual Variable Method for Solving Fluid Flow Difference Equations on Delaunay Triangulations," Copyright, 1991, with permission from Elsevier Science).

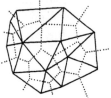

(c) Voronoi tessellation (....) and Delaunay triangulation (——) of 13 points (Reprinted from *Comput. Fluids*, Vol. 20, No. 2, C. A. Hall, J. C. Cavendish, and W. H. Frey, "The Dual Variable Method for Solving Fluid Flow Difference Equations on Delaunay Triangulations," Copyright, 1991, with permission from Elsevier Science).

Figure 11.31 Voronoi tesselation with Delaunay triangulation.

the Delaunay triangles. It is possible to construct a circle, centered at a Voronoï vertex, which passes through the three points that form a triangle.

Likewise, given the definition of Voronoï line segments and regions, no circle can contain any point; this is known as the in-circle criterion. Some of the geometric properties of this construction are shown in Fig. 11.31 Figure 11.31a (from Weatherhill et al., 1995) and Fig. 11.31b,c (Hall et al, 1991) show the combination of Voronoï tesselation with Delaunay triangulation to create a 2-D discretized domain. In three dimensions a vertex of a Voronoï diagram is at the circumcenter of a sphere that passes through four points, which then form a tetrahedron (and no other point in the construction will lie within the sphere).

Several algorithms exist for constructing Delaunay triangulation (e.g., Bowyer, 1934; Watson, 1981; Weatherhill, 1988). The most common approach is to utilize the in-circle criterion in a sequential operation. Each point is introduced into an existing Delaunay structure,

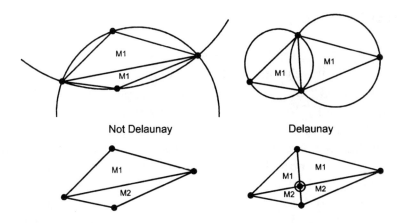

Figure 11.32 Delaunay triangulation for two triangles (from Gable et al., 1996).

which is broken and then reconnected to form a new triangulation, in general, the element connectivity matrix and data structure that relates each element to its three (or four in 3-D) neighbor elements. An example of Delaunay triangulation is shown in Fig. 11.32 for two triangles, after an example discussed by Gable et al. (1997). Four points are connected to form a Delaunay triangulation; in Fig. 11-32a both triangles are the same material so the triangulation is made Delaunay by flipping the connection. In Fig. 11.32b, the triangles are different materials; in this case, the triangulation is achieved by adding a point at the interface and making four triangles. Further details of the methodology can be found in the work by Weatherhill et al. (1995).

Points for connection by the Delaunay algorithm can be obtained in many ways. The two more popular approaches include superposition methods and points generated from independent techniques, such as structured mesh methods (see Thompson et al., 1985). The former approach yields good-quality meshes within interior regions, but quality deteriorates where tetrahedral and connections are constrained by the boundaries. The latter approach is restrictive for general geometries.

Weatherhill et al. (1995) describes the use of the Delaunay triangulation method to produce 3-D tetrahedral meshes, with an automatic point creation technique. Meshes consisting of up to 10^6 tetrahedra have been generated on a range of workstations in under 30 min (~12 min on an IBM Risc 6000; ~27 min on a SUN Sparc II; ~5 min on a Cray Y-MP).

11.5 ADAPTATION

Mesh adaptation is becoming more widely used and has begun to appear in commercial finite element codes (although primarily in structural codes). A few fluid flow codes now support adaptation – PROPHLEX (COMCO, Inc.), NEKTON (Fluent, Inc.) and GWADAPT, which is a recent code for porous media flow (Pepper and Stephenson, 1995). It has been amply demonstrated that adaptation leads to computations of better solutions, producing optimal meshes in forms of size (number of nodes), nodal positions, and element properties (e.g., shape, orthogonality). Refer to the texts by Babuska et al. (1986) and Zienkiewicz and Taylor (1989) for detailed discussions on the mathematical aspects of adaptation.

The construction of an adaptive mesh begins with a coarse mesh, although some care should be given as to defining an optimal mesh (to reduce the number of adaptations) that will yield a solution. The equations are solved and the solution at each node examined. After a choice of pertinent criteria involving the gradients of the solution, a derived field, and/or an evaluation of the interpolation error (usually a discrete evaluation), specific regions (or zones) of the initial mesh are selected for adaptation and a new, better-adapted mesh is generated. This type of adaptation may be refinement, local remeshing, unrefinement, and/or element order enhancement. In simpler terms, a *rapid change detector* examines the value at each node, and if it differs from its neighbors by more than a set threshold, the elements adjoining that node are each divided (or refined). This procedure is repeated until all the nodes have local differences below the threshold. The concept of adaptive meshes was introduced in the 1970s but did not seem to take hold until the beginning of the 1990s (due in part to the increased power and storage capabilities of computers).

11.5.1 Types of Adaptation

There are basically three types of adaptive techniques in use today: r-refinement, h-refinement, and p-refinement. In r-refinement a fixed mesh is first established; the elements within the mesh are then moved, shrunk, or expanded to accommodate regions where the solution is rapidly changing (or relatively stagnant). This technique has been shown to be effective in some cases; however, the elements can become severely distorted and eventually lead to divergent or less accurate solutions. By far the most popular methods are h- and p-refinement. In h-refinement, elements are subdivided into smaller elements; this technique creates additional nodes and elements, which must be carefully monitored through some form of bookkeeping. In p-refinement the degree of the polynomial is increased to improve the accuracy of the solution, i.e., an element that may have been originally linear is ultimately refined to a cubic, quartic, quintic, or higher order element. A smaller h and a higher p generally yield greater accuracy but slower convergence if too fine a refinement is established. Methods that adapt both h and p together are called h-p refinements. Examples of r-, h-, and p-refinements are shown in Fig. 11.33 for compressible flow over a ramp; note the distortion of the elements in the r-refinement near the shock location. Convergence of the h- and p-refinement methods in the form of error versus number of

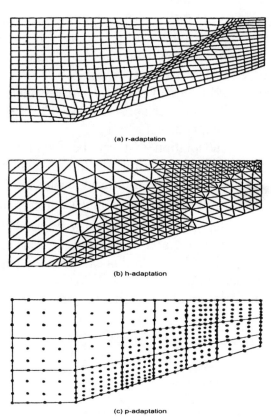

(a) r-adaptation

(b) h-adaptation

(c) p-adaptation

Figure 11.33 Types of mesh adaptation.

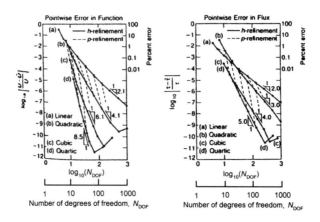

Figure 11.34 Comparison of convergence for h-, p-, and h-p adaptation.

nodes is shown in Fig. 11.34. Clearly, the h-p method is the most optimum technique but also the most complex to incorporate. The inclusion of adaptation in this text would be quite lengthy and is best left to specialized texts and reports (see Carey, 1997). Also recommended are papers by Shapiro and Murman (1988), Ramakrishnam et al. (1990), Oden et al. (1986, 1989), Pelletier and Hetu (1992), Zienkiewicz et al. (1981), and Pepper and Stephenson (1995).

Other forms of adaptation that can be found in the literature include local disenrichment (a form of h-adaptation), which removes one of several points, nested meshes and multigrid techniques (see Hackbush and Trottenberg, 1982), and local remeshing (Voronoï methods). The concept of local disenrichment states that if a vertex belongs to only three triangles (or four tetrahedrals), it can be removed. An example of this method is shown in Fig. 11.35. In this case, some adequate diagonal swappings or edge-face changes are made and an optimization step performed to prevent creation of flat (collapsed) elements.

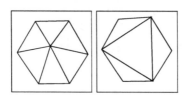

Figure 11.35 Removal of a vertex.

11.5.2 Error Estimates and Adaptation Criteria

We begin with the premise that as $h \to 0$, or $p \to \infty$, a unique exact solution exists. For example, from Taylor series,

$$\phi(x) = \phi_i + h \frac{\partial \phi_i}{\partial x} + \frac{h^2}{2!} \frac{\partial^2 \phi_i}{\partial x^2} + \cdots$$

(11.26)

If a polynomial expansion of degree p, within an element of size h, fits the Taylor expansion up to that degree, and x is of the order of magnitude h, the error in ϕ is $O(h^{p+1})$*. For a linear expansion, $p = 1$, and the convergence rate should be $O(h^2)$. Thus, halving a mesh, h to h/2, would reduce the error by 1/4, i.e., $O(h/2)^2/O(h^2) = 1/4$, yielding a monotonic convergence. At issue is the determination of the error (which is a posteriori) in the solution and how best to refine the approximation while maintaining some level of computational efficiency.

Basically, one begins by creating an original mesh, which is retained, and refining locally by either introducing new elements (h) or using the same elements but increasing the order of interpolation (p). Likewise, one can use combinations of h and p. In lieu of using local error, which may be locally infinite but globally convergent, global norms are utilized. Recall that a norm is an integral scalar quantity that avoids inaccuracies arising from singularities and extremes. For the general linear equation,

$$L\phi + Q = 0$$

(11.27)

the error in the energy norm is expressed as

$$\|e\| = \left(\int_\Omega e^T L e \, d\Omega \right)^{\frac{1}{2}} = \left(\int_\Omega (\phi - \hat{\phi})^T L (\phi - \hat{\phi}) \, d\Omega \right)^{\frac{1}{2}}$$

(11.28)

where $\hat{\phi}$ denotes the approximate solution and ϕ is the exact value. The error in the L_2 norm is defined as

---

* For quantities given by the n-th derivatives, convergence is $O\left(h^{P+1-m}\right)$.

$$\left\| e_\phi \right\|_{L_2} = \left(\int_\Omega \left(\phi - \hat{\phi} \right)^T \left(\phi - \hat{\phi} \right) d\Omega \right)^{\frac{1}{2}}$$

(11.29)

The root-mean-square error is similarly

$$\left| \Delta \phi \right| = \left(\frac{\left\| e_\phi \right\|_{L_2}^2}{\Omega} \right)^{\frac{1}{2}}$$

(11.30)

These norms are similar to the sum of the squares error that is globally computed in finite difference schemes (Long and Pepper, 1981). We note that

$$\left\| e \right\|^2 = \sum_{i=1}^m \left\| e \right\|_i^2$$

(11.31)

where i denotes the individual elements Ω_i. For well-behaved problems with no singularity, the order of convergence should also apply to the energy norm. Zienkiewicz and Taylor (1989) point out that convergence for problems with singularity[*] is

$$O(\text{DOF})^{-\frac{\min(\lambda, p)}{2}}$$

(11.32)

where DOF is the number of degrees of freedom and λ is a value associated with the intensity of the singularity.

In an optimal mesh the distribution of energy norm error ($\left\| e \right\|_i$) should be the same in all elements. We define the total permissible error requirement as (Zienkiewicz and Taylor, 1989)

---

[*] Stress analysis problem.

$$\|e\|_i < \eta \left(\frac{\|\hat{\phi}\|^2 + \|e\|^2}{m} \right)^{\frac{1}{2}} \equiv \bar{e}_m$$

(11.33)

where η is a predetermined energy norm percentage error ($\sim 5\%$) and m is the number of elements. Elements where the above is not satisfied are candidates for refinement. Letting

$$\xi_i \equiv \frac{\|e\|_i}{\bar{e}_m} > 1$$

(11.34)

only those elements where ξ is higher than a prescribed limit would be refined, e.g., halving the size of the elements at each refinement pass. This procedure is known as mesh enrichment; however, the number of trial solutions can become excessive. A more efficient procedure is to define a new mesh such that $\xi_i \leq 1$ and $\|e\|_i \propto h_i^p$, where h_i is the current element size and p is the polynomial order of approximation. To satisfy the error requirement, the element should be no larger than

$$h = \xi_i^{-1/p} h_i$$

(11.35)

In situations where the effects of singularity are felt, a more suitable form for h is

$$h = \xi_i^{-1/\lambda} h_i$$

(11.36)

where λ, the singularity strength, is $0.5 \leq \lambda \leq 1.0$, with $\lambda = 0.5$ being convenient.

A uniform mesh refinement is not necessary for problems in which a function varies rapidly along a particular coordinate direction, e.g., boundary layer flow. In such situations, correction indicators that indicate the most effective direction for refinement can be introduced (see Zienkiewicz and Taylor, 1989). Likewise, the energy norm, by itself, is not the best measure for practical adaptation. As we shall show in the next section, an effective, yet simple, measure for adaptation can be achieved using simple local error estimates.

When dealing with p-adaptive refinement, a slightly different approach is employed. Zienkiewicz et al. (1981) discuss the use of a hierarchical form of approximation that is added to the original trial approximation, i.e., $\hat{\phi} = N_i \phi_i^n + N_i^h \phi_i^h$, or

$$\begin{bmatrix} k^{nn} & k^{nh} \\ k^{hn} & k^{hh} \end{bmatrix} \begin{Bmatrix} \phi_i^n \\ \phi_i^h \end{Bmatrix} = \begin{Bmatrix} f_1 \\ f_2 \end{Bmatrix} \tag{11.37}$$

which comes from the (stiffness) matrix solution $[k]\{\phi\}=\{f\}$. The error becomes

$$\varepsilon = N_i \left(\phi_i^n - \phi_i \right) + N_i^h \phi_i^h \tag{11.38}$$

where ϕ_i is the exact solution. Letting $\phi_i^n \approx \phi_i$, one obtains

$$\phi_i^n = \left[k^{hh} \right]^{-1} \left(f_2 - \left[k^{hn} \right] \phi_i \right) \tag{11.39}$$

thus requiring only the inversion of matrix k^{hh}. At node i,

$$\left(k_{ii}^{hh} \right)^{-1} = 1 / k_{ii}^{hh}$$

Consequently, one can obtain for the energy norm

$$\| e \|_i^2 = \phi_i^{hT} k_{ii}^{hh} \phi_i^h \tag{11.40}$$

and

$$\| e \| = \left(\sum_{i=1}^{N} \| e \|_i^2 \right)^{\frac{1}{2}} \tag{11.41}$$

A correction indicator can also be calculated by introducing the residual

$$L\hat{\phi} + Q = r \quad (\text{remember } L\phi + Q = 0)$$

Thus

$$Le \equiv L(\phi - \hat{\phi}) = -r$$

$$k^{hh}\phi^h = -\int_\Omega N_i^h r d\Omega$$

Subsequently,

$$\|e\|_i^2 = \frac{\left(\int_\Omega N_i^h r d\Omega\right)^2}{k_{ii}^{hh}} = C_i^2 \tag{11.42}$$

On occasion, this indicator can cause problems, since the new shape function, N_i^h, can be orthogonal to the residual r. An improved estimator is introduced by Zienkiewicz and Taylor (1989)

$$\|e\|_i = \frac{1}{\sqrt{2}} \left[\frac{\sum\left(\int_\Omega N_i^{h^2} d\Omega \int_\Omega r^2 d\Omega + \int_I N_i^{h^2} dI \int_I J^2 dI\right)}{k_i^{hh}} \right]^{\frac{1}{2}} \tag{11.43}$$

where the Schwartz inequality has been used to replace $(\int_\Omega N_i^{h^2} d\Omega)^2$ and a *jump* variable introduced (which deals with interelement jumps in order – see Zienkiewicz et al., 1981).

A technique also employed is to use the criterion

$$C_i \geq \gamma C_{max} \tag{11.44}$$

where $0.1 \leq \gamma \leq 0.5$ and C_{max} is the maximum correction value indicator. Setting $\gamma = 0$ corresponds to complete refinement of all elements. Any reasonable value >0 tends to eventually give optimal refinement.

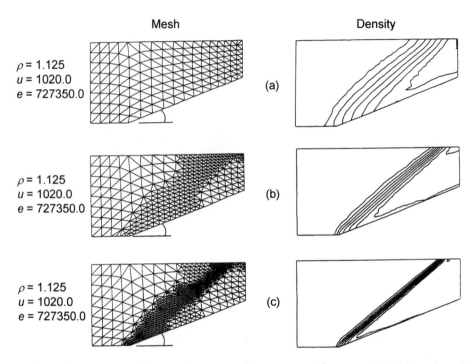

Figure 11.36 Mesh enrichment for M = 3 flow over a 20° ramp: a) initial mesh and density after 100 time steps, b) mesh after 101 time steps, density at 200 time steps, c) mesh after 201 time steps, density at 250 time steps. (Reprinted from *The Finite Element Method, Basic Formulation and Linear Problems*, Vol. 1, 4th Ed., O. C. Zienkiewicz and R. L. Taylor, with permission from McGraw-Hill Publishing Co., UK).

Obviously, there are many different norms and measures of error that can be utilized. In problems dealing with fluid flow the energy norm is not effective. An example is shown in Fig. 11.36. In this problem, compressible flow at M = 3 encounters a ramp, thereby producing a shock, and very steep gradients. In these situations the analysis focuses more on the flow variable u, as opposed to its derivatives (as in stress–type calculations). The L^2 norm is used in the error measure. An estimate of the error can be obtained as

$$e \equiv u - \hat{u} = \frac{1}{2} x_1 (h - x_1) \left| \frac{d^2 \hat{u}}{dx^2} \right|$$

(11.45)

where x_1 is the coordinate measured from the side of an element. The L^2 norm is given as

$$\|e\|_{L^2} \approx \frac{1}{8}h_i^2 \left| \frac{d^2\hat{u}}{dx^2} \right|$$

$$(11.46)$$

Notice the accuracy improvement in the density as the mesh becomes refined in the location of the shock (Fig. 11.42d–11.42f). It is relatively straightforward to extend the measure to two and three dimensions.

11.5.3 Simple h-adaptive Technique

When utilizing h-adaptation, there are basically two choices to be made, *mesh regeneration* or *element subdivision*. Mesh regeneration, or remeshing, requires completely regenerating the entire mesh, either in regions where there is high error or over the complete domain. The principal advantage of remeshing is that areas can be coarsened where the error is below an allowable amount, thus creating an optimal mesh in which every element has essentially the same level of error. However, the main disadvantage of remeshing is that a high degree of spatial flexibility is necessary when using error estimation procedures. When using element subdivision, every element that exceeds the allowable error threshold is subdivided into smaller elements. This method is particularly effective for four-node quadrilaterals and eight-node hexahedrals. However, the method produces virtual nodes (i.e., constrained midside nodes) that must be handled with care; likewise, only one level of adaptation can be performed at a time.

11.5.3.1 Mesh Regeneration

The first step one must take in performing mesh regeneration is to define a maximum permissible error, $\overline{\eta}$, which is to be obtained when the analysis is completed, i.e., all the elements in the final mesh must have the same level of error.

One can begin by calculating the square of the total error,

$$\|q\|^2 \approx \|\hat{q}\|^2 = \sum_{e=1}^{m} \int_{\Omega_e} (\nabla\hat{\phi})^T \kappa \nabla\hat{\phi} d\Omega_e \qquad (11.47)$$

or in two dimensions,

$$\|q\|^2 \approx \|\hat{q}\|^2 = \sum_{e=1}^{m} \int_{\Omega_e} \left[k_x (\frac{\partial \hat{\phi}}{\partial x})^2 + k_y (\frac{\partial \hat{\phi}}{\partial y})^2 \right] dxdy \qquad (11.48)$$

where m is total number of elements, k_x and k_y are diffusion coefficients, $\hat{q}$ is the calculated square of the error, and $\hat{\phi}$ is the calculated unknown variable. The maximum permissible error for each element is then calculated by distributing $\|q\|^2$ equally over all the elements,

$$\|\bar{e}\|_e^2 \leq \bar{\eta}(\frac{\|q\|^2}{m}) \qquad (11.49)$$

where $\bar{\eta}$ is the specified maximum value.

The error is now compared for all elements to the maximum permissible error in an element and used to modify the mesh for a second analysis. Letting ξ_e be defined as

$$\xi_e = \frac{\|e\|_e}{\|\bar{e}\|_e} \qquad (11.50)$$

where e is the approximate error obtained from smoothed values of ϕ (see Huang and Usmani, 1994),

$$\|e\|_e = \left\{ \int_{\Omega_e} k_x \left[(\frac{\partial \tilde{\phi}}{\partial x} - \frac{\partial \hat{\phi}}{\partial x})^2 + k_y (\frac{\partial \tilde{\phi}}{\partial y} - \frac{\partial \hat{\phi}}{\partial y})^2 \right] dxdy \right\}^{\frac{1}{2}} \qquad (11.51)$$

and $\tilde{\phi}$ is the smoothed value[*]. If $\xi_e > 1$, the size of element e must be reduced, thus requiring the mesh to be refined; otherwise, the size of the

---

[*] A "smoothed value" can be easily obtained by solving the relation $[M]\{\tilde{\phi}\}=\{f\}$, where $[M] = \int_{-1}^{1}\int_{-1}^{1} N_i N_j |J| d\xi d\eta$ and $\{f\} = \int_{-1}^{1}\int_{-1}^{1} N_i \hat{\phi} |J| d\xi d\eta$. Note that $\hat{\phi}$ values are obtained at the Gauss points (ξ,η) if they represent gradient values.

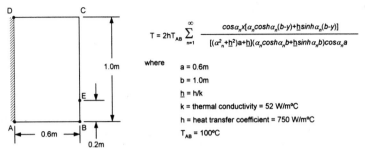

$$T = 2hT_{AB} \sum_{n=1}^{\infty} \frac{cos\alpha_n x[\alpha_n cosh\alpha_n(b-y)+\underline{h}sinh\alpha_n(b-y)]}{[(\alpha_n^2+\underline{h}^2)a+\underline{h}](\alpha_n cosh\alpha_n b+\underline{h}sinh\alpha_n b)cos\alpha_n a}$$

where

a = 0.6m

b = 1.0m

$\underline{h}$ = h/k

k = thermal conductivity = 52 W/m°C

h = heat transfer coefficient = 750 W/m°C

T_{AB} = 100°C

(a) Two dimensional heat transfer with convection

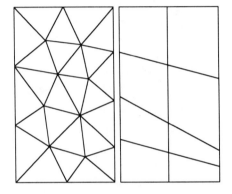 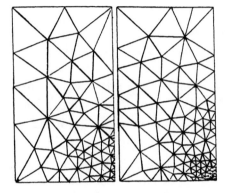

(b) Initial meshes for triangular and quadrilateral elements

(c) Set of refined meshes (three node)
based on the smoothed solution

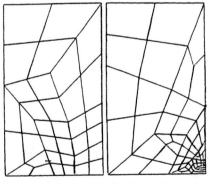

(d) Set of refined meshes (four node)
based on the smoothed solution

Figure 11.37 Remeshing example. (Reprinted from *The Finite Element Method for Heat Transfer,* A. B. Huang and A. F. Usmani, 1994, with permission from Spring-Verlag Publishing Co., UK).

element must be increased, and the mesh coarsened. The size of the new element is then calculated from the old element as (see Eq. 11.35)

$$\overline{h}_e = \frac{h_e}{\xi_e^{1/p}}$$

where $\overline{h}_e$ is the predicted size of the element, h_e is the old element, and p is of the order of the shape function. An example of remeshing for three-node triangles and four-node quadrilaterals is shown in Fig. 11.43 for the simple heat transfer problem with convective boundary conditions. This benchmark problem is illustrated in more detail in the text by Huang and Usmani (1994), which also includes a set of computer programs for generating adaptive meshes based on the remeshing principle. Figure 11.43a shows the problem domain with three different boundary conditions. The analytical solution is also given by Huang and Usmani (1994). For a uniform mesh of 30 triangular elements (Fig. 11.43b), a temperature error of −24.6% was obtained. When the adaptive procedure is applied, a first level of adaptation (Fig. 11.43c) produced a norm error of 14.3% (average temperature error of 2% overall) in the triangular mesh; a second level of adaption produced an error norm of 9.8%. In the quadrilateral case, the initial mesh consisting of eight elements (Fig. 11.43b) produced a norm error of 21.6%; the following two mesh refinements, shown in Fig. 11.43d yielded 16% and 10%, respectively. If one uses quadratic elements, the error reduces considerably but at a higher computational cost.

11.5.3.2   Element Subdivision

The starting point of the adaptation procedure is a mesh coarse enough to allow rapid convergence, yet fine enough to allow the flow details to appear. An initial solution is then computed on the crude mesh; it is not necessary to allow this solution to converge completely. The initial solution should not evolve too far before adaptation, or expensive computational time will be used needlessly since the flow features will shift location during the adaptation procedure.

Refinement indicators are computed based on the solution on the initial mesh, and elements that need to be refined are identified. All elements in the mesh that have indicators above a preset refinement threshold value are enriched, whereas those elements that have values below the unrefinement threshold value are coarsened. Refinement proceeds from the coarsest level to the finest level.

After the initial set of refinements and unrefinements are performed, the mesh is scanned for holes. A hole is defined as an element with three or more of its faces subdivided. Each of the holes is then refined to reduce the number of virtual nodes and hopefully increase the solution accuracy. The process of eliminating holes will often create new holes; therefore the process of refining holes is repeated until no holes are found.

After all the mesh changes have been made, the grid geometry is recalculated, the solution is interpolated onto the new grid, and the calculation procedure is begun again. For steady state problems the entire procedure is repeated until a "converged" mesh is obtained. A converged mesh is a mesh that no longer changes as the solution progresses. The calculation procedure continues on the converged mesh until each of the dependent variables converges to an absolute error of 10^{-4}. In transient problems the mesh is adapted as needed to properly capture high-gradient features as they evolve in time.

In order to decide which elements to refine or unrefine, an adaptation parameter must be defined. There is a great deal of literature indicating possible choices for an adaptation parameter. The two most popular refinement criteria are refinement to minimize error and refinement based on gradients. Both criteria are based upon a key variable that is representative of the solution behavior. Refinement criteria based upon the minimization of maximum errors are generally more complex and are only as accurate as the method of estimating the error.

11.5.3.3 Adaptation Parameters

The adaptation parameter is calculated for 2-D and 3-D flows as follows:

(a) In each element, calculate the absolute value of the first difference of variable ϕ. The quantity used in two dimensions is

$$A_e = \max(|\phi_2 - \phi_1|, |\phi_3 - \phi_2|, |\phi_4 - \phi_3|, |\phi_1 - \phi_4|) \qquad (11.52)$$

where the subscripts 1, 2, 3, and 4 denote the corner nodes of each quadrilateral element. The quantity in three dimensions is

$$A_e = \max \begin{Bmatrix} |\phi_1,\phi_2,\phi_3,\phi_4| - |\phi_5,\phi_6,\phi_7,\phi_8| \\ \{|\phi_2,\phi_6,\phi_7,\phi_3| - |\phi_1,\phi_5,\phi_8,\phi_4|\} \\ |\phi_2,\phi_6,\phi_5,\phi_1| - |\phi_3,\phi_7,\phi_8,\phi_4| \end{Bmatrix} \qquad (11.53)$$

where the four values in each column denote the four corner nodes on an element face.

> (b) Compute the mean and standard deviation of this quantity using the expressions

$$\overline{A}_e = \frac{1}{n}\sum_{i=1}^{n} A_e \qquad (11.54)$$

and

$$\sigma = \sqrt{\frac{\sum\limits_{i=1}^{n} A_e^2 - n\overline{A}_e^2}{n-1}} \qquad (11.55)$$

respectively, where n is the number of nodes.

> (c) Normalize this quantity by subtracting the mean and dividing by the standard deviation

$$A_e \leftarrow \frac{A_e - \overline{A}_e}{\sigma} \qquad (11.56)$$

The mesh is refined or unrefined based on the temporal evolution of any variable chosen by the user, e.g., velocity, temperature, density, or species concentration are typical examples. As the governing equations are integrated in time, the mesh is adapted as needed to properly capture the appropriate variable gradients. The adaptive refinement procedure automatically refines all elements that satisfy the criterion $A_e > R$ and unrefines all elements that satisfy $A_e < U$, where R and U are the refined and unrefined threshold values, respectively. The values of R and U are determined experimentally, based on problem geometry and flow

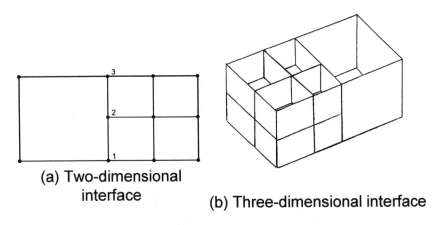

(a) Two-dimensional
interface

(b) Three-dimensional interface

Figure 11.38 Virtual node interfaces.

features. In many problems, the values of R and U vary between 0.2 and 0.4 for R and between 0.6 and 0.8 for U. Note that R and U can be varied to cause more or fewer elements to be refined or unrefined, depending upon the desired accuracy and computational resources.

The use of quadrilaterals in two dimensions results in midside nodes at the interfaces between the coarse and fine regions of the mesh. In three dimensions a face-centered node appears, which creates four quadrilateral elements on the face, resulting in eight new hexahedral elements from the original coarse element. These midside nodes, shown in Fig. 11.38, are called virtual nodes and require special treatment to obtain a stable, conservative scheme. In two dimensions an element face may have at most one additional node. Figure 11.44a shows a typical interface between a locally fine region and a coarser region. The special treatment used is to set the fluxes and the state vector at node 2 equal to the average of the fluxes and state vector at nodes 1 and 3 after each iteration. The procedure of setting node 2 to the average of the state vector of nodes 1 and 3 produces a nearly consistent, conservative scheme (errors of the order of 0.1% to 0.5% – see Shapiro and Murman, 1988). A similar procedure follows for the 3-D hexahedral element, as shown in Fig. 11.38b.

An illustration of a 3-D extension of this approach to mesh adaptation is shown in Fig. 11.39. Here, a single group of eight elements is successively refined until it consists of 13 groups totaling 98 elements.

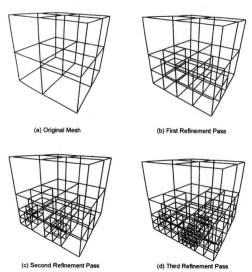

(a) Original Mesh

(b) First Refinement Pass

(c) Second Refinement Pass

(d) Third Refinement Pass

Figure 11.39 Three-dimensional adaptation of eight elements after three refinements. (Reprinted from *Comput. Methods Appl. Mech. Eng.*, Vol. 59, J. T. Oden, T. Strouboulis, and P. Devloo, "Adaptive Finite Element Methods for the Analysis of Inviscid Compressible Flow: Part I. Fast Refinement/Unrefinement and Moving Mesh Methods for Unstructured Meshes," pp. 327-362, Copyright 1986, with permission from Elsevier Science).

One can see from this figure that a maximum of one level of refinement difference exists between any two adjacent elements. While the 3-D extension of this procedure is conceptually straightforward, the additional bookkeeping is substantial.

11.5.3.4 Adaptation Rules

Several rules must be followed to successfully implement the refinement or unrefinement of a mesh. These rules are summarized below for both elements and nodal points.

Elements

1. An element may be refined only if its neighbors are at the same refinement level or higher.
2. If a neighbor element of an element to be refined is at a lower level of refinement, it must be refined first.
3. Refinement of an element results in the creation of four subelements and in the creation of one to five new nodes.
4. To be unrefined, a group of elements must not contain another group

of elements, and each element of the group to be unrefined must not be a neighbor to an element with a higher level.

Nodes

1. An embedded virtual node is common to two members of a group only.
2. An embedded node that is created along a domain boundary cannot be a virtual node.
3. If an element and its neighbor, both of which are at the same level, are connected to a third element at a lower level, then the embedded node that exists along the edge common with the third element is a virtual node.
4. If a group of elements is unrefined, then the embedded virtual node along the edge common to an element that is not a member of the group will be eliminated.
5. If a group of elements is unrefined, then the node along the edge common to this group and its neighbor group will become a virtual node.
6. If a group of elements is unrefined, all embedded nodes along a domain boundary will be eliminated.

11.5.4. **Mesh Adaptation Example**

The rules stated in the previous section can be illustrated by considering the uniform quadrilateral mesh of four elements shown in Fig. 11.40. Suppose element A is marked for refinement. By applying element rules 1 and 3, element A is divided into subelements I, II, III, and IV as shown. Application of node rules 1 and 2 shows that the nodes marked with circles are virtual nodes, and those marked with an X are not virtual nodes.

Suppose that element III is marked for further refinement. Element III cannot be refined, since one of its neighbors, B, is at a lower level. Refinement of element III before element B would violate element rule 1. Therefore element B is refined as shown in Fig. 11.40b. Note that node i is no longer a virtual node, since node rule 1 no longer applies. Node j remains a virtual node. Now that element B has been divided into elements V, VI, VII, and VIII, element rule 1 can be applied. Figure 11.40c illustrates this division.

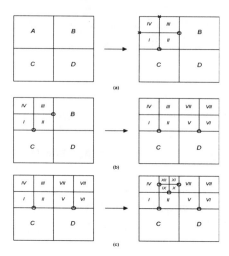

Figure 11.40 Mesh adaptation sequence.

Next, suppose that the group of elements V, VI, VII, and VIII shown in Fig. 11.40c is marked for unrefinement. This group is not eligible until the group of elements IX, X, XI, and XII has been unrefined. Element VIII has neighbors X and XI which are at a higher level, and unrefinement would violate element rule 4.

Now let the group of elements IX, X, XI, and XII be marked for unrefinement. Element rule 4 is satisfied, and elements IX, X, XI, and XII are replaced by element III. The embedded virtual nodes associated with elements IX, X, XI, and XII are eliminated by node rule 4. The embedded node along the upper domain boundary is eliminated using node rule 6.

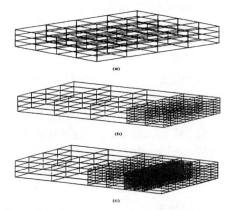

Figure 11.41 Three-dimensional example of multiple mesh adaptation.

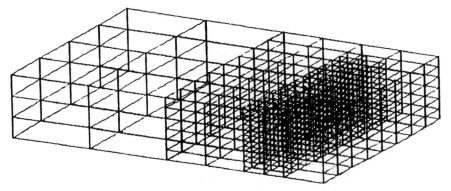

Figure 11.42 Enlarged view of 3-D adaptation after three refinements.

An example of 3-D adaptation on a group of hexahedral elements moving from a coarse mesh to higher levels of adaptation is shown in Fig. 11.41. An exploded view of a mesh after three refinement passes is shown in Fig. 11.42. Note that during a solution sequence, parts of the mesh may be undergoing refinement while other parts are being unrefined.

11.6 CLOSURE

The preprocessing operation of developing a mesh is crucial to obtaining a reliable and accurate solution. The best solver package will still yield inaccurate results if the mesh is poorly formed. Developing an optimum mesh is not straightforward and may likely involve much time in creating the final mesh. One should have patience and perseverance when generating meshes, especially when the problem domains are complicated. Remember that you are probably solving a problem that could not even be considered tractable a few years ago, and certainly could not be handled with simple numerical methods. Since mesh analysis is so crucial, and also difficult when employing large meshes, one should take advantage of graphical visualization tools; it is much easier to visually inspect a mesh than attempt to decipher a long list of numbers. Generally, the easier to display a mesh, the quicker one obtains an optimal mesh. Such graphical programs can be quite complex and are often large (many lines of code); their costs are generally well worth the price, and they are useful for examining output solution results as well.

The best numbering pattern is typically one that is easy to construct and to quickly change as necessary. It may not be the optimal mesh, but close is usually good enough. Exotic numbering patterns are not only cumbersome to create, but also difficult to analyze and examine. Fortunately, many of the commercial finite element codes now available on the market employ the latest bandwidth optimizers and frontal methods, and produce optimal (or near optimal) mesh patterns automatically.

The FEM user should have a good understanding of the physics of the problem and have some feel for the approximate behavior and locations where variables are likely to rapidly change. Such insight will allow the user to create a good mesh, limiting the number of new meshes and local refinements. An analyst with essentially no feel for the problem, or a weak background in the problem physics, will likely have a frustrating and difficult time in obtaining good results. In our teaching experience, we have seen numerous examples of students who want to use a commercial code to obtain some flow predictions, but have never had a fluid flow course, let alone a course in numerical methods or finite elements.

Skill in creating good meshes comes from experience, knowledge of the problem and expected results[*], and degree of variable complexity. Complex areas within a problem configuration generally require a finer mesh structure, or higher element order, to achieve acceptable accuracy. Likewise, there are certain areas of a problem that are typically of more interest to the modeler; refining a mesh in regions of low interest, or using a globally uniform fine mesh, although achieving desired global accuracy, may be wasteful and unnecessary. This is one of the great attractions of boundary element methods. The greater the separation of regions of interest and disinterest, the greater the tolerance of the differences in accuracy. The mesh can become coarser as one moves farther away from regions of interest, especially in those regions where the solutions are smooth and unvarying. One should begin with the coarsest mesh and the simplest geometric primitive (or combination of primitives) as possible.

---

[*] From the authors' experience, expected results differed markedly from true outcome in a series of simulations involving the bifurcation of flow experienced in natural convection within cylinders and spheres (see Pepper and Heinrich, 1992).

REFERENCES

Akin, J. E. (1993) *Finite Elements*. New York: John Wiley and Sons.

Akhras, G. and Dhatt, G. (1976) "An Automatic Node Relabelling Scheme for Minimizing a Matrix or Network Bandwidth," *Int. J. Numer. Methods Eng.*, Vol. 10, pp. 787–797.

Babuska, I. (1971) "The Rate of Convergence for the Finite Element Method," *SIAM J. Numer. Anal.*, Vol. 8, No. 2, pp. 304–315.

Babuska, I., Zienkiewicz, O. C., Gago, J. and de Oliveira, E. R. (1986) *Accuracy Estimates and Adaptive Refinements in Finite Element Computations*. New York: John Wiley and Sons.

Bowyer, A.,(1934) "Computing Dirichlet Tessellations," Computer Journal, Vol. 24, No. 2, pp. 162–166.

Burnett, D.S. (1987) *Finite Element Analysis*. Reading, Mass: Addison-Wesley.

Carey, G. F. (1997) *Computational Grids: Generation, Adaptation, and Solution Strategies*. Washington, D.C.: Taylor and Francis.

Collins, R. J. (1973) "Bandwidth Reduction by Automatic Renumbering," *Int. J. Numer. Methods Eng.*, Vol. 6, pp. 345–356.

Cuthill, E. and McKee, J. (1969) "Reducing the Bandwidth of Sparse Symmetric Matrices," *Proc. Nat. Conf. Assoc. Comput. Mach.*, 24th, pp. 157–172.

Delaunay, B. (1934) "Sur la Sphere Vide," *Bull. Acad. Sci. USSR Class. Sci. Nat.*, pp. 793–800.

Dirichlet, G. L. (1850) "Uber Die Reduction der Positiven Quadratischen Forment Mit Drei Underestimmten Ganzen Zahlen," *J. Reine Angew. Math.*, Vol. 40, No. 3, pp. 209–227.

George, P. L. (1991) *Automatic Mesh Generation*. Chichester, UK: John Wiley and Sons.

George, J. A. and Liu, J. W. (1978) "An Automatic Nested Dissection Algorithm for Irregular Finite Element Problems", *SIAM J. Numer. Anal., Vol. 15*, pp. 1053–1069.

Gibbs, N. E., Poole, W. G. and Stockmeyer, P. K. (1976) "An Algorithm for Reducing the Bandwidth and Profile of a Sparse Matrix," *SIAM J. Numer. Anal.*, Vol. 13, No. 2, pp. 236–250.

Grooms, H. R. (1972) "Algorithm for Matrix Bandwidth Reduction," *J. Struct. Div.*, Vol. 98, No. STI, pp. 203–214.

Hackbush, W. and Trottenberg, U. (1982) "Multigrid Methods," *Lect. Notes Math.*, Vol. 960.

Hall, C. A., Cavendish, J. C. and Frey, W. H. (1991) "The Dual Variable Method for Solving Fluid Flow Difference Equations on Delaunay Triangulations," *Comput. Fluids*, Vol. 20, No. 2, pp. 145–164.

Hirt, C. W. and Amsden, A. A. (1978) "The ALE Method for Fluid Dynamics." Los Alamos Scientific Laboratory Report LA-3425, Los Alamos, N.M.

Huang, A. B. and Usmani, A. F. (1994) *The Finite Element Method for Heat Transfer*. London, UK: Springer-Verlag.

Krauthammer, T. (1979) "Accuracy of the Finite Element Method near a Curved Boundary," *Int. J. Comput. Struct.*, Vol. 10, pp. 921–929.

LAPCAD User's Guide (1996). LAPCAD Engineering, San Diego, Calif.

Lipton, R. J. and Tarjan, R. E. (1979) "A Separator Theorem for Planar Graphs," *SIAM J. Appl. Math.*, Vol. 36, No. 2, pp. 177–189.

Long, P. E. and Pepper, D. W. (1981) "An Examination of Some Simple Numerical Schemes for Calculating Scalar Advection," *J. Appl. Meteorol.*, Vol. 20, No. 2, pp.146–156.

Marro, L. (1980) "Méthodes de Réduction de Largeur de Bande et de Profil Efficace des Matrices Creuses." Thèse, Université de Nice.

Melhem, R. (1987) "Toward Efficient Implementation of Preconditioned Conjugate Gradient Methods on Vector Supercomputers," *Int. J. Supercomput. Appl.*, Vol. 1.

Minkowycz, W. J., Sparrow, E, M., Schneider, G. E., and Pletcher, R. H. (Eds.) (1988) *Handbook of Numerical Heat Transfer*. New York: John Wiley and Sons.

Numerical Technologies (1996). *pcGridGen User's Guide, Version 1.1.* Tempe, Ariz.

Oden, J. T., Demkowitz, L., Rachowitz, W. and Westerman, T. A. (1989) "Toward a Universal h-p Adaptive Finite Element Strategy: Part 2. A Posteriori Error Estimation," *Comput. Methods Appl. Mech. Eng.*, Vol. 77, pp. 113–180.

Oden, J. T., Strouboulis, T. and Devloo, P. (1986) "Adaptive Finite Element Methods for the Analysis of Inviscid Compressible Flow: Part I. Fast Refinement/Unrefinement and Moving Mesh

Methods for Unstructured Meshes," *Comput. Methods Appl. Mech. Eng.*, Vol. 59, pp. 327–362.

Oran, E. S. and Boris, J. P. (1987) *Numerical Simulation of Reactive Flow*. New York: Elsevier.

Patankar, S. V. (1991) *Computation of Conduction and Duct Flow Heat Transfer*. Maple Grove, Minn.: Innovative Research.

Patankar, S. V. (1980) *Numerical Heat Transfer and Fluid Flow*. Washington, D.C.: Hemisphere (Taylor and Francis).

Pelletier, D. H. and Hetu, J. (1992) "An Adaptive Finite Element Methodology for Incompressible Viscous Flow," pp. 1–11 in *Advances In Finite Element Analysis in Fluid Dynamics*. New York: ASME Publications.

Pepper, D. W. and Long, P. E. (1978) "A Comparison of Results Using Second Order Moments with and without Width Correction to Solve the Advection Equation," *J. Appl. Meteorol.*, Vol. 17, No. 17-2, pp. 228–233.

Pepper, D. W. and Baker, A. J. (1979) "A Simple One-Dimensional Finite Element Algorithm with Multidimensional Capabilities," *Numer. Heat Transfer*, Vol. 2, pp. 81–95.

Pepper, D. W. and Cooper, R. E. (1980) "Numerical Solution of Recirculating Flow by a Simple Finite Element Recursion Relation," *Comput. Fluids*, Vol. 8, pp. 213–223.

Pepper, D. W. and Heinrich, J. C. (1992) *The Finite Element Method: Basic Concepts and Applications*, Washington, D.C.: Hemisphere (Taylor and Francis).

Pepper, D. W. and Heinrich, J. C. (1993) "Transient Natural Convection within a Sphere using a 3-D Finite Element Method," pp. 369–378 in *Finite Elements in Fluids* (K. Morgan et al, editor). UK: CIMNE/Pineridge Press.

Pepper, D. W. and Stephenson, D. E. (1995) "An Adaptive Finite-Element Model for Calculating Subsurface Transport of Contaminant," *Ground Water*, Vol. 33, No. 3, pp. 486–496.

Pointwise, Inc. (1996) *GRIDGEN User's Guide*. Fort Worth, Texas.

Ramakrishnam, R., Bey, K. S. and Thornton, E. A. (1990) "Adaptive Quadrilateral and Triangular Finite Element Scheme for Compressible Flows," *AIAA J.*, Vol. 28, No. 1, pp. 51–59.

Schneider, G. E. and Raw, M. J. (1987) "Control-Volume Finite-Element Method for Heat Transfer and Fluid Flow Using

Colocated Variables – 1: Computational Procedure," *Numer. Heat Transfer*, Vol. 11, pp. 363–390.

Segerlind, L. J. (1984) *Applied Finite Element Analysis*, 2nd ed. New York: John Wiley and Sons.

Shapiro, R. A. and Murman, E. (1988) "Adaptive Finite Element Methods for the Euler Equations." AIAA paper 88–0034.

Szabo, B. A. (1986) "Mesh Design for the p–Version of the Finite Element Method," *Comput. Methods Appl. Mech. Eng.*, Vol. 55, pp. 181–197.

Szabo, B. A. and Babuska, I. (1991) *Finite Element Analysis*. New York: John Wiley and Sons.

Thompson, J. F., Warsi, Z., U. A. and Mastin, C. W. (1985) *Numerical Grid Generation: Foundations and Applications*. Amsterdam: North-Holland.

Voronoi. G. (1908) "Novelles Applicatons des Parametre Continus a la Theorie des Formes Quadratiques: Recherches Sur les Paralleloedres Primitifs," *J. Reine Angew. Math.*, Vol. 134.

Warsi, Z. U. A. (1993) *Fluid Dynamics: Theoretical and Computational Approaches*. Boca Raton, Fla.: CRC Press.

Watson, D. F. (1981) "Computing the n-Dimensional Delaunay Tessellation with Application to Voronoi Polytopes," *Comput. J.*, Vol. 24, No. 2, pp. 167–172.

Weatherhill, N. P. (1988) "A Method for Generating Irregular Computational Grids in Multiply Connected Planar Domains," *Int. J. Numer. Methods Fluids*, Vol. 8, pp. 181–197.

Weatherhill, N. P., Hassan, O. and Marcum, D. L. (1995) "Compressible Flowfield Solutions with Unstructured Grids Generated by Delaunay Triangulation," *AIAA J.*, Vol. 33, No. 7, pp. 1196–1204.

Zienkiewicz, O. C. (1977) *The Finite Element Method*, 3rd ed. London: McGraw-Hill.

Zienkiewicz, O. C., Gago, J. and Kelly, D. W. (1981) "The Hierarchical Concept in Finite Element Analysis," *Comput. Struct.*, Vol. 16, pp. 53–65.

Zienkiewicz, O. C. and Taylor, R. L. (1989) *The Finite Element Method*, 4th ed. Vol. 1, *Basic Formulation and Linear Problems*. London, UK: McGraw-Hill.

Zlamal, M. (1973) "Curved Elements in the Finite Element Method," *SIAM J. Numer. Anal.*, Part I, Vol. 10, No. 1, pp. 229–240.

EXERCISES

11.1   Discretize the plate shown in Fig. 11.25 using four bilinear quadrilaterals with the node numbers increasing (a) horizontally and (b) vertically, beginning with the top-left node as 1. What is the difference in the half-bandwidth?

11.2   A circle exists with a radius of $r = 2$. Discretize the interior of the circle using (a) four-node bilinear quadrilaterals and (b) nine-node biquadratic quadrilaterals. Is there a minimum number of elements to discretize the interior region? What is the bandwidth of each mesh?

11.3   A simple rectangular domain is shown below. Discretize the problem domain using three-node triangles with (a) right-running hypotenuse and (b) left-running hypotenuse, and solve for the steady state temperature distribution using 20 elements. What can be said about the accuracy of the left -versus right-running triangular meshes?

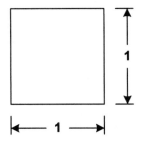

11.4   Solve the example problem described in Fig. 11.37 using (a) 30 triangular elements, (b) 60 triangular elements, and (c) 120 triangular elements, all utilizing three-node linear elements, and compare results.

11.5   Plot accuracy versus node number for Exercise 11.4 using a log-log scale.

FURTHER APPLICATIONS

12.1 OVERVIEW

In the previous chapters, we introduced a variety of examples of physical situations that can be modeled using the finite element method. The main objective has been to show simple situations in which the appropriate physics can be readily understood. In all the examples, we had simple rectangular geometries, and only steady state solutions were sought. In this chapter, we will look at some more complicated problems and geometries.

12.2 TIME-DEPENDENT FLOWS AND FLOWS IN ROTATING SYSTEMS

There is an interesting analogy between convective flows and Taylor-Couette flows that we will exploit in the following examples. To this purpose, we rewrite the governing equations in the form

$$\frac{\partial u_i}{\partial x_i} = 0 \tag{12.1}$$

$$\frac{\partial u_i}{\partial t} + u_j \frac{\partial u_i}{\partial x_j} = -\frac{\partial p}{\partial x_i} + \varepsilon_1 \frac{\partial^2 u_i}{\partial x_j \partial x_j} + f_i \tag{12.2}$$

$$\frac{\partial \theta}{\partial t} + u_j \frac{\partial \theta}{\partial x_j} = \varepsilon_2 \frac{\partial^2 \theta}{\partial x_j \partial x_j} \tag{12.3}$$

REMARKS

1. In the case of an isothermal flow without body forces, $f_i = 0$ and Eq. (12.2) is either irrelevant or it may model the transport of a species that does not affect the fluid motion. In this case, we set $\varepsilon_1 = 1/Re$ in Eq. (12.2) to recover Eqs. (10.12) and (10.13).

2. If a Boussinesq fluid is being modeled, Eq. (12.3) becomes the energy equation and θ the temperature. Now, we may have two distinct cases:

(i) In the case of natural convection the appropriate parameters are the Prandtl and Rayleigh numbers [Eqs. (6.73) and (6.74)]. We set $\varepsilon_1 = Pr$, $\varepsilon_2 = 1.0$, $f_1 = 0$, and $f_2 = PrRa\theta$.

(ii) For free and forced convection, we characterize the flow by the Reynolds, Péclet, and Grashof or Froude numbers [Eqs. (10.14), (10.70), and (10.71)]. The Froude number is related to the Grashof and Reynolds numbers by the expression

$$Fr = \frac{Re^2}{Gr} = \frac{\overline{U}^2}{\beta g |\Delta T| L} \tag{12.4}$$

Therefore we set $\varepsilon_1 = 1/Re$, $\varepsilon_2 = 1/Pe$, $f_1 = 0$, and $f_2 = \theta/Fr$.

3. Finally, if we are modeling an isothermal flow in a rotating system (of the Taylor-Couette type), the only relevant parameter is the Taylor number, Ta, given by

$$Ta = \frac{\Omega^2 R d^3}{v^2} \tag{12.5}$$

where Ω is the angular velocity, R is the outer radius, and d the characteristic thickness of the fluid layer. In this case we set $\varepsilon_1 = \varepsilon_2 = 1.0$, $f_1 = 0$, and $f_2 = Ta\theta^2$. Equation (12.3) becomes the equation for the circumferential component of the velocity.

It is important to note how a single finite element model can be made to accommodate a variety of widely different flows by simply introducing the appropriate scaling. We will show examples of each of the cases given above, but first we establish a general algorithm for the time-dependent case. This is based on the work of Hughes et al. (1979) and is a variation of the Newmark method.

In general, a Petrov-Galerkin discretization will be necessary to guarantee stability of the algorithm, as discussed in section 10.3.1. We use isoparametric bilinear elements and the penalty formulation with penalty parameter λ. The weak form of Eqs. (12.1)–(12.3) become

$$
\int_\Omega N_i \left[\frac{\partial u}{\partial t} + \varepsilon_1 \left(\frac{\partial N_i}{\partial x} \frac{\partial u}{\partial x} + \frac{\partial N_i}{\partial y} \frac{\partial u}{\partial y} \right) \right] d\Omega + \int_\Omega \lambda \frac{\partial N_i}{\partial x} \left(\frac{\partial u}{\partial x} + \frac{\partial v}{\partial y} \right) d\Omega
$$

$$
= \int_\Omega (N_i + P_{i1}) \left(u \frac{\partial u}{\partial x} + v \frac{\partial u}{\partial y} \right) d\Omega - \int_{\Gamma_0} N_i p n_x \, d\Gamma \tag{12.6}
$$

$$
\int_\Omega N_i \left[\frac{\partial v}{\partial t} + \varepsilon_1 \left(\frac{\partial N_i}{\partial x} \frac{\partial v}{\partial x} + \frac{\partial N_i}{\partial y} \frac{\partial v}{\partial y} \right) \right] d\Omega + \int_\Omega \lambda \frac{\partial N_i}{\partial y} \left(\frac{\partial u}{\partial x} + \frac{\partial v}{\partial y} \right) d\Omega
$$

$$
= \int_\Omega \left[(N_i + P_{i1}) \left(u \frac{\partial v}{\partial x} + v \frac{\partial v}{\partial y} \right) + N_i f_2 \right] d\Omega - \int_{\Gamma_0} N_i p n_y \, d\Gamma \tag{12.7}
$$

$$
\int_\Omega \left[N_i \frac{\partial \theta}{\partial t} + \varepsilon_2 \left(\frac{\partial N_i}{\partial x} \frac{\partial \theta}{\partial x} + \frac{\partial N_i}{\partial y} \frac{\partial \theta}{\partial y} \right) \right] d\Omega
$$

$$
= \int_\Omega (N_i + P_{i2}) \left(u \frac{\partial \theta}{\partial x} + v \frac{\partial \theta}{\partial y} \right) d\Omega + \int_{\Gamma_1} N_i q \, d\Gamma \tag{12.8}
$$

Here N_i denotes the bilinear shape functions and P_{ij} are the Petrov-Galerkin perturbations applied to the convective terms only, which are given by

$$P_{ij} = k_j \left(u \frac{\partial N_i}{\partial x} + v \frac{\partial N_i}{\partial y} \right) \qquad j = 1, 2 \qquad (12.9)$$

$$k_j = \frac{\alpha_j \overline{h}}{\|V\|} \qquad (12.10)$$

$$\alpha_j = \coth \frac{\gamma_j}{2} - \frac{2}{\gamma_j} \qquad (12.11)$$

$$\gamma_j = \frac{\|V\| \overline{h}}{\varepsilon_j} \qquad (12.12)$$

as derived in Chapter 8. The open portions of the boundary are denoted by Γ_0, and those for which a heat flux q is prescribed, by Γ_2.

The spatial discretization leads to the system of ordinary differential equations

$$M\dot{d} + Kd = F(\theta) - N(d) \qquad (12.13)$$

$$C\dot{\theta} + D\theta = -G(d, \theta) \qquad (12.14)$$

M and **C** are the mass matrices for the momentum and energy equations, respectively; **d** is the vector of velocity degrees of freedom; **K** contains the contributions of the viscous and penalty terms; **N** is the vector containing the nonlinear convective terms in the momentum equation; **F** contains the contributions of the body forces and the prescribed boundary conditions; **D** is the heat diffusion matrix; and **G** contains the convective terms and contributions from the boundary conditions in the energy equation.

The time-stepping algorithm proceeds as follows.
1. At time $t = t_n, d_n$ and θ_n are known.
2. Calculate predictors for the temperatures and velocities; set the iteration index $i = 0$:

$$\theta_{n+1}^{(0)} = \theta_n + (1 - \gamma)\Delta t \dot{\theta}_n \qquad (12.15)$$

$$d_{n+1}^{(0)} = d_n + (1 - \gamma)\Delta t \dot{d}_n \qquad (12.16)$$

3. Calculate a corrected temperature from

$$(\mathbf{C} + \gamma \Delta t \mathbf{D})\theta_{n+1}^{(i+1)} = \mathbf{C}\theta_{n+1}^{(0)} - \gamma \Delta t \mathbf{G}\left(d_n^{(i)}, \theta_n^{(i)}\right) \qquad (12.17)$$

4. Calculate corrected velocities from

$$(\mathbf{M} + \gamma \Delta t \mathbf{K})\,d_{n+1}^{(i+1)} = \mathbf{M}d_{n+1}^{(0)} + \gamma \Delta t\left[\mathbf{F}\left(\theta_{n+1}^{(i)}\right) - \mathbf{N}\left(d_{n+1}^{(i)}\right)\right] \qquad (12.18)$$

5. Once θ_{n+1} and d_{n+1} have been calculated, we have

$$\dot{\theta}_{n+1}^{(i+1)} = \left(\theta_{n+1}^{(i+1)} - \theta_{n+1}^{(0)}\right)/(\gamma \Delta t) \qquad (12.19)$$

and

$$\dot{d}_{n+1}^{(i+1)} = \left(d_{n+1}^{(i+1)} - d_{n+1}^{(0)}\right)/(\gamma \Delta t) \qquad (12.20)$$

REMARKS

1. If $\gamma = 1$ and only one correction is performed to obtain the values at t_{n+1}, i.e., $\theta_{n+1} = \theta_{n+1}^{(1)}$ and $d_{n+1} = d_{n+1}^{(1)}$, it can be shown (Hughes et al., 1979) that the algorithm is first order in time. If, on the other hand, $\gamma = 1/2$ and two corrections are performed, i.e., $\theta_{n+1} = \theta_{n+1}^{(2)}$ and $d_{n+1} = d_{n+1}^{(2)}$, the algorithm is second order in time.

2. The nonlinear convective terms and the body force terms are calculated explicitly in the right-hand side using the latest available values of the dependent variables. This has the advantage that the system matrices $(\mathbf{M} + \gamma \Delta t \mathbf{K})$ and $(\mathbf{C} + \gamma \Delta t \mathbf{D})$ on the left-hand side are constant throughout the calculation, provided the time step remains constant. Therefore the matrices can be factored at the beginning of the calculation

using an LU decomposition method. To advance the solution in time after the first time step, it suffices to update the right-hand side and perform a forward and backward substitution, thereby reducing the computational time significantly. Furthermore, these matrices are symmetric, further enhancing computational efficiency.

3. The fact that the convective and body force terms are treated explicitly imposes a stability limitation on the time-step size. The convective terms require the Courant number to be less than 1 everywhere in the mesh and to be estimated using Eq. (9.53). The body force term introduces a limitation of the form

$$\Delta t \leq \frac{1}{\text{Pr Ra}}$$

in the case of natural convection and

$$\Delta t \leq \frac{1}{\text{Ta}}$$

for rotating flows. The most restrictive of these bounds must always be satisfied.

4. When Boussinesq fluids are considered, it is often necessary to use lumped mass matrices in order to avoid oscillation that may destabilize the calculations. This is discussed in Chapter 7.

We now show examples of calculations performed using the algorithm given above, or very similar, modified versions of it.

12.2.1 Isothermal Flow Past a Circular Cylinder

Flow past a circular cylinder is a well-defined example of a time-dependent flow. When Re exceeds about 60, vortices are shed from the two re-circulating cells behind the cylinder. Brooks and Hughes (1982) performed calculations at Re = 100 based on the cylinder diameter. The domain and mesh used in their work are shown in Fig. 12.1. The mesh contains 1510 nodes and 1436 isoparametric bilinear elements; the time step was fixed at $\Delta t = 0.03$. The simulation was started from zero velocities, and a unit horizontal inlet velocity was applied at t = 0. A

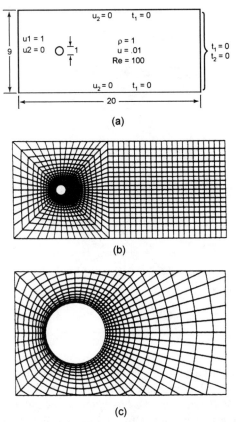

Figure 12.1 Flow past a circular cylinder at Re = 100: (a) Domain and boundary conditions, (b) finite element mesh, and (c) detail of the mesh next to the cylinder. (Reprinted from *Computer Methods in Applied Mechanics and Engineering*, Vol. 32, A. Brooks and T. J. R. Hughes, Streamline Upwind/Petrov-Galerkin Formulations for Convection Dominated Flows with Particular Emphasis on the Incompressible Navier-Stokes Equations, pp. 199-259, with permission from Elsevier Science).

velocities, and a unit horizontal inlet velocity was applied at t = 0. A symmetric steady state solution is obtained unless the flow is perturbed (Anderson et al., 1990). A perturbation was accomplished by imposing small forces of magnitude 0.0001 to the boundary layer nodes between time t = 54 and t = 58.5. The perturbation effectively destabilized the flow, and vortex shedding began at time t = 96. Stream function plots at times t = 102 and t = 144 are shown in Fig. 12.2, together with a pressure plot in the neighborhood of the cylinder at time t = 144.

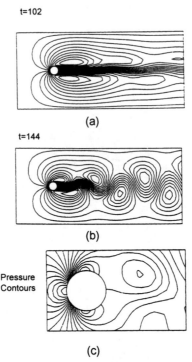

t=102

(a)

t=144

(b)

Pressure Contours

(c)

Figure 12.2 Calculated flow past a circular cylinder at Re = 100: (a) Streamlines at $t = 10^2$, (b) streamlines at $t = 144$, and (c) pressure contours close to the cylinder at $t = 144$. (Reprinted from *Computer Methods in Applied Mechanics and Engineering*, Vol. 32, A. Brooks and T. J. R. Hughes, Streamline Upwind/Petrov-Galerkin Formulations for Convection Dominated Flows with Particular Emphasis on the Incompressible Navier-Stokes Equations, pp. 199-259, with permission from Elsevier Science).

12.2.2 Natural Convection in a Horizontal Circular Cylinder

This is an example of a flow driven by density differences. We assume a circular cylinder filled with fluid is aligned horizontally, perpendicular to the direction of the gravity force. The cylinder is infinitely long, and two-dimensional flow is obtained at any cross section. Figure 12.3 shows the geometry and the finite element mesh used in the calculations. The initial conditions are those of a fluid at rest, at constant temperature. The boundary conditions for the velocities are no slip (u = v = 0) along the cylinder boundary and, at time t = 0, the imposed temperature distribution is hot at the bottom and cold at the top, as shown in Fig. 12.4. Calculations for Pr = 1 and $0 < Ra \leq 5 \times 10^4$ were reported by

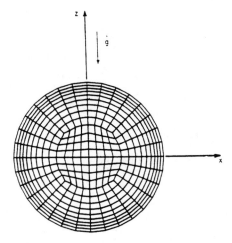

Figure 12.3 Domain and finite element mesh for natural convection in a horizontal circular cylinder.

Heinrich and Yu (1988). Two bifurcation points were identified where the mode of circulation changes, one at approximately Ra = 400 and a second at about Ra = 4800. For Ra > 4800, the flow no longer reached a steady state but oscillated with a fixed frequency. Calculations at Pr = 1 and Ra = 50,000 were performed in the mesh of Fig. 12.3 using bilinear elements and the penalty formulation. The time step was $\Delta t = 10^{-4}$, and

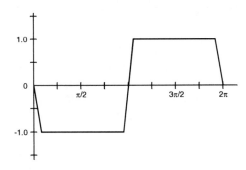

Figure 12.4 Temperature boundary conditions for a horizontal circular cylinder heated from below and cooled from above.

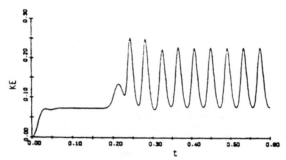

Figure 12.5 Total kinetic energy versus time; flow in a horizontal circular cylinder at Pr = 1 and Re = 10,000.

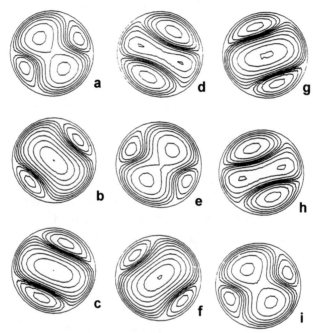

Figure 12.6 Streamline plots for natural convection in a horizontal circular cylinder at Pr = 1 and Re = 10,000: (a) t = 0.4354, ψ = -60.0, -40.0, -25.0, -15.0, -7.5, -2.5, 2.5, 7.5, and 15.0, (b) t = 0.4438, ψ = -120.0, -100.0, -80.0, -60.0, -40.0, -25.0, -15.0, -7.5, -2.5, 2.5, 7.5, and 15.0, (c)t = 0.4564, ψ =, -100.0, -80.0, -60.0, -40.0, -25.0, -15.0, -7.5, -2.5, 2.5, 7.5, 15.0, and 25.0, (d) t = 0.4672, ψ = -32.0, -25.0, -15.0, -7.5, -2.5, 2.5, 7.5, 15.0, 25.0, and 40.0, (e) t = 0.4774, ψ = -15.0, -7.5, -2.5, 2.5, 7.5, 15.0, 25.0, and 40.0, (f) t = 0.4858, ψ = -15.0, -7.5, -2.5, 2.5, 7.5, 15.0, 25.0, 40.0, 60.0, 80.0, 100.0, and 120.0, (g) t = 0.4978, ψ = -25.0, -15.0, -7.5, -2.5, 2.7, 7.5, 15.0, 40.0, 60.0, 80.0, and 100.0, (h) t = 0.5086, ψ = -40.0, -25.0, -15.0, -7.5, -2.5, 2.5, 7.5, 15.0, 25.0, and 33.0, and (i) t = 0.5188, ψ = -60.0, -40.0, -25.0, -15.0, -7.5, -2.5, 2.5, 7.5, and 15.0

the calculation was terminated at t = 0.6. The transient was monitored through changes in the total kinetic energy of the fluid, shown in Fig. 12.5. Observe how the kinetic energy rises monotonically at the beginning, reaches a plateau where it remains almost constant for a while, and eventually breaks into an oscillatory mode with a fixed frequency. Figure 12.6 shows the streamlines at various stages of the calculation, starting at t = 0.4354, when the kinetic energy in Fig. 12.5 is at a minimum, and ending at t = 0.5188, after the flow has undergone a full cycle and is back to the initial configuration.

12.2.3 Lubricant Flow in a Microgap

Lubrication problems are usually very challenging because they involve drastically different length scales. Here we consider flow in the microgap between a lip seal and a rotating shaft. An axisymmetric section of an idealized model is shown in Fig. 12.7. The dimension of the gap is 10 μm. The finite element mesh used in the discretization is shown in Fig. 12.8; it contains 2035 nodes and 1864 bilinear isoparametric elements. The steady state solution for Ta = 15, based on the length $d = 200$ μm, is shown in Fig. 12.9. In this problem it is critical that the pressure along the open boundaries be modified. The solution in Fig. 12.9 was obtained by retaining the line integrals of the pressure in the formulation as discussed in Section 10.4.3. If this had not been done, we would have obtained a completely erroneous solution in which the flow in the gap actually goes in the wrong direction. In Fig. 12.10, we show the solution obtained using the natural stress boundary conditions, $\sigma_{ij}n_j = 0$, along

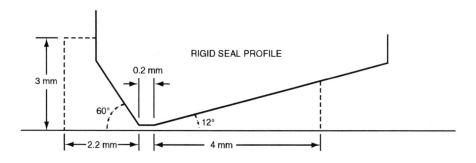

SHAFT

Figure 12.7 Cross section of a seal-shaft configuration.

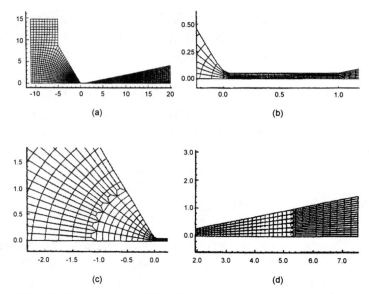

Figure 12.8 Finite element mesh for flow through a microgap.

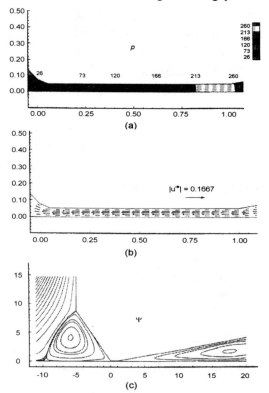

Figure 12.9 Flow through a microgap at Ta = 15: (a) pressure, (b) velocity vectors, and (c) stream functions

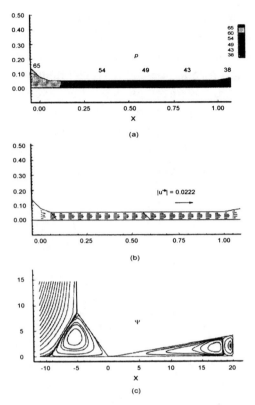

Figure 12.10 Flow through a microgap obtained using incorrect open boundary conditions: (a-b) pressure gradient and the flow going in the wrong direction, and (c) spurious recirculation at the right open boundary.

the open boundaries. Here the flow goes from left to right, which is qualitatively wrong. Notice that the streamline plot shows a recirculating cell in the right open boundary, very similar to those that appeared in the stratified flow examples of Chapter 10, and that a pressure gradient an order of magnitude smaller is obtained in the microgap. The dynamics of the flow have been significantly modified by the wrong boundary conditions.

12.3 TURBULENT FLOW

So far we have only considered laminar flows in the development of our numerical methods because the concepts related to the numerics are easier to explain in this context and are essentially the same as those needed in the numerical treatment of turbulent flows. We know that most real-life flows are turbulent; however, none of the currently available approaches to turbulence can give us an answer to the question of how to find instantaneous turbulent velocities, even if we assume that they satisfy the momentum equations. This is a formidable problem, which can be somewhat simplified if we concentrate on the mean flow and the mean turbulence (Arpaci and Larsen, 1984; Munson et al., 1994; Warsi, 1993). We define the mean (or time-averaged) values as

$$\bar{f} = \frac{1}{T} \int_{t_0}^{t_0 + T} f(x, y, t) \, dt \tag{12.21}$$

where f is a variable such as the velocity components or the pressure and T is a time interval that must be large when compared to the period of the fluctuations in the velocity but small when compared to unsteadiness in the average velocity of the fluid.

The velocities and pressure can then be written as

$$u_i = \bar{u}_i + u_i' \tag{12.22}$$

$$p = \bar{p} + p' \tag{12.23}$$

where u_i' and p' are the velocity and pressure fluctuations about the mean. Substitution into the Navier-Stokes equations yields the equations in terms of the averaged velocities:

$$\frac{\partial \bar{u}_j}{\partial x_j} = 0 \tag{12.24}$$

$$\frac{\partial \bar{u}_i}{\partial t} + \bar{u}_j \frac{\partial \bar{u}_i}{\partial x_j} = -\frac{1}{\rho} \frac{\partial \bar{p}}{\partial x_i} + \nu \frac{\partial^2 \bar{u}_i}{\partial x_j \partial x_j} - \frac{\partial}{\partial x_j} (\overline{u_i' u_j'}) \tag{12.25}$$

The last term in Eq. (12.25) is called the Reynolds stress term and is defined by

$$\tau_{ij} = -\rho \overline{u_i' u_j'}$$

(12.26)

Equations (12.24) and (12.25) are now incomplete because we have introduced the new unknown functions τ_{ij}. We must therefore produce a turbulence model for τ_{ij}. This is a very difficult task, and many models have been constructed to deal with different types of flow, ranging from simple algebraic models (Prandtl's mixing length) to multiple-equations model. The most popular are the two-equation models, also known as k-ε models. A great variety of turbulence models are presented and discussed by Launder and Spalding (1972) and in the review by Rodi (1980).

In a k-ε model, k is the turbulence kinetic energy

$$k = \frac{1}{2} \overline{u_i' u_i'}$$

(12.27)

and ε is the kinematic dissipation rate of the turbulence energy

$$\varepsilon = \nu \left(\frac{\overline{\partial u_i'}}{\partial x_j} \right)^2$$

(12.28)

The use of turbulence models is difficult and usually becomes the dominant aspect of the numerical computations. To illustrate the complexity of turbulent flow calculations, we present an example of turbulent flow over a backward-facing step from Sohn (1988). The k-ε turbulence model is taken form Launder and Spalding (1974), where the equations for k and ε are given by

$$\frac{\partial k}{\partial t} + \bar{u}_i \frac{\partial k}{\partial x_i} = \frac{1}{\rho} \frac{\partial}{\partial x_i} \left(\frac{\mu_t}{\sigma_k} \frac{\partial k}{\partial x_i} \right) + \frac{\mu_t}{\rho} \left(\frac{\partial \bar{u}_i}{\partial x_j} + \frac{\partial \bar{u}_j}{\partial x_i} \right) \frac{\partial \bar{u}_i}{\partial x_j} - \varepsilon$$

(12.29)

and

$$\frac{\partial \varepsilon}{\partial t} + \bar{u}_i \frac{\partial \varepsilon}{\partial x_i} = \frac{1}{\rho} \frac{\partial}{\partial x_i}\left(\frac{\mu_t}{\sigma_\varepsilon} \frac{\partial \varepsilon}{\partial x_i}\right) + \frac{c_1 \mu_t}{\rho} \frac{\varepsilon}{k}\left(\frac{\partial \bar{u}_i}{\partial x_j} + \frac{\partial \bar{u}_j}{\partial x_i}\right)\frac{\partial \bar{u}_i}{\partial x_j} - c_2 \frac{\varepsilon^2}{k} \quad (12.30)$$

The Reynolds stresses are given by

$$\tau_{ij} = \mu_t\left(\frac{\partial \bar{u}_i}{\partial x_j} + \frac{\partial \bar{u}_j}{\partial x_i}\right) \quad (12.31)$$

Here μ_t is the turbulent eddy viscosity obtained from

$$\mu_t = \rho c_\mu \frac{k^2}{\varepsilon} \quad (12.32)$$

and σ_k, σ_ε, c_1, c_2, and c_μ are empirical constants. These equations can be solved using the finite element methodology already available, provided that we can obtain values for the constants. The fact that we have to solve an extra two equations just makes the calculations more expensive.

The model was applied to the benchmark problem of flow over a backward-facing step by Sohn (1988). The geometry and boundary conditions are shown in Fig. 12.11. The Reynolds number based on the inlet velocity and the height of the step was chosen to be 69,610 in order to match the experiments of Kim et al. (1980). The finite element mesh consisted of 22 × 16 nine-noded biquadratic elements refined near the walls.

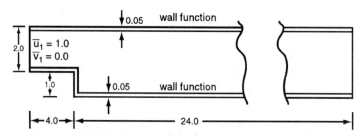

Figure 12.11 Domain and boundary conditions for the turbulent flow over a backward-facing step. (From Evaluation of FIDAP on Some Classical Laminar and Turbulent Benchmarks, by J. L. Sohn. Copyright John Wiley and Sons Ltd. Reproduced with permission).

A further complication with the use of turbulence models is that they do not always do a good job over the entire domain. In this case, it is known that the k-ε model is not valid near the walls, where laminar sublayers may develop (Warsi, 1993). To correct for this deficiency, the computational domain was truncated at a distance from the wall and a "wall function" was used to describe the velocity in the immediate vicinity of the walls, as shown in Fig. 12.11. The tangential velocity, u_t, near the walls was calculated as

$$u_t = \begin{cases} wz & 0 < z < 30 \\ \dfrac{w}{\kappa} \ln(Ez) & 30 < z < 100 \end{cases} \tag{12.33}$$

where z and w are a nondimensional distance from the wall and a frictional velocity, respectively, and are given by

$$z = \frac{wy}{v} \tag{12.34}$$

$$w = \sqrt{v \frac{\partial u_t}{\partial y}} \tag{12.35}$$

Here κ is the von Karmàn constant, set at $\kappa = 0.41$; E is a roughness parameter equal to 9.0. The normal velocity is zero here.

The rest of the empirical constants in the turbulence model are taken from Launder and Spalding (1974) and are $\sigma_k = 1.0$, $\sigma_\varepsilon = 1.3$, $c_1 = 1.44$, $c_2 = 1.92$, and $c_\mu = 0.09$. Finally, k and ε are prescribed at the inlet as $k = 0.003\,\bar{u}_1^2$ and $\varepsilon = 100\,c_\mu k^{1.5}$.

Steady state results for the mean horizontal velocity are shown in Fig. 12.12, and the calculated pressure coefficient is shown in Fig. 12.13. The experimental results of Kim et al. (1980) are presented for comparison. We see that after all this work, the model does a good job only in certain parts of the domain. The flow is not predicted well in the recirculating zone, and the pressure coefficient is only captured properly near the exit of the region.

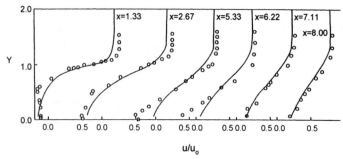

Figure 12.12 Numerical prediction and experimental measurements of the averaged horizontal velocity, $\bar{u}$, for turbulent flow over a backward-facing step. (From Evaluation of FIDAP on Some Classical Laminar and Turbulent Benchmarks, by J. L. Sohn. Copyright John Wiley and Sons Ltd. Reproduced with permission).

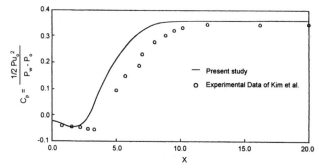

Figure 12.13 Numerically calculated and experimentally measured pressure coefficient for turbulent flow over a backward-facing step (from Sohn, 1988). (From Evaluation of FIDAP on Some Classical Laminar and Turbulent Benchmarks, by J. L. Sohn. Copyright John Wiley and Sons Ltd. Reproduced with permission).

This example should underscore the difficulties that we face in predicting turbulent flows. Turbulence modeling becomes the dominant issue in these problems. The numerical techniques, on the other hand, remain the same as discussed in the context of laminar flows throughout this book.

12.4 COMPRESSIBLE FLOW

A great amount of effort in computational fluid dynamics has been devoted to the calculation of high-speed flows in aerodynamics. This is in fact such a vast area of research that entire books have been devoted to its computational aspects. We will restrict ourselves, here, to showing how the methods developed in this book can be extended to perform compressible flow calculations. For comprehensive reference lists on the use of finite element methods to model high-speed flows, we refer the readers to books by Carey and Oden (1986) and Zienkiewicz and Taylor (1991).

We will write the governing equation using the nondimensional form given by Anderson (1989), and we will use the internal energy form, which is simpler, even though this means that the equations will not be written in conservation form (Brueckner, 1991). The governing equations are as follows

Continuity

$$\frac{\partial \rho}{\partial t} + u_j \frac{\partial \rho}{\partial x_j} + \rho \frac{\partial u_j}{\partial x_j} = 0 \tag{12.36}$$

Momentum

$$\rho \left(\frac{\partial u_i}{\partial t} + u_j \frac{\partial u_i}{\partial x_j} \right) = -\frac{\partial}{\partial x_i} \left(\frac{1}{\gamma M^2} p - \frac{1}{3} \frac{\mu}{Re} \frac{\partial u_j}{\partial x_j} \right) + \frac{\mu}{Re} \frac{\partial^2 u_i}{\partial x_j \partial x_j} \tag{12.37}$$

Energy

$$\rho \left(\frac{\partial e}{\partial t} + u_j \frac{\partial e}{\partial x_j} \right) = \frac{\gamma}{Re\, Pr} \frac{\partial}{\partial x_j} \mu \frac{\partial e}{\partial x_j} - \left(\frac{1}{\gamma M^2} p + \frac{2}{3} \frac{\mu}{Re} \frac{\partial u_j}{\partial x_j} \right) \frac{\partial u_j}{\partial x_j}$$
$$+ \frac{\mu}{Re} \left(\frac{\partial u_i}{\partial x_j} + \frac{\partial u_j}{\partial x_i} \right) \frac{\partial u_i}{\partial x_j} \tag{12.38}$$

State

$$p = \rho e \tag{12.39}$$

In Eq. (12.37), we have kept the divergence term in the first parentheses on the right-hand side separate from the viscous terms (the second term on the right-hand side, even though they are of the same form and are usually combined into one term. Combining them is misleading, however, because the divergence term may behave very differently, especially when the flow approaches conditions of constant density. Also, they must be treated differently from a numerical point of view. In fact, to avoid pressure oscillations in regions of stagnation or where the flow is locally of constant density, we need to treat such terms in a manner consistent with an incompressible flow.

In the equations given above, the nondimensionalization utilizes the following variables:

$$\rho = \frac{\rho'}{\rho_\infty} \quad \mathbf{u} = \frac{\mathbf{u}'}{U_\infty} \quad T = \frac{T'}{e_\infty} \quad p = \frac{p'}{T_\infty} \quad e = \frac{e'}{e_\infty}$$

$$\mathbf{x} = \frac{x'}{L} \quad t = \frac{t'U_\infty}{L} \quad \mu = \frac{\mu'}{\mu_\infty} \quad k = \frac{k'}{k_\infty}$$

where the prime denotes a dimensional quantity and $e = T$, where e is the internal energy. The viscosity is considered temperature dependent and is calculated using Sutherland's law (Anderson, 1989)

$$\mu = T^{3/2} \frac{1+S}{T+S} \tag{12.40}$$

where S is a constant. The quantities ρ_∞, U_∞, e_∞, T_∞, L, μ_∞, and k_∞ are reference values, and the nondimensional parameters are:

Mach number

$$M = \frac{U_\infty}{\sqrt{\gamma \dfrac{p_\infty}{\rho_\infty}}} \tag{12.41}$$

Reynolds number

$$Re = \frac{\rho_\infty U_\infty L}{\mu_\infty} \tag{12.42}$$

Prandtl number

$$Pr = \frac{\mu_\infty c_p}{k_\infty} \tag{12.43}$$

Ratio of specific heats

$$\gamma = \frac{c_p}{c_v} \tag{12.44}$$

A stable, consistent Petrov-Galerkin weak formulation using bilinear elements to approximate the density, velocities and temperature, and piecewise-constant pressure can be obtained as follows (Dyne and Heinrich, 1993):

Continuity

$$\oint_\Omega N_i \frac{\partial \rho}{\partial t} d\Omega = -\int_\Omega (W_\rho)_i \left(u\frac{\partial \rho}{\partial x} + v\frac{\partial \rho}{\partial y} \right) d\Omega - \oint_\Omega N_i \rho \left(\frac{\partial u}{\partial x} + \frac{\partial v}{\partial y} \right) d\Omega \tag{12.45}$$

x-momentum

$$\int_\Omega \rho N_i \frac{\partial u}{\partial t} d\Omega = -\int_\Omega \left[(W_u)_i \left[u\frac{\partial u}{\partial x} + v\frac{\partial u}{\partial y} \right] + \frac{\mu}{Re} \left(\frac{\partial N_i}{\partial x} \frac{\partial u}{\partial x} + \frac{\partial N_i}{\partial y} \frac{\partial u}{\partial y} \right) \right] d\Omega$$

$$+ \oint_\Omega \frac{\partial N_i}{\partial x} \left[\frac{1}{\gamma M^2} p - \frac{1}{3}\frac{\mu}{Re} \left(\frac{\partial u}{\partial x} + \frac{\partial v}{\partial y} \right) \right] d\Omega$$

$$+ \oint_\Gamma N_i \left[-\frac{1}{\gamma M^2} p - \frac{1}{3}\frac{\mu}{Re} \left(\frac{\partial u}{\partial x} + \frac{\partial v}{\partial y} \right) \right] n_x d\Gamma$$

$$+ \int_\Gamma N_i \frac{\mu}{Re} \left[2\frac{\partial u}{\partial x} n_x + \left(\frac{\partial u}{\partial y} + \frac{\partial v}{\partial x} \right) n_y \right] d\Gamma \tag{12.46}$$

y-momentum

$$\int_\Omega \rho N_i \, \frac{\partial v}{\partial t} \, d\Omega = -\int_\Omega \left[(W_v)_i \left(u\frac{\partial v}{\partial x} + v\frac{\partial v}{\partial y} \right) + \frac{\mu}{Re}\left(\frac{\partial N_i}{\partial x}\frac{\partial v}{\partial x} + \frac{\partial N_i}{\partial y}\frac{\partial v}{\partial y} \right) \right] d\Omega$$

$$+ R\oint_\Omega \frac{\partial N_i}{\partial y}\left[\frac{1}{\gamma M^2}p - \frac{1}{3}\frac{\mu}{Re}\left(\frac{\partial u}{\partial x} + \frac{\partial v}{\partial y} \right) \right] d\Omega \qquad (12.47)$$

$$+ R\oint_\Gamma N_i\left[-\frac{1}{\gamma M^2}p + \frac{1}{3}\frac{\mu}{Re}\left(\frac{\partial u}{\partial x} + \frac{\partial v}{\partial y} \right) \right] n_y \, d\Gamma$$

$$+ \int_\Gamma N_i\frac{\mu}{Re}\left[\left(\frac{\partial u}{\partial y} + \frac{\partial v}{\partial x} \right)n_x + 2\frac{\partial v}{\partial y}n_y \right] d\Gamma$$

Energy

$$\int_\Omega \rho N_i \frac{\partial e}{\partial t}d\Omega = -\int_\Omega \left[(W_e)_i\left(u\frac{\partial e}{\partial x} + v\frac{\partial e}{\partial y} \right) + \frac{\gamma\mu}{Re\,Pr}\left(\frac{\partial N_i}{\partial x}\frac{\partial e}{\partial x} + \frac{\partial N_i}{\partial y}\frac{\partial e}{\partial y} \right) \right] d\Omega$$

$$+ R\oint_\Omega N_i\left[\frac{1}{\gamma M^2}p + \frac{2}{3}\frac{\mu}{Re}\left(\frac{\partial u}{\partial x} + \frac{\partial v}{\partial y} \right) \right]\left(\frac{\partial u}{\partial x} + \frac{\partial v}{\partial y} \right)d\Omega \qquad (12.48)$$

$$+ \int_\Gamma N_i\frac{\mu}{Re}\left[2\left(\frac{\partial u}{\partial x} \right)^2 + 2\left(\frac{\partial v^2}{\partial y} \right) + \frac{\partial u}{\partial y} + \frac{\partial v}{\partial x} \right]d\Omega$$

$$+ \int_\Gamma N_i\frac{\gamma\mu}{Re\,Pr}\left(\frac{\partial e}{\partial x}n_x + \frac{\partial e}{\partial y}n_y \right)d\Gamma$$

State

$$\oint_\Omega M_i p \, d\Omega = \oint_\Omega \rho M_i e \, d\Omega \qquad (12.49)$$

In the equations given above, N_i are the bilinear shape functions and M_i are piecewise-constant shape functions such that $M_i = 1$ over element e_i and $= 0$ otherwise.

The Petrov-Galerkin weighting functions, W_ρ, W_u, W_v, and W_e are given by

$$(W_a)_i = N_i + \alpha_a \frac{\overline{h}}{2\|\mathbf{V}\|} \mathbf{V} \cdot \nabla N_i \tag{12.50}$$

where $a = \rho, u, v$, and e. The parameter α_a is given by

$$\alpha_a = \coth \frac{\gamma_a}{2} - \frac{2}{\gamma_a} \tag{12.51}$$

and the local numbers, γ_a, are

$$\gamma_\rho = \infty \qquad (\text{i.e., } \alpha_\rho = 1) \tag{12.52}$$

$$\gamma_u = \frac{\rho \operatorname{Re}\|\mathbf{V}\|\overline{h}}{\mu\left(1 + \dfrac{u^2}{3\|\mathbf{V}\|^2}\right)} \tag{12.53}$$

$$\gamma_v = \frac{\rho \operatorname{Re}\|\mathbf{V}\|\overline{h}}{\mu\left(1 + \dfrac{v^2}{3\|\mathbf{V}\|^2}\right)} \tag{12.54}$$

$$\gamma_e = \frac{\rho \operatorname{Pr} \operatorname{Re}\|\mathbf{V}\|\overline{h}}{\mu\gamma} \tag{12.55}$$

Details on the derivation of these Petrov-Galerkin weights are found in the work by Brueckner (1991).

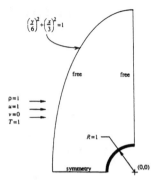

Figure 12.14 Geometry and boundary conditions for viscous flow impinging on a cylinder of unit radius.

We now look at two examples of the application of the algorithms given above. In both cases we seek steady state solutions; these were obtained using mass lumping in Eqs.(14.44)–(14.47) and a second-order Runge-Kutta method to integrate in time.

12.4.1 Supersonic Flow Impinging on a Cylinder

The computational domain and boundary conditions for this example are shown in Fig. 12.14. The cylinder has unit radius, the free stream Mach number is $M = 2$, and we set $Pr = 0.72$, $\gamma = 1.4$, and $Re = 1000$ based on the radius of the cylinder. In this example the viscosity is assumed to be constant. The finite element mesh is shown in Fig. 12.15 and consists of a 40×25 node mesh clustered near the cylinder surface.

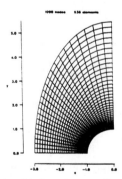

Figure 12.15 Computational grid for viscous flow impinging on a cylinder of unit radius.

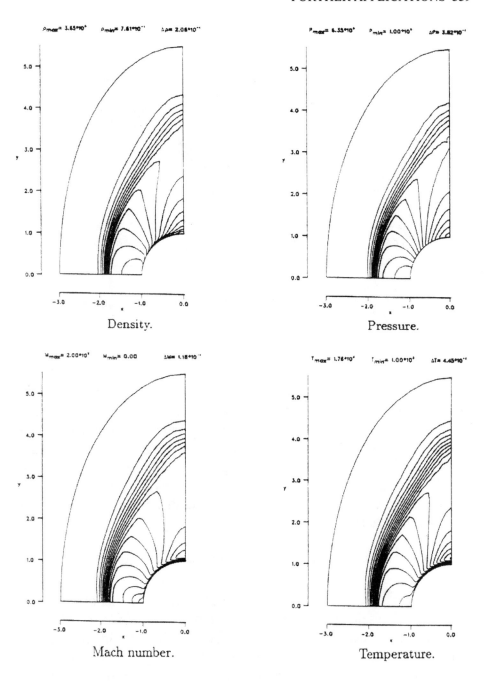

Figure 12.16 Results for viscous flow impinging on a cylinder at $M = 2$ and Re = 1000.

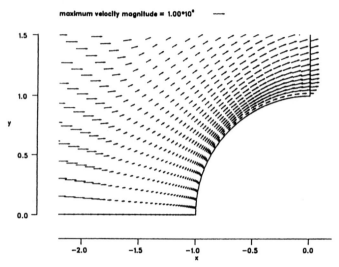

maximum velocity magnitude = 1.00*10°   —

Figure 12.17 Detail of the velocity field for viscous flow impinging on a cylinder of unit radius at $M = 2$ and Re = 1000.

Figure 12.16 shows contours of density, pressure, local Mach number, and temperature calculated at steady state. The detached bow shock is evident; three to four elements are needed to accommodate it along the line of symmetry. The formation and development of the boundary layer can be observed in the plots of local Mach number and temperature; a further illustration is given in Fig. 12.17, where the velocity vectors near the cylinder surface are shown. No recirculation is predicted near the stagnation point. Also note that no oscillation can be observed in the subsonic flow region.

12.4.2 Chemically Reacting Supersonic Flow

Our next example involves supersonic flow with combustion in a ram accelerator (Hertzberg et al., 1988). The ram accelerator is a method of accelerating a projectile in a tube using gasdynamic techniques. Here we consider the detonation wave mode illustrated in Fig. 12.18. The nose of the projectile creates a cone-shaped bow shock, which reflects off the tube wall. The fluid properties and geometry are calibrated so that the bow shock does not initiate combustion, but the reflected shock does.

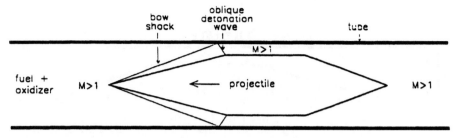

Figure 12.18 Oblique detonation ram accelerator.

Optimal performance is obtained when the detonation wave strikes the projectile at the shoulder.

Here we present an inviscid calculation with finite rate chemistry for a mixture of hydrogen and oxygen. The governing equations are as follows.

Continuity

$$\frac{\partial \rho}{\partial t} + \rho \nabla \cdot V = 0 \qquad (12.56)$$

$$\rho \frac{Dc_i}{Dt} = \dot{w}_i \qquad i = 1, ns-1 \qquad (12.57)$$

$$\rho \frac{DV}{Dt} = -\nabla p \qquad (12.58)$$

$$\rho \frac{De}{Dt} = -\nabla \cdot V \qquad (12.59)$$

where c_i is the mass fraction of species i, $\dot{w}_i$ is the rate of change in ρ due to the chemical reaction, and ns is the number of species. We calculate the mass fraction of species ns from the conservation equation

$$\sum_{i=1}^{ns} c_i = 1 \qquad (12.60)$$

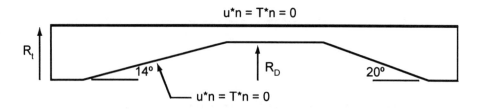

$u^*n = T^*n = 0$

R_t

$14°$

R_D

$20°$

$u^*n = T^*n = 0$

Figure 12.19 Computational geometry for the ram accelerator.

and the equation of state remains unchanged. The combustion model involves eight reactions and six species. Calculations proceed in the following fashion.

1. Find a steady state solution for flow without chemical reactions.

2. Turn on the chemistry: At each timestep, first advance the flow, freezing the chemistry; then allow the reactions to take place with the flow fixed.

The chemistry calculations were performed using the program CFDK of Radhakrishnan and Pratt (1988). The present calculations were reported in detail by Dyne and Heinrich (1996).

Calculations were performed in axisymmetric coordinates on a configuration with a nose cone angle of $14°$ and a nozzle angle of $20°$, as shown in Fig. 12.19. The projectile-to-tube-radius ratio is $r = R_p/R_t = 0.9$. The mesh of bilinear isoparametric elements consists of 35 equally spaced nodes in the y-direction and 250 nodes in the x-direction, clustered near the detonation region. The gas mixture is $O_2 + 2H_2$ at a fill pressure of 50 atm and $M = 8$. The calculated temperature and pressure are shown in Fig. 12.20 and show that the detonation wave has been well resolved. Figure 12.21 shows a detail of the velocity field near the detonation region.

We close by reminding the reader that this calculation was performed using the algorithm explained earlier in this chapter (with viscous effects suppressed) to obtain the density, velocity, internal energy, pressure, and species concentrations. The CFDK program was used for the chemistry calculation.

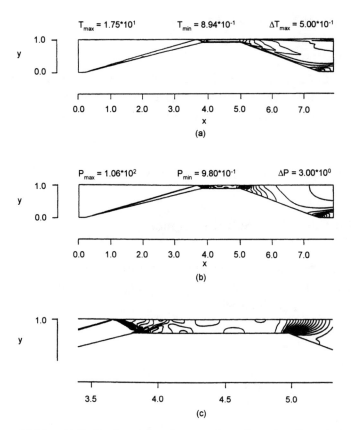

Figure 12.20 Calculated results for reacting flow in the ram accelerator: (a) temperature, (b) pressure, and (c) expanded view of the pressure in the detonation region.

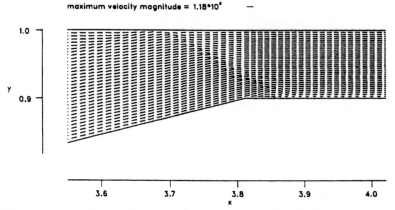

Figure 12.21 Velocity field near the detonation region for flow at Mach 8 through the ram accelerator.

12.5 THREE-DIMENSIONAL FLOW

Throughout this book, we have indicated that the finite element algorithms developed here are useful in both two and three dimensions. So far, however, we have restricted our examples to one- and two-dimensional situations. We now present examples of application to three-dimensional analysis. We hope that these will also show that, in three dimensions, the processes of mesh generation and results visualization are greatly magnified and become serious problems in themselves.

12.5.1 Natural Convection Within a Sphere

This is similar to the case of convection in a horizontal circular cylinder considered in Section 12.2.2, and the same form of the Navier-Stokes equations is used in three dimensions:

Continuity

$$\frac{\partial u}{\partial x} + \frac{\partial v}{\partial y} + \frac{\partial w}{\partial z} = 0 \tag{12.61}$$

Momentum

$$\frac{\partial u}{\partial t} + u\frac{\partial u}{\partial x} + v\frac{\partial u}{\partial y} + w\frac{\partial u}{\partial z} = -\frac{\partial p}{\partial x} + Pr\left(\frac{\partial^2 u}{\partial x^2} + \frac{\partial^2 u}{\partial y^2} + \frac{\partial^2 u}{\partial z^2}\right) \tag{12.62}$$

$$\frac{\partial v}{\partial t} + u\frac{\partial v}{\partial x} + v\frac{\partial v}{\partial y} + w\frac{\partial v}{\partial z} = -\frac{\partial p}{\partial y} + Pr\left(\frac{\partial^2 v}{\partial x^2} + \frac{\partial^2 v}{\partial y^2} + \frac{\partial^2 v}{\partial z^2}\right) \tag{12.63}$$

$$\frac{\partial w}{\partial t} + u\frac{\partial w}{\partial x} + v\frac{\partial w}{\partial y} + w\frac{\partial w}{\partial z} = -\frac{\partial p}{\partial z} + Pr\left(\frac{\partial^2 w}{\partial x^2} + \frac{\partial^2 w}{\partial y^2} + \frac{\partial^2 w}{\partial z^2}\right) + Pr\,Ra\,T \tag{12.64}$$

Energy

$$\frac{\partial T}{\partial t} + u\frac{\partial T}{\partial x} + v\frac{\partial T}{\partial y} + w\frac{\partial T}{\partial z} = \frac{\partial^2 T}{\partial x^2} + \frac{\partial^2 T}{\partial y^2} + \frac{\partial^2 T}{\partial z^2} \tag{12.65}$$

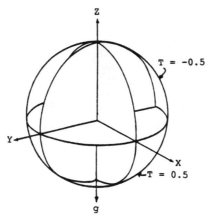

Figure 12.22 Spherical problem domain.

The boundary conditions are no slip, $u = v = w = 0$, on the sphere's surface, which is kept hot at the bottom and cold at the top, as shown in Fig. 12.22. The initial conditions are $u = v = w = T = 0$ at $t = 0$. The algorithm described in Section 12.2 is used to integrate in time using eight-noded brick elements and the penalty and Petrov-Galerkin formulations.

The mesh consists of 4600 trilinear isoparametric elements and 5000 nodes and is depicted in Fig. 12.23. Results for $Re = 5 \times 10^4$ and $Pr =1$ are shown in Figs. 12.24 and 12.25 at time $t = 2$ (nondimensional). A very complex cellular pattern can be observed that makes the flow very difficult to interpret. A discussion on the visualization aspects of this problem is given by Heinrich and Pepper (1989).

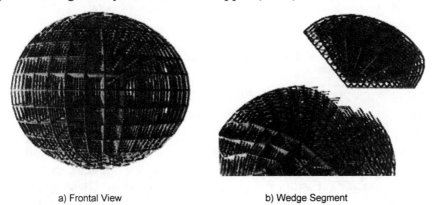

a) Frontal View   b) Wedge Segment

Figure 12.23 Mesh for natural convection within a sphere.

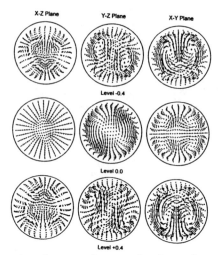

Figure 12.24 Velocity plots for natural convection in a sphere.

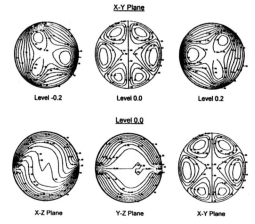

Figure 12.25 Temperature contours for natural convection in a sphere.

12.5.2 Transonic Flow Through a Rectangular Nozzle

This is an example of a three-dimensional compressible flow moving through a nozzle configuration that changes from annular to rectangular studied by Brueckner and Pepper (1995). Their study was motivated by an interest in thrust vectoring. The geometry and computational mesh are shown in Fig. 12.26. In order to reduce the model by one half, symmetry conditions are applied along the x-y plane. The mesh consists of 71,512 trilinear elements with 77,395 nodes. The reference length was taken to be the half-width in the y-direction. The free stream Mach number is 0.22, the Reynolds number is 500, and the Prandtl number is 0.72. The reference temperature is 865°K, $\gamma = 1.4$, and the pressure ratio p_{inlet}/p_{outlet} is 1.83. The incoming flow is subject to a 20° swirl angle. A plot of the transverse velocity vector at the nozzle exit plane is given in Fig. 12.27, where axial vortices that form along the sidewalls of the nozzle as the fluid moves down the transition region can be observed. Figure 12.28 shows temperature contours at various cross sections through the computed plane downstream of the nozzle exit, showing the effect of swirl on the plume.

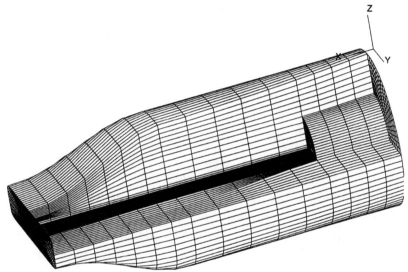

Figure 12.26 Computational mesh for viscous transonic flow through a nozzle.

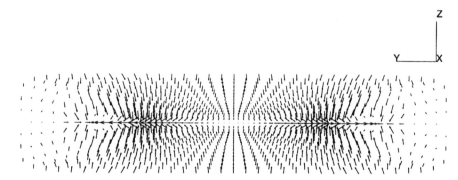

Figure 12.27 Transverse velocity vector field at the nozzle exit plane.

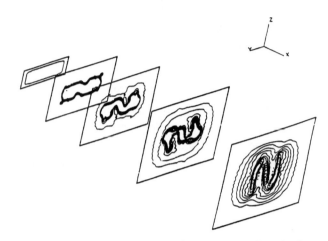

Figure 12.28 Temperature contours at various cross sections in the nozzle plume.

12.6 CLOSURE

In this chapter, we have looked at some rather complex applications of the finite element method to problems that are time dependent, have difficult geometries, are three dimensional, or require extensive modeling. We have covered a wide cross section of applications, which should give a good idea of the power and versatility of the methods presented in this book. We have followed a fairly straight line in the sense that we have always used the same type of algorithm, no matter what the problem at hand. There are, of course, many other ways to approach numerical modeling. Our hope is that readers who have been exposed to this material will have reached a level of understanding of the issues involved in numerical modeling that will allow them to move on to their particular problems and choose the appropriate numerical tools to solve them.

REFERENCES

Anderson, J. D. (1989) *Hypersonic and High Temperature Gas Dynamics*. New York: McGraw-Hill.

Anderson, C. R., Greengard, C., Greengard, L. and Rokhlin, V. (1990) "On the Accurate Calculation of Vortex Shedding," *Phys. Fluids A,* Vol. 2, pp. 883–885.

Arpaci, V. S. and Larsen, P. S. (1984) *Convection Heat Transfer*. Englewood Cliffs, N.J: Prentice-Hall.

Brooks, A. and Hughes, T. J. R. (1982) "Streamline Upwind/Petrov-Galerkin Formulations for Convection Dominated Flows with Particular Emphasis on the Incompressible Navier-Stokes Equations," *Comput. Methods Appl. Mech. Eng.,* Vol. 32, pp. 199–259.

Brueckner, F. P. (1991) "Finite Element Analysis of High-Speed Flows with Application to the Ram Accelerator Concept" Ph.D. thesis, Univ. of Arizona, Tucson.

Brueckner, F. P. and Pepper, D. W. (1995) "Parallel Finite Element Algorithm for Three-Dimensional Inviscid and Viscous Flow," *J. Thermophys. Heat Transfer,* Vol. 9, pp. 240–246.

Carey, G. F. and Oden, J. T. (1986) *Finite Elements*, Vol. 6, *Fluid Mechanics*. Englewood Cliffs, N.J: Prentice-Hall.

Dyne, B. R. and Heinrich, J. C. (1993) "Physically Correct Penalty-Like Formulations for Accurate Pressure Calculation in Finite Element Algorithms of the Navier-Stokes Equations," *Int. J. Numer. Methods Eng.,* Vol. 36, pp. 3883–3902.

Dyne, B. R. and Heinrich, J. C. (1996) "Numerical Analysis of the Scramaccelerator with Hydrogen-Oxygen Combustion," *J. Propul. Power,* Vol. 12, pp. 336–340.

Heinrich, J. C. and Pepper, D. W. (1989) "Flow Visualization of Natural Convection in a Differentially Heated Sphere," pp. 135–141 in *Flow Visualization* FED Vol. 85, (B. Khalighi, M. J. Braun, and C. J. Freitas, editors). New York: ASME.

Heinrich, J. C. and Yu, C.-C. (1988) "Finite Element Simulation of Buoyancy-Driven Flows with Emphasis on Natural Convection in a Horizontal Circular Cylinder," *Comput. Methods Appl. Mech. Eng.,* Vol. 69, pp. 1–27.

Hertzberg, A., Bruckner, A. P. and Burgdanoff, D. W. (1988) "Ram Accelerator: A New Chemical Method for Accelerating Projectiles to Ultrahigh Velocities," *AIAA J.,* Vol. 26, pp. 195–203.

Hughes, T. J. R., Liu, W. K. and Brooks, A. (1979) "Finite Element Analysis of Incompressible Viscous Flows by the Penalty Function Formulation," *J. Comput. Phys.,* Vol. 30, pp. 1–60.

Kim, J., Kline, S. J. and Johnson, J. P. (1980) "Investigation of a Reattaching Turbulent Shear Layer: Flow over a Backward Facing Step," *J. Fluid Eng., ASME Trans.,* Vol. 102, pp. 302–308.

Launder, B. E. and Spalding, D. B. (1972) *Lectures in Mathematical Models of Turbulence.* London: Academic Press.

Launder, B. E. and Spalding, D. B. (1974) "The Numerical Computation of Turbulent Flows," *Comput. Methods Appl. Mech. Eng.,* Vol. 3, pp. 269–289.

Munson, B. R., Young, D. F. and Okiishi, T. H. (1994) *Fundamentals of Fluid Mechanics.* New York: John Wiley and Sons.

Radhakrishnan, K. and Pratt, D. T. (1988) "Fast Algorithm for Calculating Chemical Kinetics in Turbulent Reacting Flow," *Combust. Sci. Technol.,* Vol. 58, pp. 155–176.

Rodi, W. (1980) *Turbulence Models and Their Application in Hydraulics.* Delft, The Netherlands: International Association for Hydraulic Research Secretariat.

Sohn, J. L. (1988) "Evaluation of FIDAP on Some Classical Laminar and Turbulent Benchmarks," *Int. J. Numer. Methods Fluids,* Vol. 8, pp. 1469–1490.

Warsi, Z. U. A. (1993) *Fluid Dynamics: Theoretical and Computational Aspects.* Boca Raton, Fla.: CRC Press.

Zienkiewicz, O. C. and Taylor, R. L. (1991) *The Finite Element Method,* Vol. 2, 4th ed. London: McGraw-Hill.

LINEAR OPERATIONS

A.1 LINEAR VECTOR SPACES

The spaces of functions considered in this book fall under the general category of *linear vector spaces*. A linear vector space S is a collection or set of vectors that satisfies the following conditions:

1) An operation called *addition* is defined in S such that if $\mathbf{x}$ and $\mathbf{y}$ are elements of S, then $\mathbf{x} + \mathbf{y}$ is also an element of S, i.e., S is *closed* under addition. Furthermore, it satisfies the following properties:

i) $\mathbf{x} + \mathbf{y} = \mathbf{y} + \mathbf{x}$
ii) $(\mathbf{x} + \mathbf{y}) + \mathbf{z} = \mathbf{x} + (\mathbf{y} + \mathbf{z})$
iii) S contains a "zero vector" $\mathbf{O}$, such that if $\mathbf{x}$ is in S, then $\mathbf{x} + \mathbf{O} = \mathbf{x}$.
iv) for each $\mathbf{x}$ in S there is a vector "$-\mathbf{x}$" in S such that $\mathbf{x} + (-\mathbf{x} = 0)$

2) A second operation called *scalar multiplication* is defined such that if $\mathbf{x}$ is in S and α is a scalar, then $\alpha\mathbf{x}$ is also in S, i.e., S is *closed* under scalar multiplication. In addition, it satisfies the following properties:

i) $\alpha\,(\beta\mathbf{x}) = (\alpha\beta)\,\mathbf{x}$
ii) $(\alpha + \beta)\,\mathbf{x} = \alpha\,\mathbf{x} + \beta\mathbf{x}$
iii) $\alpha\,(\mathbf{x} + \mathbf{y}) = \alpha\,\mathbf{x} + \alpha\,\mathbf{y}$
iv) $1 \cdot \mathbf{x} = \mathbf{x}$

v) $\mathbf{O} \cdot \mathbf{x} = \mathbf{O}$

If the scalars are complex numbers, S is a complex linear vector space; we restrict ourselves to real scalars only, and we refer to S simply as a *vector space*.

A set of vectors $\mathbf{x}_1, \mathbf{x}_2, \dots, \mathbf{x}_n$ in S is said to be *linearly dependent*, if there exists a set of scalars $\alpha_1, \alpha_2, \dots, \alpha_n$ *not all zero* such that

$$\alpha_1 \mathbf{x}_1 + \alpha_2 \mathbf{x}_2 + \dots + \alpha_n \mathbf{x}_n = 0 \tag{A.1}$$

If (A.1) is satisfied only when $\alpha_1 = \alpha_2 = \dots = \alpha_n = 0$, the vectors are *linearly independent*. A vector $\mathbf{x}$ such that $\mathbf{x} = \alpha_1 \mathbf{x}_1 + \alpha_2 \mathbf{x}_2 + \dots + \alpha_n \mathbf{x}_n$ is called a *linear combination* of $\mathbf{x}_1, \mathbf{x}_2, \dots, \mathbf{x}_n$.

A vector space S is *n-dimensional* if it contains a set of n linearly independent vectors, but any set of $n + 1$ vectors is linearly dependent. The space S is *∞-dimensional* if, for any positive integer k, we can find k independent vectors.

To measure length in a vector space, we introduce the *norm* defined for any vector $\mathbf{x}$ as $\|\mathbf{x}\|$. The norm satisfies the following properties:

i) $\|\alpha \mathbf{x}\| = |\alpha| \|\mathbf{x}\|$

ii) $\begin{cases} \|\mathbf{x}\| > 0 & \text{if } \mathbf{x} \neq 0 \\ \|\mathbf{x}\| = 0 & \text{if } \mathbf{x} = 0 \end{cases}$

iii) $\|\mathbf{x} + \mathbf{y}\| \geq \|\mathbf{x}\| + \|\mathbf{y}\|$

A measure of the angle between any two vectors $\mathbf{x}, \mathbf{y}$ can be obtained by introducing the *inner product* between them. The inner product is defined as a real number $(\mathbf{x}, \mathbf{y})$ that satisfies:

i) $(\mathbf{x}, \mathbf{y}) = (\mathbf{y}, \mathbf{x})$

ii) $(\alpha \mathbf{x} + \beta \mathbf{y}, \mathbf{z}) = \alpha (\mathbf{x}, \mathbf{z}) + \beta (\mathbf{y}, \mathbf{z})$

iii) $(\mathbf{x}, \mathbf{x}) > 0$ if $\mathbf{x} \neq 0$ and $(\mathbf{x}, \mathbf{x}) = 0$ if $\mathbf{x} = 0$

Notice that an inner product induces a norm defined as $\|\mathbf{x}\|^2 = (\mathbf{x}, \mathbf{x})$; hence, an inner product space is always a normed space, but the opposite is not true.

A subset M of S is a *subspace* of S if, given **x** and **y** in M and any two scalars α and β, $\alpha\mathbf{x} + \beta\mathbf{y}$ is also in M. Notice that **O** must be in M for it to be a subspace.

The space of vectors with n components constitutes an example of a finite dimensional vector space, with dimension n. The inner product and norm are defined respectively by

$$(x,y) = y^T x = \sum_{i=1}^{n} x_i y_i \tag{A.2}$$

and

$$\|\mathbf{x}\| = (\sum_{i=1}^{n} x_i^2)^{\frac{1}{2}} \tag{A.3}$$

The former equation is usually referred to as the *dot product*, and the latter is usually referred to as the *Euclidean norm*.

Examples of infinite dimensional vector spaces used in this book include the following:

i)   The spaces of functions $C^n(\Omega)$ $n = -1,0,1,2,...$, where $C^{-1}(\Omega)$ is the space of piecewise continuous function in Ω. $C^n(\Omega)$ $n = 0,1,2,...$ denotes the space of functions that are continuous and have continuous derivatives up to order n in Ω. If Ω is a domain in two-dimensional space, then

$$C^n(\Omega) = \{f = f(x,y) /$$

$$f, \frac{\partial f}{\partial x}, \frac{\partial f}{\partial y}, \frac{\partial^2 f}{\partial x^2}, \frac{\partial^2 f}{\partial x \partial y}, \frac{\partial^2 f}{\partial y^2}, ..., \frac{\partial^n f}{\partial x^n}, \frac{\partial^n f}{\partial x^{n-1} \partial y}, ..., \frac{\partial^n f}{\partial x \partial y^{n-1}}, \frac{\partial^n f}{\partial y^n}$$

are all continuous in $\Omega\}$.

ii)   The space $L^2(\Omega)$ is defined as the space of functions defined over Ω for which the integral over Ω of the square of the function exists, i.e.,

$$L^2(\Omega) = \left\{ f / \int_\Omega f^2 d\Omega < \infty \right\}$$

Notice that the function in $L^2(\Omega)$ need not be continuous and may even contain a weak singularity in Ω.

iii) The Sobolev space $H^n(\Omega)$ in the space of functions f defined in Ω such that f and all of its partial derivatives of order less than or equal to n are in $L^2(\Omega)$. In two space dimensions we have

$$H^n(\Omega) = \left\{ f / f, \frac{\partial f}{\partial x}, \frac{\partial f}{\partial y}, \frac{\partial^2 f}{\partial x^2}, \frac{\partial^2 f}{\partial x \partial y}, ..., \frac{\partial^n f}{\partial y^n} \text{ are in } L^2(\Omega) \right\}$$

In particular, in three space dimensions the space $H^1(\Omega)$ is characterized as

$$H^1(\Omega) = \{ f / \int_\Omega [f^2 + (\frac{\partial f}{\partial x})^2 + (\frac{\partial f}{\partial y})^2 + (\frac{\partial f}{\partial z})^2] d\Omega < \infty \}$$

Notice that the subset $H^1_0(\Omega)$ of functions in $H^1(\Omega)$ that vanish on the boundary Γ of Ω is a subspace of $H^1(\Omega)$.

The inner product and norm for the $L^2(\Omega)$ space are defined respectively as

$$(f,g) = \int_\Omega fg \, d\Omega \tag{A.4}$$

and

$$\| f \|_0 = \left(\int_\Omega f^2 d\Omega \right)^{\frac{1}{2}} \tag{A.5}$$

This norm and inner product are also commonly used when we work in the $H^1(\Omega)$ space. It is sometimes convenient, however, to use the 1-norm in $H^1(\Omega)$. In three space dimensions this is given by

$$\|f\|_{1} = \left[\int_{\Omega} \left\{ f^2 + \left(\frac{\partial f}{\partial x}\right)^2 + \left(\frac{\partial f}{\partial y}\right)^2 + \left(\frac{\partial f}{\partial z}\right)^2 \right\} d\Omega \right]^{\frac{1}{2}} \qquad (A.6)$$

There are a variety of other norms, inner products, and seminorms that can be defined in Sobolev space, however, they are not relevant for our purposes.

There is still one more norm that is important in error analysis. This is the *infinite norm*. If f is a bounded function defined over a domain Ω, then

$$\|f\|_{\infty} = \sup_{\mathbf{x} \in \Omega} |f(\mathbf{x})| \qquad (A.7)$$

If Ω is a closed domain, $\|f\|_{\infty}$ can be expressed as the form used in the text:

$$\|f\|_{\infty} = \max_{\mathbf{x} \in \Omega} |f(\mathbf{x})| \qquad (A.8)$$

A.2   LINEAR OPERATORS

A *linear operator* or *transformation* L is a mapping from a vector space S^1 to a vector space S^2, such that if $\mathbf{x}$ and $\mathbf{y}$ are in S^1 and α and β are scalars, then

$$L(\alpha \mathbf{x} + \beta \mathbf{y}) = \alpha L(\mathbf{x}) + \beta L(\mathbf{y})$$

The norm of an operator is defined as

$$\|L\| = \sup_{\substack{\mathbf{x} \in S^1 \\ \mathbf{x} \neq 0}} \frac{\|L\mathbf{x}\|_{S^2}}{\|\mathbf{x}\|_{S^1}} \qquad (A.9)$$

where $\|\cdot\|_S^1$ and $\|\cdot\|_S^2$ denote norms in S^1 and S^2, respectively. The norm of L is a measure of the largest amplification of a vector $\mathbf{x}$ in S^1 by L.

A linear operator is said to be *bounded* if there exists a scalar $c > 0$, such that

$$\| L \mathbf{x} \|_{S^2} \leq c \| \mathbf{x} \|_{S^1}$$

Note that differential operators are *unbounded* while matrix operators are always bounded.

The adjoint L^* of an operator L is an operator associated with L such that $(L\mathbf{x},\mathbf{y}) = (\mathbf{x},L^*\mathbf{y})$ for all $\mathbf{x}$ in S^1 and $\mathbf{y}$ in S^2. If $S^1 = S^2$ and $L = L^*$, then L is *self adjoint*. If $L = L^*$ but $S^1 \neq S^2$, the operator L is called *formally self adjoint*. The difference between S^1 and S^2 is normally due to the boundary conditions.

If L is a linear operator, the equation $L\mathbf{x} = \mathbf{y}$ has a solution only if $(\mathbf{y}, \mathbf{z}) = 0$ for all $\mathbf{z}$ such that $L^* \mathbf{z} = 0$. Moreover, if the only solution of $L\mathbf{x} = 0$ is $\mathbf{x} = 0$, then the solution is unique.

Another important result related to the adjoint operator pertains to the eigenvalues of a linear operator: Let $\mathbf{x}_i$ and λ_i denote the eigenvectors and eigenvalues of L, i.e., $\mathbf{x}_i$ and λ_i satisfy $L\mathbf{x}_i = \lambda_i \mathbf{x}_i$. Then if L is self adjoint, all the eigenvalues λ_i are real and eigenvectors corresponding to different eigenvalues are orthogonal.

UNITS

Quantity	Symbol	SI	English	Conversion
Area	A	m^2	ft^2	$1\ m^2 = 10.7639\ ft^2$
Convection heat transfer coefficient	h	$W/m^2\text{-}^\circ C$	$Btu/hr\text{-}ft^2\text{-}^\circ F$	$1\ W/m^2\text{-}^\circ C = 0.1761\ Btu/hr\text{-}ft^2\text{-}^\circ F$
Density	ρ	kg/m^3	lb_m/ft^3	$1\ kg/m^3 = 0.06243\ lb_m/ft^3$
Energy (heat)	e	kJ	Btu	$1\ kJ = 0.94783\ Btu$
Force	F	N	lb_f	$1\ N = 0.2248\ lb_f$
Gravity	g	m/s^2	ft/s^2	$9.8\ m/s^2 = 32.2\ ft/s^2$
Heat flux (per unit area)	q/A	W/m^2	$Btu/hr\text{-}ft^2$	$1\ W/m^2 = 0.317\ Btu/hr\text{-}ft^2$
Heat flux (per unit length)	q/L	W/m	Btu/hr-ft	$1\ W/m = 1.0403\ Btu/hr\text{-}ft$
Heat generation (per unit volume)	Q	W/m^3	$Btu/hr\text{-}ft^3$	$1\ W/m^3 = 0.096623\ Btu/hr\text{-}ft^3$
Length	L	m	ft	$1\ m = 3.2808\ ft$
Mass	m	kg	lb_m	$1\ kg = 2.20462\ lb_m$

Pressure	p	N/m^2	lbf/in^2	1 N/m^2 = 1.45038 x 10^{-4} lbf/in^2
Specific heat	c_p	kJ/kg-°C	Btu/lbm-°F	1 kJ/kg-°C = 0.23884 Btu/lbm-°F
Thermal conductivity	κ	W/m-°C	Btu/hr-ft-°F	1 W/m-°C = 0.5778 Btu/hr-ft-°F
Thermal diffusivity	α(κ/ρc_p)	m^2/s	ft^2/s	1 m^2/s = 10.7639 ft^2/s
Velocity	v	m/s	ft/s	1 m/s = 3.2808 ft/s
Viscosity (dynamic)	μ	kg/m-s	lbm/ft-s	1 kg/m-s = 0.672 lbm/ft-s
Viscosity (kinematic)	ν(μ/ρ)	m^2/s	ft^2/s	1 m^2/s = 10.7639 ft^2/s
Volume	V	m^3	ft^3	1 m^3 = 35.3134 ft^3

NOMENCLATURE

Only those principal symbols that are used in this book are listed below. However, all symbols are defined within the chapters as they are introduced. Occasionally situations occur where duplication of symbols are intentionally used, with different meanings for the same notation clearly identified.

A	area
A	advection matrix
B	body force
$B_{x,y}$	body force term components
$B_{1/2}$	half bandwidth
c	mass fraction of material; Courant number
C	mass (capacity) matrix
$C^{0,\ldots n}$	space of functions
c_i	unknown coefficient (weighted residual)
c_n	coefficient (wave number)
c_v	specific heat at constant volume
d	viscous damping coefficient; thickness of fluid
d	degrees of freedom
D	mass diffusion coefficient; material derivative
D'	anisotropic balancing diffusion
D_T	thermal diffusivity
e	internal energy per unit mass
E	error
$e_{1,2,3}$	unit coordinate vectors
$\|e\|$	energy norm error
$\|e_\phi\|_{L2}$	L_2 norm error
f	volumetric heat source/sink; load vector
$\bar{f}$	time-averaged (mean) value

F	forcing term (wave equation)
f_n	function
Fr	Froude number
g	gravity
G	mean surface curvature
G	mass flux vector
g(x)	functional
Gr	Grashof number
h	convection (film) coefficient; height (free surface)
$\bar{h}$	element length
H^n	Sobolev space
I( )	integral
i,j	node numbers
i,j,k	unit vectors in the x, y, and z directions
I_m	imaginary
J	Jacobian matrix
k	wave number; turbulence kinetic energy
K	conductance (stiffness) matrix
L	linear operator; reference length; wave length
$L_{1,2,3}$	natural coordinate system (triangles; tetrahedrals)
m	mass; total number of elements
M	Mach number
M(x,y)	differential function
n	total number of nodes
n	outward unit vector normal to the boundary
N	one-row matrix of shape functions
N_{max}	largest node number (bandwidth)
N_{min}	smallest node number (bandwidth)
N(x,y)	differential function
N(x,y,z)	shape function
Nu	Nusselt number
O	zero vector
p	pressure
Pe	Peclet number
P_{ij}	Petrov-Galerkin perturbation
Pr	Prandtl number
P_s	static component of pressure
$P_{x,y}$	geometric projection vector

P	vector prescribed flow rates at the boundary
Pe	Peclet number (= Pr·Re)
q	heat transfer flux
$\|q\|^2$	square of total error
Q	source/sink term (heat)
r	radial direction
r	number of nodes in an element
r	vector of nodal flow rates at the external boundary
R	residual; universal gas constant; outer radius
Ra	Rayleigh number
Re	Reynolds number; real
s	convection-diffusion coordinate aligned with velocity
S	source/sink term for concentration
$S_{1,2}$	interface (jump)
t	time; convection-diffusion coordinate (tangent)
t	tangent to surface
T	temperature
Ta	Taylor number
u	horizontal velocity (x); unknown variable
u_t	tangential velocity
v	lateral velocity (y)
V_k	Voronoi regions
w	vertical velocity (z)
w(x)	weight
x	horizontal Cartesian coordinate
X	iterate value
y	lateral Cartesian coordinate
y_c	characteristic length
z	vertical Cartesian coordinate
α	Petrov-Galerkin optimal value
α_n	wave number
β	coefficient of thermal expansion; stability parameter
γ	Petrov-Galerkin stability parameter; specific heat ratio; cel Peclet number
Γ	boundary
Δ	interval (time or space)
δ_{ij}	Kronecker delta
ε	diffusivity (Burgers equation); turbulence dissipation rate

ζ	natural (nondimensional) coordinate (z)
η	natural (nondimensional) coordinate (y)
θ	relaxation parameter
Θ	phase error
κ	thermal conductivity; von Karman's constant
λ	second coefficient of viscosity; penalty parameter
μ	dynamic viscosity
μ_t	turbulent eddy viscosity
ν	kinematic viscosity
ϕ	function; viscous heat dissipation
ϕ	approximate function
Φ_n	analytical amplification factor
τ	phase angle
τ_{ij}	Reynolds stress term
φ	variable
ρ	density
σ	standard deviation
σ_{ij}	stress tensor
ξ	nondimensional distance (x)
ξ_i	mesh enrichment error parameter
ψ	streamfunction
Ω	domain (area or volume); angular velocity
Σ	summation
∇	divergence operator
$\cup$	union
$\cap$	intersection

Subscripts

i,j	node numbers; column-row reference in vectors
i,j,k	unit vectors in the x, y, and z directions
u,v,w	velocity components in the x,y,z directions
x,y,z	x,y,z coordinate directions
o	reference (free stream)
*	exact value

Superscripts

e	element
n	previous time level
n+1	new time level
p	order of shape function
$'$	dimensional value; perturbation
$\cdot$	time dependence
*	approximate

INDEX

Acceleration 284, 298
Accuracy 129, 146, 192
Acoustic waves 433
Adaptation (mesh) 468, 469, 492,
        508, 509, 510, 511, 514, 517,
        520, 523
Adaptation parameter 521
Adjoint 576
Advancing front 473
Advection 383, 388
Aerodynamics 552
Akhras and Dhatt method (mesh) 500
Aliasing 472
Amplification factor 377, 401, 575
        analytical (AAF) 362
        numerical (NAF) 363, 366, 367
Amplitude 357, 358, 360, 362, 364,
        366, 375, 378, 388, 395
Angular velocity 536
Anisotropic
        balancing diffusion 333
        diffusion 354
ANSYS 4
Approximating function 5
Area
        coordinates 118, 164
        element 446
        integral 95

Artificial numerical diffusion 320, 321
Assembly process 57, 241, 262
Atmospheric modeling 489
Attractions 481, 482
Aubin-Nitsche 128
Axisymmetric
        coordinates 562
        heat conduction 101

Backward difference 260, 265, 399, 417
Backward facing step 241, 405, 418,
        419, 420, 449, 450, 451, 452,
        453, 549, 550, 552
Bandwidth 114, 247, 466, 495, 496, 501,
        503, 533
        half 495, 496, 497, 502, 533
Barotropic 432
Basis functions 48
Bifurcation points 344, 345, 543
Blasius equation 251
Blending function interpolation 109, 127,
        147, 150, 151, 153, 154, 160, 174
Body force 13, 404, 409, 426, 436, 437,
        536, 538, 539, 540
Boolean sum 153, 154, 165
Boundary conditions

application of 23
concave 483
cyclic 406
Dirichlet 23, 28, 53, 84, 97, 160,
  187, 213, 238, 405, 406,
  407, 413, 465, 490, 505
  flux 37, 45, 62, 63, 72, 79, 83,
  92, 196, 202, 314, 439, 452,
  494
  free 458
  inflow 456
  mixed 406
  natural 441, 448
  Neumann 24, 160, 406, 407,
  465, 490
  no-slip 18, 245, 405, 565, 405,
  443, 565
  open 238, 406, 441, 448, 458
  outflow 405, 439, 443, 448,
  449, 451
  parallel 442
Boundary element method (BEM)
  528
Boundary fitted coordinates (BFC)
  2, 465, 467, 471, 478
Boundary layer 253, 314, 343
Boundary value problem 101
Boussinesq approximation 243, 435,
  536, 540
Bouyancy forces 243
Bubble function 282
Burgers equation 212, 221, 224,
  278, 300, 303, 304, 342,
  343, 346, 354, 394, 402

CAD/CAM 466
Cartesian 12, 101, 331, 488
Cartwheel (nodal hub) 505
Cavity-driven flow 419
CDFK (chemistry code) 562
Cell Reynolds number 417
Chain rule 133, 479
Characteristic

length scale 410, 455
  equation 350
  thickness 536
  velocity 410, 455
Chebyshev interpolation 111
Checkerboard mode 419, 420, 421, 422,
  427, 428
Chemical reaction 561
Chill plate 22
Classical solution 29
Closed (operation) 571
Coefficient
  bulk viscosity 13
  dynamic viscosity 13
  mass diffusion 17
  second viscosity 13, 424
  thermal expansion 244, 435
  viscous damping 18, 258
  volumetric expansion 435
Combination method (mesh) 473
Combustion 560, 562
Commercial
  aircraft 7
  FEM codes 7
Computational errors
  numerical differentiation 5
  numerical integration 5
  round-off 5
Computational Fluid Dynamics (CFD) 8
Conduction, heat 77, 97, 185
Conductivity 225, 226
  matrix 58
  thermal 16, 92, 225
Conformal mapping 143
Conformity 76
Connected region 34
Consistent 409
Continuity of fluxes 72
Convection 315
  diffusion equation 6, 321, 326, 327,
  331, 339, 346, 350, 354, 392, 402
  heat transfer coefficient 24, 208
Rayleigh-Bernard 443, 444
  transport 313, 358

Convergence 61, 66, 87, 103, 144, 192, 193, 252, 292, 383, 444, 511, 512, 521
Convex hull 505
Coordinates
  area 164
  boundary fitted 2
  natural 164
  parent 135, 176
  transformation 353
  volume 164
Correction indicator 515
Correctors 292, 298, 300, 311
Cosine hill 335
Couette flow 239, 282, 283, 443, 444
Courant-Friedrich-Levy property 393
Courant number 364, 365, 366, 367, 375, 386, 388, 393, 416, 540
Crack propagation 112
Crank-Nicolson scheme 265
Crank-Nicolson-Galerkin 265, 272, 358, 363, 364, 365, 376, 388, 397, 398, 399, 400, 402
Critical
  depth 455
  time step 272
  value 322
Cuthill-McKee algorithm (mesh) 500, 505

Damping 363, 365, 378
Decomposition
  geometries (C, H, O) 483
  LU 540
Degree of precision 176
Delaunay triangulation 505, 506, 507
Delaunay-Voronoi 473
Density 283, 404
Dependent variables 410

Derivative operator 11, 12
Deviation (mean, standard) 522
Diagonal dominance 85
Diffusion 353, 371, 372, 383
  anisotropic 354, 384
  balancing 321, 324, 325, 327, 328, 333, 372, 401
  coefficient 314, 381, 386
  cross-flow 335
  equation 308, 313, 354, 397
  heat 244, 257, 443, 538
  length scale 318
  matrix 332
  mass 17
Dilatation 13
Dirichlet boundary condition 23, 28, 53, 84, 97, 160, 187, 213, 238, 405, 406, 407, 413, 465, 490, 505
Discontinuity capturing algorithms 338
Dispersion terms 374
Displacement 66, 284
Divergence term 553, 554
Do loops 467, 476
Dot product 573
Double precision 430
Dufour and Soret effects 17

Element 5, 51
  bilinear 116, 140, 195, 332, 537
  bicubic 152
  biquadratic 152, 195, 308
  brick 95, 565
  connectivity 490
  cubic 110, 113, 114, 120, 123, 140, 157, 169, 173, 327, 393
  curved 146
  distorted 490
  equations 56
  exponential 115
  Hermite 393
  hexahedral 341, 468, 472, 477, 489, 517, 523, 527
  isoparametric 130, 140, 143, 166,

176, 192, 204, 230, 271,
338, 485, 494, 537, 562,
565

Lagrangian 95, 112, 114, 116,
124, 152, 166, 172, 308,
485, 486, 488

linear 110, 113, 114

macro 157, 472, 490

mixed 493, 498

parent 133, 136, 137, 172

prism 489

quadratic 111, 113, 114, 140,
157, 282, 327, 520

quadrilateral 109, 116, 271,
468, 471, 472, 490, 493,
496, 502, 504, 517, 522,
523

rectangular 124, 132, 147, 195,
270, 332

space-time 309, 359, 371, 397,
400

tetrahedral 468, 487, 506, 507

transition 158, 164, 491, 493

triangular 116, 117, 120, 123,
142, 147, 282, 338, 468,
471, 486, 489, 493, 496,
497, 502, 504, 520

trilinear 97, 341, 565, 566

tricubic 163

triquadratic 163

Eigenvalue 265, 576

Energy, internal 16

Equation
Burgers 212, 221, 224, 278,
300, 303, 304, 342, 343,
346, 354, 394, 402

concentration 393

continuity 12, 422, 425, 432,
461, 553, 561, 564

diffusion 308, 397

elliptic 479

energy 15, 244, 443

heat 19, 399

hyperbolic 257, 284, 291, 305,

307, 314, 479, 483

Laplace 19

Momentum 553, 564

Navier-Stokes 13, 196, 209, 225,
238, 244, 248, 250, 296, 297,
305, 313, 342, 343, 344, 346,
403, 404, 405, 408, 415, 428,
443, 456, 548, 564

parabolic 18, 310, 479, 483

pendulum 310

Poisson 420, 446, 451, 452, 470

state 15, 244, 432, 435, 553, 562

wave 19

Error 87, 97, 178, 219, 363, 428, 491,
511

approximate 518

local 514

maximum permissible 517, 518

relative 382

root mean square 512

sum of the squares 512

total permissible 512, 517, 518

Essential boundary condition 23, 36

Euler
equation 392

formula 364

Lagrange 393

scheme 265, 275, 276, 281, 307, 422

Euler-Galerkin 265, 266, 268, 269, 307,
308, 398, 399

Euler Taylor Galerkin 392, 401

Exact solution 30, 33, 86, 199, 286, 289,
290, 296, 307, 328, 352, 439,
514

Explicit scheme 257, 268, 275, 281, 422,
493, 495

Fick's Law 17, 314

Finite difference 2, 319, 350, 422, 465,
471, 473, 474, 475, 476, 492,
512

Finite rate chemistry 561

Finite strip 46

Finite volume 2, 473, 474, 484
First difference 469, 521
First law of thermodynamics 15
Flow
    capillary 456
    cavity-driven 419
    circular cylinder 540, 542, 558, 559
    compressible 275, 309, 392, 432, 469, 516, 552, 553, 566
    Couette 239, 282, 283, 443, 444
    free surface 444, 452, 454
    Hagen-Poiseuille 444
    incompressible 425, 428, 433
    inviscid 561
    isothermal 448, 452, 463, 536, 540
    Poiseuille 415, 430, 431, 438, 440, 442, 444, 463
    Stokes 449
    stratified 426, 434, 438, 440, 442, 448, 449, 450, 451, 452, 453
    Taylor-Couette 535, 536
    temperature dependent 435
    transonic 566
    turbulent 547, 548, 552
    with finite elements 4
FLOW2D 7
Fluid
    Boussinesq 536, 540
    calorically perfect 16
    incompressible 244, 403, 404
    Newtonian 405
    viscous 403, 404
Flux 37, 45, 62, 63, 72, 79, 83, 92, 196, 202, 314, 439, 452, 494
FORTRAN 7
Four-color theorem (mesh) 501
Fourier
    components 362, 363, 364, 368, 378
    series 361
Fourier's law 15
Fractional step method 462

Fracture mechanics 112
Free surface 444, 452, 454, 456
Frictional velocity 551
Frontal solver 498, 501
Froude number 455, 536

Galerkin
    formulation 259
    least squares (GLS) 338, 339, 340, 354
    method 37, 38, 39
Gamma function 162
Gas constant 15
Gateaux derivative 211
Gauss
    cone 389, 390
    integration 182, 197, 428
    -Legendre 181
    points 198, 200, 202, 203, 270
    puff/plume 469
    quadrature 180, 181, 184, 188, 190, 194, 195, 204
    Theorem 97, 233, 255, 448
    weights 191
General Galerkin Least Squares (GGLS) 340
Geometric primitive 528
Gibb's method (mesh) 499
Global 58, 220
    interpolation function 43, 45
    transformation 143
Gradient 196, 440, 455, 516, 521
Grashof number 438, 439, 449, 450, 451, 536
Gravity 244, 310, 435, 437, 452, 456
Green's Theorem 34, 100, 335
Grid
    multiple 474
    telescoping 474
Groom's method (mesh) 500
Groundwater 470, 489
GWADAPT 508

h-method 113, 493, 509, 510, 517
h-p method 113, 509, 510
Half angle formulae 352
Heat
    conduction 21, 77, 96, 97, 111
    diffusion 258, 443, 538
    diffusivity 262
    equation 399
    flux 24, 439, 452
    heptadiagonal 114
    source/sink 16
    transfer 196, 199, 504
Hermite polynomials 115, 150, 169,
    393
Hole (virtual node) 521
Homogeneous conditions 213
HVAC 489
Hypergeometric function 162

Ideal gas equation 15
Idempotence 148
Ill-conditioned 425, 450
Implicit scheme 257, 274, 444, 495
Incompressibility condition 403,
    424, 426, 433
Indicial notation 12, 14, 404
Initial conditions 258, 542
Inner product 572, 573, 574
Integration by parts 30, 68
Interelement boundaries 144, 491
Interface 66, 69, 72, 88, 89, 97, 454
Internal energy 553, 554
Interpolation
    bilinear 152
    equal-order 424
    linear 70
    mixed 411
Intersection 24, 77
Inverse system 481
Isoparametric
    elements 130, 140, 143, 166,
        176, 192, 204, 230, 271,

        338, 485, 494, 537, 562, 565
    transformation 76, 124, 132, 133,
        135, 144, 164, 171, 172, 465
Isotherm 86
Isotropic material 332
Iteration 289, 293
    direct 210, 220, 226, 227, 243, 249,
        254, 342, 343
    explicit 221
    index 538
    nonlinear 255, 342
    Newton 224
    Picard 224, 225

Jacobian
    determinant 271
    matrix 134, 164, 211
    transformation 132, 137, 138, 147,
        480
Jump operator 67, 72, 89, 91, 94, 515

k-ε model 549, 551
Kinematic dissipation rate 549
Kinetic energy 544, 545
Kronecker delta 13

Ladyzhenkaya-Babuska-Brezzi condition
    (LBB) 403, 409, 410, 412, 424,
        427, 429
Lagrangian
    elements 112, 114, 116, 124, 485,
        486, 488
    formula 111, 120, 121, 177
    quasi- 469
    techniques 469
LAPCAD 472
Laplace
    equation 101, 135, 208
    operator 138, 171, 414
Least squares 198, 200, 201, 338, 339,
    340, 421, 448
Legendre polynomials 181

Line
   integral 35, 95, 238, 239, 413,
      427, 448, 451
   tangent 210
Linear
   combination 572
   operator 575, 576
      bounded 576
      unbounded 576
   system of algebraic equations 6,
      38
Linearly
   dependent 572
   independent 572
Lip seal 545
Lipton-Trajan method (mesh) 500
Local functions 45
Locking 427
Lubrication, 225, 545

Mach number 433, 554, 558, 559,
   560, 567
Manual generation (mesh) 470
Mass 310
   conservation 12
   diffusion coefficient 17
   flux 17
   fraction of material 17, 561
   lumping 201, 268, 270, 271,
      275, 277, 279, 367, 377,
      422, 436, 462, 558
   matrix 257, 270, 291, 413, 421,
      538
   system 258
Matrix algebra 4
Matrix
   asymmetric 449, 495
   banded 495
   capacitance 259, 260
   connectivity 507
   consistent mass 259, 260, 269,
      271, 275, 308, 377, 436
   determinant 138

ill-conditioned 450, 493
   inverse 272
   Jacobian 211
   lumped mass 257, 269, 271, 274,
      307, 308, 365, 377, 398, 399
   mass 257, 265, 365, 413, 421, 538
   nonsingular 123, 134
   sparse 495
   stiffness 58, 94, 114, 208, 223, 235,
      238, 274, 315, 350, 391, 413,
      450, 493, 496, 514
   symmetric 495, 540
   tangent 211, 213, 291, 295, 417
Mean
   surface curvature 454
   value (time average) 548
Mesh 6
   adapting 458, 508, 523
   canonical 478
   coarse 87, 370, 472, 474, 508, 527,
      528
   enrichment 513
   fine 494, 528
   generation 6
   gradation 492
   irregular 370
   nested 510
   nonuniform 476
   orthogonal 471, 478, 483
   refinement 508, 511, 513, 514, 520,
      521, 524, 527
   regeneration 517
   structured 7, 465, 466, 473, 474,
      476, 477, 479, 490, 507
   tetrahedral 507
   uniform 242, 325, 358, 363, 369,
      492
   unrefinement 508, 524, 526
   unstructured 7, 466, 467, 468, 484,
      490
Method of moments 393
Microgap 545, 546, 547
Midpoint rule 182, 206
Mixed method 428, 433, 439, 449
Moment of inertia 310

Monotonic 227, 267, 511

Multigrid techniques 510

NASTRAN 4

Natural boundary condition 24, 36, 37, 242, 282, 439, 441, 448, 463

Natural convection 243, 248, 255, 528, 536, 540, 542, 543, 544, 564, 565, 566

Natural coordinates 119, 122, 123, 164, 239

Navier-Stokes equations 13, 196, 209, 225, 238, 244, 248, 250, 296, 297, 305, 313, 342, 343, 344, 346, 403, 404, 405, 408, 415, 428, 443, 456, 548, 564

NEKTON 508

Neumann 24, 160, 406, 407, 465, 490

Newmark algorithm 283, 286, 287, 296, 297, 302, 305, 311, 312, 537

Newton-Cotes formula 178, 180, 181, 187, 188, 206, 270

Newton-Raphson 210, 212, 218, 220, 223, 224, 225, 228, 243, 245, 230, 247, 249, 251, 252, 253, 254, 255, 287, 342, 343, 344, 346, 354, 416, 417, 444

Newton's
    method 224
    second law 13

Newtonian
    fluid 13, 405
    laminar flow 11

Node 51
    numbering 466, 498, 499
    points 465
    virtual 517, 523, 525

Noise 272, 359

Nonconformity 145

Nondimensional 244, 409, 410, 428, 438, 443, 455, 544

Nonlinear 258, 291, 495
    convection 538
    equations 209, 210, 281, 289, 299, 300, 305
    iteration 342, 416
    operator 252

Norm 59,60, 511, 520, 572, 573, 574, 575
    energy 511, 512, 513, 514, 516
    Euclidean 215, 224, 293, 573
    k^{th} 128
    L^2 60, 87, 216, 511, 516
    infinity 219, 575
    maximum 112, 293
    mean square 60

Normal derivative 26

No slip 18, 245, 405, 565, 405, 443, 565

Nozzle 566, 567, 568

Numerical
    damping 357, 358, 361
    diffusion 320, 321, 323, 324, 325
    dispersion 359, 360, 388, 391
    dissipation 297
    integration 175, 176, 204

Nusselt number
    average 249
    local 249

Optimal
    numbering 500
    value 324

Ordinary differential equation 260, 281, 291, 538

Orthogonal 43, 169, 177

Oscillations 87, 227, 267, 271, 272, 273, 274, 303, 316, 336, 343, 365, 394, 395, 419, 420, 427, 428, 436, 540, 545, 554, 560

Outer radius 536

Overlapping deformation (mesh) 472

Overshoot 395

p-method 113, 493, 509, 514
Pade approximation 316
Parabolic 257, 297, 305
Parallel
    plates 415
    processing 281
Partial differential equation 471
Partition 64, 473, 483
Pascal's triangle 116, 118, 120, 128,
    206
Patch test 76
PATRAN 472, 473, 489
Peclet number 316, 343, 345, 438,
    439, 449, 450, 536
Penalty function 196, 424, 425, 426,
    427, 428, 430, 432, 433,
    434, 436, 437, 439, 440,
    442, 444, 450, 453, 537,
    538, 543
Pentadiagonal 114
Petrov-Galerkin method 6, 37, 51,
    239, 242, 321, 325, 326,
    328, 329, 330, 335, 338,
    340, 342, 343, 344, 345,
    346, 352, 354, 355, 357,
    365, 370, 377, 378, 379,
    380, 383, 384, 391, 392,
    393, 394, 395, 400, 401,
    417, 418, 436, 462, 537,
    555, 557, 565
Phase
    angle 368, 369, 375
    error 368, 369, 378
    lag 357, 358, 360
Piecewise
    constant 56, 230, 233, 416, 418,
        433, 446, 448, 555, 557
    functions 46, 50, 557
    linear approximation 3, 47, 52,
        59, 158
    polynomial 112

Plate problem 503
Poiseuille flow 415, 430, 431, 438, 440,
    442, 444, 463
Poisson equation 420, 424, 446, 451,
    452, 471
Polynomial trial functions 43, 45
Pontriagin-Vitt equation 160
Positive definite 58
Prandtl mixing length 549
Prandtl number 244, 248, 443, 536, 542,
    543, 544, 555, 558, 567
Predictors 285, 287, 291, 294, 298, 299,
    538
Preprocessor 474, 491
Pressure 232, 233, 238, 239, 240, 243,
    282, 299, 404, 406, 408, 412,
    416, 418, 419, 421, 422, 427,
    428, 433, 439, 440, 446, 448,
    449, 454, 463, 541, 545, 547,
    548, 551, 555, 563
    absolute 456
    dynamic 441
    hydrostatic 441, 448, 451, 456
    mean 14, 424
    modified 437, 442, 443, 444, 446,
        448, 449, 451, 456
    static 424, 426, 437, 442, 443, 448,
        451, 455
    thermodynamic 13, 424
    total 441, 443, 444, 446, 451
Product
    approximation 294, 297, 311
    operator 151
Projection error 5
Projector 148, 153, 337
Pseudo code 476

Quadrature 177
    formula 175, 189, 191, 206
    Gauss 180, 184, 188, 190, 194, 195,
        204
    points 176, 190
    weights 176, 178, 191
Quadrilateral 73, 520, 522, 523

594 INDEX

Quasi-linearization 219

Radiation heat transfer 24, 209
Radius 220
Ram accelerator 560, 561, 562, 563
Rapid change detector 508
Ratio of specific heats 555
Rayleigh-Bernard convection 443, 444
Rayleigh number 244, 248, 417, 443, 536, 542, 543
Rayleigh-Ritz method 5
Reactive term 216, 331
Reduced integration 6, 175, 194, 195, 198, 199, 201, 202, 203, 204, 428, 438, 446
Reference
    density 244, 404, 435, 441
    depth 452
    pressure 229, 410, 432, 443, 449
    temperature 244
    time 410
Refinement (h, h-p, p, r) 508, 509, 513
Regions 472
Relaxation parameter 260
Remeshing 520
Renumbering sequence 502, 503
Residual function 28
Reynolds number 228, 229, 238, 242, 248, 344, 410, 417, 418, 428, 430, 431, 438, 439, 441, 449, 450, 451, 455, 462, 536, 542, 544, 554, 558, 559, 560, 567
Reynolds stress term 549, 550
Right parallelepipeds 95
Robins boundary condition 24
Roots 254, 269
Rotating system 536, 540
Runge-Kutta 275, 277, 278, 279, 281, 282, 300, 305, 309, 310

SAP 4
Scalar multiplication 571
Schwartz-Christofol 471
Second-order 88, 97
    operators 75
    partial differential equations 11, 18, 59, 144, 165, 305
    time stepping 375
Self-adjoint operator 146, 198, 313, 576
Semi-automatic generation (mesh) 470
Semi-discrete Galerkin method 259
Separation of variables 183, 206
Serendipity element 126, 144, 157, 166, 173, 472, 485, 486, 488, 489
Shape function 38, 48, 52, 53, 109, 125, 159, 172, 186, 238, 271, 356, 359, 384, 485, 487, 520, 557
    bilinear 76
    global 77
    linear 148
    space-time 384, 385
    trilinear 96, 341, 384, 385
Sharp fronts 314, 394
Shock 560
Single-step method 371
Simpson's rule 179, 182, 188, 205, 271, 308
Sine-Gordon wave equation 253, 293, 296
Singularity 512, 513
Singular matrix 187, 189, 219
Sobolev space 28, 38, 100, 574, 575
Soliton 253
Source/Sink 257, 314, 328, 381
    heat 23
    mass 17
    volumetric 16
Space
    n-dimensional 572
    vector 572, 575
    ∞-dimensional 572
Spatial coordinates 12

Specific heat, constant volume 16
Spectral density 161
Species concentration 225, 314, 434,
        469, 504, 561, 562
Speed of sound 404, 432, 433
Sphere 564, 565, 566
Splines 115
Stability
    analysis 265
    limit 267, 268, 269, 281, 307,
        386, 540
    von Neumann 367, 368, 399
Stefan-Bolzman constant
Stiffness matrix 58, 94, 114, 208,
        223, 235, 238, 274, 315,
        350, 391, 413, 450, 493,
        496, 514
Stochastic equation 160
Stokes
    equation 408, 409, 426
    flow 449
    hypothesis 14, 424
Stratified flow 426, 434, 438, 442,
        448, 449, 450, 451, 452,
        4453
Streamfunction 242, 418, 419
Streamline 248
Streamline Upwind Petrov-Galerkin
        (SUPG) 335
Stress 87, 196, 225, 405, 454, 516,
        545
    normal 14
    tensor 13, 404
Structural mechanics 487, 504
Sublayer 551
Suborbital aircraft 8
Subparametric 145
Subspace 573
Superconvergence 51, 65, 199, 324
Superparametric 145
Superposition methods 507
Surface tension 454, 456
Sutherland's law 554
Swirl 567

Symmetric matrix 58
Symmetry boundary conditions 407

Tangent operator 249, 299
Taylor Galerkin 391, 392, 402
Taylor number 536, 545, 546
Taylor series 177, 210, 218, 276, 284,
        310, 322, 346, 350, 352, 391,
        475
Tensor 448
    product 124
Tesselation 505
Test function 38, 40, 385
Tetrahedral 487
Theta method 260, 265, 268, 299, 305,
        307, 308, 309, 312, 361, 365,
        367, 397, 398, 399
Threshold values (adaptation) 522
Thrust vectoring 566
Time-dependent 250, 257, 258, 283, 296,
        303, 305, 307, 339, 340, 344,
        346, 347, 358, 367, 370, 395,
        416, 417, 422, 450, 535, 537
Time-independent 259, 331, 338
Total
    derivative 435
    head 455, 456
    potential energy 3
Transcendental functions 45, 150, 209,
        219
Transfinite interpolant 149
Transformation
    geometric 473, 575
    parametric 487
    topologic 473
Transport mapping (mesh) 471
Trapezoidal rule 179, 182, 187, 188, 205,
        270
Trial function 38, 40, 48
Triangles 73, 117, 142, 486, 487, 497,
        506, 507, 520
    linear 192, 412
    Pascal's 116, 118, 120, 128
Tridiagonal 58, 114

Truncation error 265, 320, 322,
      323, 346, 371, 374, 375,
      376
Turbulent
      eddy viscosity 550
      flow 547, 548, 552
Turbulence kinetic energy 549
Two point boundary value 39, 42

Unconditionally stable method
      265, 367
Underdiffused 321, 323, 343, 351
Underrelaxation 227
Undershoot 395
Union 24, 51, 77, 482
Unit normal vector 12, 26, 35, 89,
      234, 454
Unit vector 12
Universal limiter 395
Upwind 319, 320, 321, 342, 350,
      353

Vector
      computers 501
      linear 571
      outward normal 454
      residual (out-of-balance) 288,
            291, 295, 298, 299
      space 572, 575
      state 522
      unit 12
      unit normal 12, 26, 35, 89,
            234, 454
      unit tangent 454
      zero 571
Vectorization 281
Velocity 284, 311, 318, 325, 332,
      381, 420, 422, 433, 538,
      548, 563
      characteristic 410, 455
      components 12
Vibration 258

Viscosity 225, 283, 409, 424, 436, 554,
      558
      bulk 425
      dynamic 229
      kinematic 229, 404, 436
      second coefficient 424, 425
Viscous heat dissipation 16
Volume
      coordinates 164
      fluid particle 13
      integral 95
von Karman constant 551
von Neumann stability analysis 367, 368,
      399
Voronoi 505, 506, 510
Vortex 335, 541

Wall function 551
Wave
      acoustic 433
      detonation 561
      equation 19
      front 303, 495
      Gaussian 360, 365, 383, 393
      gravity 452, 455
      -length 364, 366, 369
      number 362, 364
      propagation 17, 258
      reflection 370
Weak formulation 30, 33, 37, 54, 97,
            135, 212, 314, 326, 345, 407,
            447, 448, 451, 537, 555
Weighted residual formulation 29
Weighting
      bilinear 377, 380
      functions 53, 54, 326
      space-time 377, 380
Well-posed problem 6
Weir 454, 455, 457
Words (storage) 247, 430

Zeroth moment of time 161

DATE DUE

APR 2 4 2001			
11/01/01			
OCT 1 8 2001			
APR 0 7 2003			
MAR 2 4 2003			
08/25/03			
12/10/03			
MAY 0 1 2008			
JUN 1 2 2008			
GAYLORD			PRINTED IN U.S.A.